PLUMBING

4th Edition

R. D. Treloar

Colchester Institute

WILEY-BLACKWELL

A John Wiley & Sons, Ltd., Publication

This edition first published 2012
© 1994, 2000, 2006, 2012 by Blackwell Publishing Ltd

Wiley-Blackwell is an imprint of John Wiley & Sons, formed by the merger of Wiley's global Scientific, Technical and Medical business with Blackwell Publishing.

Registered office:
John Wiley & Sons, Ltd, The Atrium, Southern Gate, Chichester, West Sussex, PO19 8SQ, UK

Editorial offices:
9600 Garsington Road, Oxford, OX4 2DQ, UK
The Atrium, Southern Gate, Chichester, West Sussex, PO19 8SQ, UK
2121 State Avenue, Ames, Iowa 50014-8300, USA

For details of our global editorial offices, for customer services and for information about how to apply for permission to reuse the copyright material in this book please see our website at www.wiley.com/wiley-blackwell.

Extracts from British Standards are reproduced with the permission of BSI. Complete copies can be obtained by post from BSI Customer Services, 389 Chiswick High Road, London W4 4AL.

Library of Congress Cataloging-in-Publication Data
Treloar, Roy.
    Plumbing / R.D. Treloar. – 4th ed.
        p. cm.
    Includes bibliographical references and index.
    ISBN 978-1-4051-8959-0 (pbk.)
1. Plumbing–Great Britain–Examinations–Study guides.   2. Plumbing–Problems, exercises, etc.   3. Plumbers–Certification–Great Britain.   I. Title.
    TH6123.T74 2011
    696'.1–dc23
                                   2011035002

A catalogue record for this book is available from the British Library.

This book is published in the following electronic formats: ePDF 9781444398908; ePub 9781444398915; Mobi 9781444398922

Set in 10/12 pt Sabon by Toppan Best-set Premedia Limited
Printed and bound in Singapore by Markono Print Media Pte Ltd

1   2012

Contents

Introduction	x
List of Abbreviations	xi
Acknowledgements	xii
A Qualification in Plumbing	xiii
Workplace Evidence for NVQ Achievement	xiv
Part 1 Key Principles, Safety and Common Plumbing Processes	1
The Plumbing Industry	2
Customer Relations and Communication Skills	4
Functional Skills	6
Employment Rights and Responsibilities	8
Effective Working Relationships	10
Health and Safety Legislation	12
Safe Working Practices	14
Personal Protective Equipment	16
Safety Signs	18
Control of Substances Hazardous to Health	20
Access Equipment and Working at Heights	22
Safe Working with Hand Tools and Machinery	24
Fire Safety	26
Regulations Governing Plumbing Work	28
Identification of Pipework	30
The International Metric System	32
Areas, Volumes and Capacities	34
Mass, Weight, Force, Pressure and Density	36
Water Pressure	38
Physical Properties of Materials	40
Heat and its Effects	42
Corrosion	44
Plumbers' Tool Kit	46
Specialist Hand Tools for Pipework	48
Specialist Hand Tools for Sheetwork	50
Plastic Pipe and Fittings	52
Low Carbon Steel Pipe and Fittings	54
Copper Pipe and Fittings	56
Low Carbon Steel Pipe Bending	58
Copper Pipe Bending	60
Pipe Bending Using Heat	62
Soldering	64
Welding Equipment and Safety	66

Welding Processes 68
Bronze Welding 70
Lead Welding (Lead Burning) 72
In-Line Valves 74
Terminal Valves 76
Float-operated Valves 78

Part 2   Hot and Cold Water Supplies 81
Classification of Water 82
Cold Water to the Consumer 84
Backflow Prevention 86
Assessing Water Efficiency in New Dwellings 88
Cold Water Systems 90
Cold Water Storage 92
Water Treatment 94
Boosted Water Supplies 96
Fire-fighting Systems 98
Domestic Sprinkler Systems 100
Hot Water Systems (Design Considerations) 102
Direct Hot Water Supply (Centralised) 104
Indirect Hot Water Supply 106
Unvented Domestic Hot Water Supply 108
Hot Distribution Pipework 110
Heat Recovery Period 112
Instantaneous Domestic Hot Water Supplies (Centralised) 114
Localised Hot Water Heaters 116
Connections to Hot and Cold Pipework 118
Installation of Pipework 1 120
Installation of Pipework 2 122
Sizing of Hot and Cold Pipework 124
Noise Transmission in Pipework 128
Commissioning of Hot and Cold Supplies 130
Disinfection of Hot and Cold Water Systems 132
Maintenance and Servicing Schedule 134

Part 3   Central Heating 137
Domestic Central Heating 138
Wet Central Heating Systems 140
Central Heating Components 1 142
Central Heating Components 2 144
Heating Controls 146
Fully Pumped System 148
Sealed (Closed) Systems 150
Boilers 152
Combination Boiler (Combi) 154
High Efficiency or Condensing Boiler 156

Domestic Heating and the Building Regulations   158
Domestic Boilers Requirements   160
Central Heating System Protection   162
Advanced Central Heating Control   164
Radiator and Boiler Sizing   166
Whole House Boiler Sizing Method   170
Pipe and Pump Sizing   172
Radiant Heating   176
Underfloor Central Heating Sizing   178
Warm Air Heating   180
Commissioning of Wet Central Heating Systems   182

Part 4   Gas Supplies   185
Properties and Combustion of Natural Gas   186
The Law Relating to Gas Installation Work   188
Gas Supply to the Consumer   190
Sizing of Domestic Gas Pipework   192
Domestic Tightness Testing and Purging   196
Pressure and Flow   198
Gas Controls 1   200
Gas Controls 2   202
Flame Supervision Devices (Flame Failure Devices)   204
Open Flued Appliances (Conventional Flue)   206
Terminal Location for Open Flues   208
Materials and Construction of Open Flues   210
Installation of Gas Fires   212
Installation of Gas Cookers   214
Room Sealed Appliances 1 (Balanced Flue)   216
Room Sealed Appliances 2   218
Ventilation Requirements up to 70 kW Net Input   220
Other Flueing Systems   222
Liquefied Petroleum Gas Installations 1   224
Liquefied Petroleum Gas Installations 2   226
Combustion Analysis   228
Flue Efficiency in Gas Appliances   230
Commissioning of Gas Installations   232
Maintenance and Servicing   234
Carbon Monoxide Detection   236
Unsafe Gas Installations   238

Part 5   Oil Supplies   241
Properties and Combustion of Fuel Oils   242
Oil Storage   244
Oil Feed Pipework   246
Controls Used on Oil Feed Pipework   248
Pressure Jet Burners 1   250

Pressure Jet Burners 2   252
Pressure Jet Burners 3   254
Open Flued Appliances   256
Room Sealed Appliances   258
Vaporising Burners   260
Combustion Efficiency Testing   262
Commissioning and Fault Diagnosis   264

Part 6   Electrical Work   267
Basic Electrical Theory   268
Electrical Current   270
Electrical Supply   272
Electrical Safety   274
Earth Continuity   276
Domestic House Wiring 1   278
Domestic House Wiring 2   280
Installation Practices   282
Electrical Components 1   284
Electrical Components 2   286
Central Heating Wiring Systems   288
Inspection and Testing of Electrical Work   290

Part 7   Sanitation   293
Sanitary Accommodation   294
Sanitary Appliances 1   296
Sanitary Appliances 2   298
Flushing Cisterns 1   300
Flushing Cisterns 2   302
Waste Pipe Connections   304
Sanitary Pipework   306
Primary Ventilated Stack System   308
Ventilation of Sanitary Pipework   310
Trap Seal Loss   312
Mechanical Disposal Units   314
Sizing of Sanitary Pipework   316
Testing of Sanitary Pipework   318
Maintenance and Periodic Inspection   320

Part 8   Drainage   323
Below Ground Drainage   324
Protection of Pipework   326
Gullies and Traps   328
Provision for Access   330
Connections to Existing Systems   332
Determining Drainage Levels   334
Sizing of Drainage Pipework   336

Eaves Guttering   338
Rainwater Pipes   340
Soakaways, Cesspools and Septic Tanks   342
Soundness Testing of Drainage Systems   344

Part 9   Sheet Weathering   347
Lead Sheet   348
Lead Bossing   350
Sheet Fixing   352
Lead Roof Coverings and Wall Cladding   354
Expansion Joints for Lead Roofs   356
Abutment Flashings in Lead   358
Chimney Flashings in Lead   360
Lead Slates and Pitched Valley Gutters   362
Gutter Linings   364
Dormer Windows   366

Part 10   Energy Conservation & Sustainability   369
Environmental Awareness   370
Energy Efficiency in Domestic Dwellings   372
Renewable Energy   374
Energy Costs & Payback Period   376
Solar Energy   378
Solar Hot Water Heating Systems 1   380
Solar Hot Water Heating Systems 2   382
Solar Power   384
The Heat Pump   386
Installation of a Heat Pump   388
Biomass   390
Biomass/Wood Burners 1   392
Biomass/Wood Burners 2   394
Wind & Micro Hydro (Water) Power   396
Combined Heat & Power (CHP)   398
Rainwater Harvesting 1   400
Rainwater Harvesting 2 (System Design)   402
Grey Water Recovery   404

Part 11   Assessing Your Knowledge   407
Assessing Your Knowledge   408
Self Assessment   409
Supplementary Assessment   424
Problem Solving   464
Answers to Problems 1–10   474
Answers to Multiple Choice Questions   476

Index   477

Introduction

This book is designed to provide easily accessible information on a wide range of subjects relating to Mechanical Engineering Services. For the purpose of simplification, I have used the words *plumber/plumbing* when relating to the various aspects of work in the MES industry, to include job titles such as heating engineer, sanitation engineer, gas installer, sheet lead worker, etc. Few books could possibly hope to have all the answers to questions relating to plumbing skills; I have, however, endeavoured to cover as many topics as possible in the hope that the book will be a source of useful information both for the student with no knowledge of the subject, and for the trained plumber seeking guidance in particular areas of study. This new edition includes a whole new chapter on energy conservation. Some of the topics covered are beyond the scope of the plumber; however, in order to grasp the whole concept of renewable fuels and this aspect of plumbing works I have included an outline of several future technologies.

The book covers topics found in the technical certificates and NVQs/SNVQs at levels 2 and 3 which are currently standard for plumbers in the UK; additional skills are identified which no respectable plumber can afford to ignore.

The book is in eleven parts. Parts 1 to 10 offer a programme of training and information while Part 11 is designed to allow you to assess your level of knowledge. There are sections for self- and supplementary assessments and a few typical plumbing problems.

Broadly speaking, the subject matter dealt with in the supplementary assessment questions is introduced in the same order as the subject matter of the book itself; hence these sections can be used as self-learning packages.

Further useful information can be found in the preliminary pages which follow. The nature of the NVQ is, for example, identified, as is a guide showing how to complete the scheme.

In addition to the new chapter in relation to energy conservation, this new edition also considers the changes to the Building Regulations to include approved documents: Part F – Ventilation; Part G – Sanitation, hot water safety and water efficiency; Part H – Drainage and waste disposal; Part L – Conservation of fuel and power; and Part P – Electrical safety.

In order to further assist your studies, the following website should be visited (www.blackwellpublishing.com/treloar), where the answers to the supplementary questions given on pages 424–463 will be found.

Special note to trainers and lecturers
To facilitate your delivery of the subject, all of the illustrations and photos found within this book can be accessed at the above website for your use in the classroom.

List of Abbreviations

a.c.	alternating current
BS	British Standard
BSP	British Standard pipe
Btu	British thermal unit
c.h.	central heating
cpc	circuit protective conductor
d.c.	direct current
dhw	domestic hot water
doc	drain off cock
dpc	damp proof course
dpm	damp proof membrane
emf	electromotive force
f & e	feed and expansion
f & r	flow and return
f.w.g.	foul water gully
GRP	glass-reinforced plastic
HSE	Health & Safety Executive
HT	high tension
i.d.	inside diameter
IET	Institution of Engineering and Technology
IGEM	Institution of Gas Engineers and Managers
LCD	liquid crystal display
LDF	leak detection fluid
LPG	liquefied petroleum gas
MCB	miniature circuit breaker
MES	Mechanical Engineering Services
o.d.	outside diameter
Pa	pascal (unit of pressure)
PAS	Product Assessment Specification
PE	polyethylene
PTFE	polytetrafluoroethylene
PVC	polyvinyl chloride
RCD	residual current device
RIDDOR	Reporting of Injuries, Diseases and Dangerous Occurrences Regulations
r.w.p.	rainwater pipe
s.v.p.	soil vent pipe
TRV	thermostatic radiator valve
wg	water gauge

Acknowledgements

I would like to thank the following organisations for permission to use extracts from their material:

British Standards Institution
389 Chiswick High Road, London. W4 4AL.
Telephone: 020 8996 9000

Institute of Plumbing and Heating Engineering
64 Station Lane, Hornchurch, Essex. RM12 6NB.
Telephone: 01708 472791

Chartered Institution of Building Services Engineers
222 Balham High Rd, Balham, London. SW12 9BS.
Telephone: 020 8675 5211

Energy Saving Trust
21 Dartmouth Street, London. SW1H 9BP.
Telephone: 020 7222 0101

I would also like to thank the following organisations and companies for permission to reproduce photographs:

Green roof: Green Roof Consultancy Ltd (livingroofs.org), by bere:architects

Images of wood chips, pellets etc: Euroheat Natural Energy Company

Biomass plant in Wales: Aalborg Energie Technik (AET)

PV cells on roof: EvoEnergy

Radiant tubes on roof: Sky Flair Ltd

Wave generator: Pelamis Wave Power

Wind turbine: RenewableUK

A Qualification in Plumbing

The current starting point to help the learner progress as a plumber is to obtain the NVQ Diploma in Plumbing & Domestic Heating. This is achieved by the individual obtaining employment, ideally through an apprenticeship where they attend a college or training centre, typically for one day a week.

The National Vocational Qualification (NVQ)

The NVQ Diploma is a competency-based scheme, where there is a requirement to undertake some compulsory specific tasks within the workplace, thereby demonstrating that the individual has carried out certain specific tasks themselves under real working conditions. The NVQ has two levels, being Level 2 and Level 3.

- At Level 2 the learner learns basic skills to include: Effective Working Relationships; Safety; Functional Skills; Key Principles; Common Plumbing processes; Hot & Cold water supplies; Sanitation & Drainage; Central Heating pipe work; and Environmental Awareness. In addition the candidate may choose to study two additional units to include sheet weathering and Warm air heating.
- At Level 3 the above skill base is developed to include a greater understanding and in-depth study, fault diagnosis and maintenance. At Level 3 the learner also has to choose one of the following four options: Gas; Oil; Solid Fuel or Environmental Technologies.

Additional qualifications

Where a trainee is enrolled onto an apprenticeship programme there is also a requirement to undertake the following mandatory subjects:

- Functional Skills (unless the trainee has already achieved a good GCSE grade [A–C] within the last 5 years)
- Employment Rights and Responsibilities.

Timescales

The length of the apprenticeship is approximately 4 years. It is not time-serving and is completed only when the apprentice has achieved the specific outcomes. Apprentices will have individual learning plans drawn up at the start of their programme to identify specific goals with agreed and anticipated achievement dates. These plans are reviewed on a quarterly basis, thereby reviewing the progress and setting new targets.

Whilst at college the candidate will be continually monitored as to their ability and will undertake numerous theoretical phase tests and be judged against practical competencies set by the lead body; plus, in addition, there is a requirement to achieve workplace evidence as identified over.

Workplace Evidence for NVQ Achievement

In order to gain the full NVQ achievement, as previously stated, evidence is required from the workplace. This evidence is logged by the candidate into a portfolio of evidence. This consists of several stages, to include: planning, questioning, evidence gathering for the completed task, and witnessing by a workplace recorder and/or assessor.

What is regarded as evidence

Evidence can range from a whole collection of gathered material, such as: photographs, installation drawing or plans, copies of completion certificates or forms and, of course, a written description of the actual work completed. The evidence also needs to be authenticated by the person who commissioned the work, such as the employer, and this will need to be assessed by a qualified assessor for the plumbing programme.

It is a competency-based programme and is only concerned with what actual work the operative has undertaken successfully and whether it is up to the required standard.

The portfolio is issued to the candidate by the awarding body prior to commencement of the programme. Throughout this period there will be specific times when actual work activities are witnessed by the NVQ assessor, thus demonstrating compliance with the scheme. Shown here is a small section taken from a completed portfolio, identifying how and where the candidate has produced the evidence required.

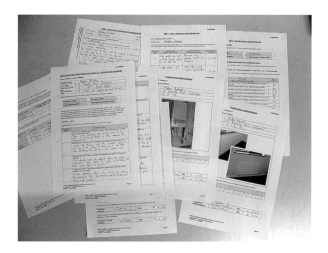

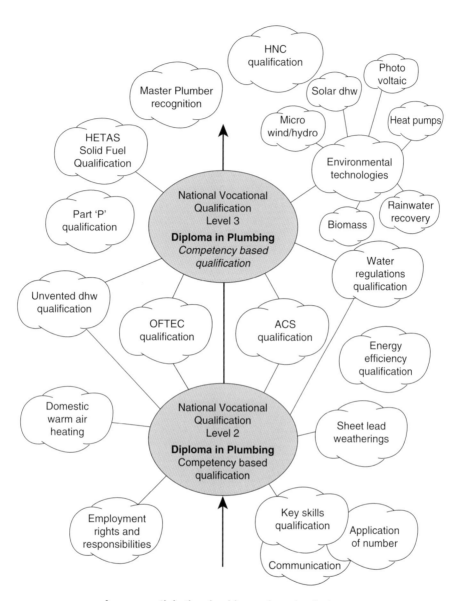

A career path in the plumbing and mechanical engineering services industry

Part 1

Key Principles, Safety and Common Plumbing Processes

Plumbing, 4th Edition. R. D. Treloar.
© 2012 Blackwell Publishing Ltd. Published 2012 by Blackwell Publishing Ltd.

The Plumbing Industry

The role of the plumber

In the eyes of many lay people, the plumber is simply someone who joins pipes together, running the water from one appliance to another. Few people appear to give much consideration to the depth of knowledge required to carry out the basic plumbing skills, or the trained professional's need to have a full understanding of related activities – or their ability to design installations and identify likely problems before they occur.

Unfortunately, many students come to college and find it difficult to take on board skills associated with tasks which their own company does not undertake. They seem to think that skills they are not currently making use of are irrelevant. I would respond by saying that if you want to have greater freedom and be in high demand – and not only by your present employer – then, yes, these extra skills are very relevant indeed.

Trained plumbers are able to turn their hands to many, if not all, the following skills, although levels of knowledge will vary from one individual to another:

- The supply and distribution of cold and hot water for drinking purposes, sanitation, heating and fire fighting, etc., and the connection of associated equipment and controls.
- The removal of water from the building via a suitable system of drainage, to include foul water from soil and waste appliances and surface water from roofs and paved areas.
- Consider the installation of energy conservation systems, to include solar power and water recovery systems.
- The weathering of roof penetrations, etc., in metallic sheet materials.
- The supply and provision of fuels, including gas, oil, solid fuel and electricity, to various appliances and components, such as those for heating or cooking, and the provision of such ventilation as is necessary for combustion.
- The removal of the combustion products from appliances by way of a safe and effective flue system.
- Designing and estimating the cost of any of the above installations, in a domestic situation, giving efficient and effective usage.
- The knowledge to identify, rectify and service any of the said installations.

In addition, plumbers are well served by a number of skills normally associated with other operatives, such as carpenters and bricklayers – skills that might enable them, for example, to make good a hole or remove/replace a floorboard, as part of the job.

The impressions you create are very important; going to work in dirty clothes and trainers gives the same impression as submitting an estimate for a job on a piece of paper taken from a school notebook. Laying a dustsheet down, even for the smallest job, takes very little time, yet it gives a lasting effect. A positive image projected today is likely to bring you work for tomorrow.

Plumbing organisations (addresses correct at time of going to press)

There are many organisations which have connections with the plumbing industry. The following list identifies some of those representing the Mechanical Engineering Services, Plumbing sector:

Association of Plumbing & Heating Contractors (APHC)

Address:	12 The Pavilions Cranmore Drive Solihull B90 4SB	Role:	The employer organisation for the MES Plumbing Industry in England and Wales.
Tel:	0121 711 5030		www.aphc.co.uk

Scottish & Northern Ireland Plumbing Employers Federation (SNIPEF)

Address:	Bellevue House 22 Hopetown Street Edinburgh EH7 4GH	Role:	The employer organisation for the MES Plumbing Industry in Scotland and Northern Ireland.
Tel:	0131 225 2255		

www.snipef.org

SummitSkills

Address:	Vega House, Opal Drive Fox Milne Milton Keynes MK15 0BF		
Tel:	01908 303960		www.summitskills.org.uk

The Joint Industry Board for Plumbing MES in England & Wales (JIB-PMES)

Address:	Brook House Brook Street, St Neots PE19 2HW	Role:	Deals with the grading of plumbing operatives and registration of apprentices. Also agrees terms and conditions of employment.
Tel:	01480 476925		

www.jib-pmes.org.uk

Chartered Institute of Plumbing and Heating Engineering (IPHE)

Address:	64 Station Lane Hornchurch, Essex RM12 6NB	Role:	An independent body having the prime objective of promoting better plumbing practices. The IPHE operates a plumbing registration scheme.
Tel:	01708 472791		

www.iphe.org.uk

Gas Safe Register

Address:	PO Box 6804 Basingstoke RG24 4NB	Role:	To register and verify the competence of those working within the gas industry.
Tel:	0800 408 5500		

Institution of Gas Engineers and Managers (IGEM)

Address:	IGEM House High Street Kegworth, Derbyshire DE74 2DA	Role:	The body that promotes, develops and identifies good practices for use in the gas industry.
Tel:	01509 678198		www.igem.org.uk

Customer Relations and Communication Skills

Customer relations

In order to gain the respect you deserve as a professional in the field of plumbing, you have to show respect for the views of others. Ear and body piercing, tattoos or outrageous clothes may reflect your personality; however, this image may not be in keeping with the views of those you are working for. In an ideal world, this should make no difference, providing that you are demonstrating good plumbing skills. However, in practice, first impressions count and it is the first impression that invariably leads to work. Dress like a tramp and expect to be treated as such; dress cleanly and respectably and it is surprising how you are accepted. Rubbish it may be, but it is a fact.

When arranging to meet someone, be there on time. The person you are to meet can soon become disheartened if you are late, as time is precious to everyone. Finally, when working on a property, remember that it belongs to someone. That person has possibly been working hard to maintain it, it is something which they are proud of and falls within their standards. Therefore, do not treat it as a tip; lay out dust sheets and clean up after your work. Give respect and you will be respected – and get recommended.

Communication

Communicating information to others is something which needs to be right first time if delays are to be avoided. Communication needs to be clear and precise. A recipient can only act upon the information they receive, which, if wrong, can result in (1) incorrect interpretation of data and (2) loss of respect and confidence in the person supplying the information. Looking at the two lists of materials shown opposite: which one indicates accurately to the supplier the items required?

Presentation of information can also give an impression of the knowledge, skill and time you have for a specific job. Look at the two estimates shown opposite: both say much the same thing, yet one indicates a much more desirable company – which do you think it is? The effort required to produce the printed version may look time consuming, but with this modern age of computers, it requires the least amount of effort – much of the data could be standard text and used in all estimates. Note also that this version was produced very promptly (the next day), which can often win the contract for small works and gives the customer confidence in your attention. Also, the customer knows what the final cost will be. The customer of the hand-written version on the other hand does not know whether VAT is included and what, if any, charges for materials there will be. Time spent promoting your company image is time well spent.

1 – 22 mm x 22 mm x 15 mm end feed tee
4 – 15 mm equal end feed tees
2 – 15 mm solder ring elbows
2 – 15 mm compression stopcocks
3 m – 15 mm half hard (R250) copper tube
1.5 m – 22 mm half hard (R250) copper tube

Example of a good material list identifying exactly what is required

Poor example of material list.
Is compression or solder fitting required?
How much copper tube is needed?

1   22 – 22 – 15   T
4   15 t's
2   15 elbows
2   15 stopcocks
     copper tube   15 mm
     "   "   22 mm

RDT PLUMBING

Attention of: R.D. Treloar
Mrs B Brown   Plumbing Contractors
16 Harcourt Road   Priory Court
Chelmsford   Colchester
Essex CM1 1PJ   Essex CO3 1LP
                                  Phone: 01206 318904

Dear Mrs Brown

RE: Proposed replacement WC Suite

Thank you for the opportunity of allowing me to estimate for the replacement WC suite yesterday.

I have pleasure in enclosing the estimate as promised, which I hope you find agreeable, and within your budget.

I have investigated the price and availability and can confirm that the suite is still available and can be made ready for delivery with two days notice.

Taking into account the removal of all rubbish and unforeseen items I estimate the total price for the work to be £487, inclusive of VAT. This price includes the cost of the WC Suite as discussed and any additional materials necessary.

Yours sincerely
R. Treloar
R.Treloar

RDT PLUMBING

R.D. Treloar
Plumbing Contractors
Priory Court
Colchester
Essex CO3 1LP
Phone: 01206 318904

Dear Madam
    The price to install the "WC" we discussed last week will be £487.

If you want me to do the job please contact me on my mobile number 01620 81432
            Thanks
                R.Treloar

P.S. The price quoted is dependent upon the availability of the toilet you want.

Examples of good and bad estimates

Communication skills

Functional Skills

Functional Skills is the name given to the basic communication and numeracy skills that you use in everyday life, at home and whilst at work. There is nothing new about functional skills; you are using them all the time. You have the opportunity to demonstrate your level of competence by completing a qualification to show how good you are. This can then be used by an employer to judge exactly what you can do.

There are three functional skills, which are:

1. **English**
2. **Maths**
3. **Information and Communication Technology (ICT)**

For each of these 3 skills there are several levels that may be achieved:

Entry Level: This basically is a lower-level equivalent to a GSCE
Level 1: Equivalent to GCSE grade D–F, foundation, or NVQ level 1
Level 2: Equivalent to GCSE grade A–C, intermediate, or NVQ level 2

Should you be undertaking an apprenticeship programme, it is a requirement that you complete as part of your course English and Maths at level 2 as a minimum.

Does everyone have to do a functional skills programme?

Trainees undertaking an apprenticeship will need to complete the functional skills elements, and they usually form part of a course where you are aged between 16 and 18. Some courses, however, do not require the completion of a functional skills qualification. Further guidance in relation to this can be obtained from your training provider.

The point of functional skills is that you apply them in your work and life skills all the time, e.g. writing to the bank. For example, you will have learned how to add up, subtract, multiply and divide by doing sums; that is mathematics. But you will often have a real problem to tackle, or a question to answer, requiring a transference of this knowledge in a meaningful way.

For example, you might become involved in pricing up for the installation of a new plumbing or heating system. You might need to calculate the radiator sizes and price up the job to do, considering what profit you need to make or even how you would break even. You will have to estimate how long it will take to do the job, plus many other details, thereby applying your skills to help you to solve a particular problem.

Similarly, you will have learned how to write grammatically and to spell: that's English. You might have to write down the details that identify specific contractual requirements, or you may write a letter that determines whether you get the job that you really want or whether you win the contract for a company. It is all about deciding the best way to deal with a specific problem and demonstrating these skills.

How are functional skills assessed?

Currently the assessment of functional skills is by end testing. There are no multiple choice papers and the use of a computer is restricted, thereby confirming your ability to spell and write grammatically correct English.

Are they worth it?

Achieving this additional qualification can all too often seem a burden. It does, however, provide the trainee with the necessary transferable skills required in today's workplace, where competition is high. Surveys of employers have indicated that they are often looking to employ applicants with:

- Good communication skills, both in speaking and writing
- A fair ability of understanding simple calculations
- A willingness to learn.

Functional skills demonstrate these qualities.

Employment Rights and Responsibilities

Contracts of employment

The relationship between the employer and employee is governed by terms and conditions that are contained within what is referred to as a 'Contract of Employment'. Under the Employment Rights Act all employees are entitled to a written statement, identifying the key terms and conditions of employment, such as the date your salary is paid into the bank, etc. This is a legally binding document and should be issued within the first two months of employment. It is designed to serve and protect both parties. It should be noted that even if no formal document has been issued the contract of employment is still in force, even as soon as the offer of employment has been made and accepted, such as at an interview. Employment contracts may be either open-ended, such as for permanent posts, or fixed-term for temporary positions. Termination of an employment contract is governed by specific rules, as identified in the contract, and is designed to protect the employee and employer from unfair treatment. Also, this part of the contract would identify the period of notice that would be given should you wish to leave, or be asked to leave. Once a contract of employment has been issued it cannot be altered, however procedures that allow consultation with the employee need to be provided to ensure no unfair treatment. Should you have a grievance with your employer there should be a procedure that is followed to help resolve matters. Employees who believe they have been unfairly dismissed or otherwise treated have the right to take their case to an independent employment tribunal, who may award damages as appropriate. However, certain rules need to be met, such as length of service. Self-employed operatives have different rights and responsibilities with regard to their entitlements of statutory benefits. They enter into a different kind of contract, which is governed by different legislation.

Anti-discrimination

You have a legal right not to be discriminated against on the grounds of gender, race or disability from the first day of your employment and it also applies to the recruitment process for the position. There are several Acts covering discrimination:

- **Sex Discrimination Act:** Under this law you must not be refused employment based upon your gender. The law does, however, cover specific situations, such as where carers may be employed, allowing for a specific gender only.
- **Equal Pay Act:** Under this Act you have a right to equal terms and conditions, to include pay, whether you are a man or woman.
- **Race Relations Act:** This Act is in place to ensure you are not discriminated against on the basis of your skin colour, religion or ethnic background. There are a few exceptions, but these cover only a few situations, based mainly outside the UK.
- **Disability Discrimination Act:** This Act applies to all employers with 15 or more employees. The definition of a disability covers a wide range of conditions and

the Act states that a disabled person is one who has a physical or mental impairment, which has been substantial over a long term, affecting their ability to perform normal day-to-day activities. Employers have a responsibility to make reasonable adjustments to the working place or practices so that the disabled employee is not at a disadvantage. This may include altering desk arrangements, etc., to allow wheelchair access. Where someone develops a back problem or heart condition it may be necessary to reallocate heavy work duties to another employee. Unlike discrimination on the grounds of sex or race it may be possible to justify discrimination against a disabled person on the grounds that the disability may lead to safety being compromised.

There is currently no Law in relation to age discrimination in the UK, however certain work activities require a person to have reached a minimum age, such as with using a particular piece of equipment or plant.

Working hours and holiday entitlement

The *Working Time Regulations* identify the amount of time that employees can work for, and the amount of rest time that should be provided; this includes the time spent off-the-job undertaking training for your employer. In addition, there is a specific provision that limits the number of hours young workers, aged 16–17, can work. There are some types of work, for example the fire service, where work activities cannot be interrupted on technical grounds, thus, as always, there may be exceptions to the rule. Both employers and employees have legal rights and responsibilities in relation to the amount and timing of holidays from work; this would be clearly identified within the contract of employment. In addition to holiday entitlement there are specific rights that apply in the case of maternity and parental leave. Time off work for public duties may be granted, although these may not be paid for by the employer.

Absence and sickness

There is a statutory prescribed level of sick pay that employers must provide, however some employers go beyond this and pay additional entitlements when employees are sick. The rules about who can claim statutory sick pay relate to the age of the employee, how much they earn and whether they are claiming any other form of benefit. Employees are allowed to 'self-certificate' for the first few days of sickness, without the need to get a doctor's certificate.

Data protection and access to personal information

Legislation exists to protect individuals in relation to information held in manual or computerised filing systems. It governs what sort of data is acceptable to collect, how it is processed and who should have access to it. There is a clear requirement to ensure protection and that no unauthorised personnel can gain access.

Effective Working Relationships

Plumbing is just a small part of the construction industry. It falls within a sector called Building Engineering Services, which includes: water, heating, ventilation, refrigeration, air conditioning, electrical, gas, telecommunications, plus escalators and lifts.

There are over 160 000 companies working within the construction industry. About 4% of these companies employ between 14 and 299 people, and 1% employ over 300. The remaining 95% employ between 1 and 13 people. Small companies tend to concentrate on extensions, maintenance work and sub-contract work for the larger builder. They may undertake the occasional one-off building. The larger company undertakes major construction projects, usually running over a long period of time. The legal status of a company is dependent upon its size:

- **Sole trader:** One person owns the business, possibly employing one to two operatives. All profits go to the owner.
- **Partnership:** Two or more people own the company and, therefore, share the profits.
- **Limited company:** A company that has grown in size and the individual(s) running the company would not be liable if the business fails. These companies sometimes have a board of directors.
- **Public limited company (PLC):** Same as a limited company, however they trade their shares on the stock market.

The construction team

It is essential that you know about the other people involved within the industry so you can communicate effectively with them, knowing their position in the chain. The following list identifies some of the key people:

Client: The person or organisation for whom the work is being undertaken.

Architect: These people plan and design buildings. They usually use the services of structural and civil engineers or consultants to provide specific details.

Project manager or clerk of works: This person acts as the client's or architect's representative on site, ensuring the requirements are met.

Building control officer: The person who works for the local authority. Their job is to ensure that the Building Regulations are complied with.

Building services engineer: The person responsible for the design of all water and electrical services plus any other services as listed above.

Contracts manager: The person responsible for the overall running of the contract. Their job is to ensure that the job runs to programme and within cost. A contracts manager may be responsible for several sites.

Site manager or site agent: The person leading the project on site. Their job is to solve problems before they happen.

Planner: An individual, or team of people, who works closely with the site agent and estimator to plan the most effective use of all resources. They will reschedule the work if necessary to bring a project back on target.

Co-ordinator (safety officer): The person who manages the overall site safety throughout all stages of the construction project.

Quantity surveyor: These people advise and monitor the cost of the project.

Buyer: The person who purchases the materials needed for the contract. They negotiate with suppliers in order to obtain the best prices.

Estimator: The estimator calculates how much a project is likely to cost. They consider all aspects of the contract, to include materials, plant and labour. This forms a basis upon which a tender is made for the contract.

Building surveyor: The person who provides advice on building matters, organises and undertakes building surveys and prepares plans and specifications.

Site engineer: This person ensures that various technical aspects of the construction project are correct. They confirm that things are built correctly and to the correct specification.

Site supervisor or foreman: The person responsible for the supervision of the trade supervisors and the day-to-day running of the contract.

Trade supervisor or charge hand: The person responsible for the specific supervision of the trade craft operatives within their specialist field of activity.

Building craft operatives: The specialist construction workers. They include: carpenters, plumbers, bricklayers, scaffolders, electricians, plasterers, wall tilers, roof tilers, painters and decorators, etc.

Plant operators: Operatives who undertake the role of operating heavy machinery from cranes to diggers and earth movers or power access equipment.

Groundworkers and labourers: Employed on large sites to undertake the less specialised tasks, to include digging holes and demolition works.

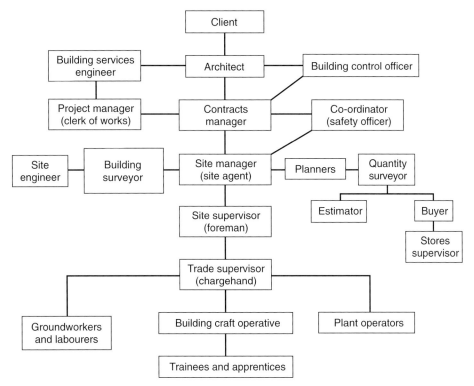

Health and Safety Legislation

Health and Safety at Work Act 1974

This is the Act of Parliament which provides the framework for all subsequent safety legislation. The Act involves everyone: employers, employees, self-employed, managers, representatives, manufacturers, etc., in matters of health and safety. Failure to meet the requirements of this Act, together with other health and safety regulations carried out under the provisions of this Act, constitutes a criminal offence and may lead to prosecution by the Health and Safety Executive (HSE).

One of the main features of this Act is to make everyone responsible for site safety, thus ensuring the safety of themselves and others. Employers must provide and maintain the plant, and make arrangements for safety, information, instruction, training and supervision, etc., as necessary, to ensure the health and safety of their employees. Where a company employs more than five staff it would be a requirement that a safety policy was produced, to include specific risk assessments for work activities, with periodic reviews, thus ensuring control measures are put in place to ensure safety.

There are 50–60 Acts and over 400 Regulations relevant to health and safety at work. It is these regulations which identify the minimum standard for safety required by law. The following are just a few of the regulations applicable:

Workplace (Health, Safety and Welfare) Regulations 1992 These Regulations cover the working environment and the facilities in which you work. They are not applicable to domestic premises. Specific points include:

- A minimum inside working temperature should be maintained.
- All areas must be kept clean and tidy, with no risk to health and safety.
- There must be sufficient space available to work safely.
- Measures must be put in place to ensure no object or items can fall.
- For outside working, protection must be provided from adverse weather.
- Appropriate emergency stop buttons are located and readily identifiable.
- Sufficient and suitable toilets, with hot and cold water, soap and towel need to be provided, as does a supply of potable water, for drinking purposes.

Construction (Design and Management) Regulations 2007 (CDM) The CDM Regulations have placed the duty for site safety firmly in the lap of the client, designer and main contractor. It is these people who can ultimately contribute to health and safety. Their aim is to radically improve the industry safety record by promoting an integrated team, working from the beginning to the end of the project, thus improving planning and management from the very start. These Regulations have created the role of a duty holder, referred to as the 'Safety co-ordinator', whose job it is to oversee the overall site safety management throughout all stages of the construction project (under the previous CDM Regulations this role was referred to as the planning supervisor). If no appropriate safety co-ordinator appointment is made, the client will be deemed to have appointed themself and be directly accountable in matters of site safety. The Regulations require the contract to maintain a health and safety file, which includes the various health and safety plans and risk assessments. The new CDM Regulations not only replace the old CDM Regulations, but also encompass the old Construction (Health, Safety and Welfare) Regulations of 1996. These new Regulations also strengthen links to about 21 other health and safety legislations, including the Management of Health and Safety at Work Regulations 1999. The Regulations apply to most of the larger construction contracts. They are not, however, applicable to demolition contracts or domestic projects and small works lasting no longer than 30 days or involving no more than four people. It is a requirement that employees must undertake the precautions as stipulated by their employer and, as stated earlier, they have a duty to carry out their work activities in a safe manner. This includes co-operating with others on matters relating to health and safety, and reporting any situations that might present a serious risk.

Manual Handling Operation Regulations 1992 These Regulations cover the human effort involved in handling manual loads. This includes direct and indirect (by means of a lever or through straining on a rope) lifting. Basically, these Regulations give guidance for employers and employees on the correct and safest methods in manual handling operations to avoid injury, usually to the back. The maximum load a reasonably fit person can be expected to lift is 20–25 kg, however this load may need to be substantially reduced for women and smaller-framed men, or where the load is awkward in shape or manoeuvreability is restricted. It is essential to maintain a straight back at all times when lifting. See the following section for an example of the correct posture to adopt when lifting any object.

Reporting of Injuries, Diseases and Dangerous Occurrence Regulations 1995 RIDDOR, as it is commonly known, requires notification of all major injuries resulting from work activities and dangerous practices to be made to the relevant enforcement authority (e.g. HSE). Failure to do so would be a criminal offence.

Typical site entry sign

Safe Working Practices

The vast majority of safe working practices are common sense. All too often the operative knows they are taking a risk, saying to themselves 'it will only take a minute', or 'I've only got this little bit to do'. Unfortunately, this action may result in an accident, causing damage to the property, personal injury or injury to others. Many people today suffer as a result of such actions, including back problems, breathing problems and an array of disfigurements and disabilities. Once the accident has occurred it is too late!

Lifting and carrying (manual handling)

You should never lift a load that is too heavy for you. Even light loads can cause damage where insufficient room is available to manoeuvre sufficiently and a twisting action is used. When lifting, you should be able to approach the load squarely and be able to lift whilst keeping the back straight, using the legs, not the back, to do the work. The load should be close to the body; never attempt a load with the arms stretched out. In carrying a load, ensure you have good visibility and can see where you are going. Above all, do not try to act macho and demonstrate how strong you are, or long term damage of your back is something you may have to live with. See the note on page 13 in relation to the Manual Handling Operation Regulations 1992.

Trips and other hazards

Tools and materials left lying about, including trailing extension leads, welding hoses or spillages of oil, etc., can cause people to trip or fall over. Likewise, cardboard packaging, oily rags, etc., may cause a fire hazard. These hazards can be avoided and it is your duty to maintain your work area in a clean and tidy condition. It also leads to better working relations with colleagues.

Excavations

Two types of accidents can occur with excavations: either the trench itself collapses, or people fall into the trench. Therefore, always make sure the trench is well supported, keeping heavy loads away from its edges, and erecting barriers around the excavation where necessary. When using liquefied petroleum gas (LPG), such as propane, never leave the bottles in or around the trench, because if the gas were to leak, it would fall (being heavier than air), filling the trench, and an explosion may result.

Most people think accidents will not happen to them or that they are not at fault. Anyone who sees a hazard and does nothing about it is making an accident more likely. Make safe, or report dangerous situations.

Risk assessment

A risk assessment is a method employed where, prior to undertaking a specific task, the likelihood of an accident occurring and the consequences of such an accident are recorded. It would list the control measures put in place and identify all those at risk. Most, if not all, activities involve some risk; the significance of the risk determines the level of risk assessment completed. Where a major accident occurs, these risk assessments are used to determine if everything that could have been done to prevent the accident was done. Failure to complete a suitable risk assessment may result in prosecution.

incorrect
(lower back strain)

correct
(legs bent and
straight back)

Lifting heavy loads

Completed risk assessment forms

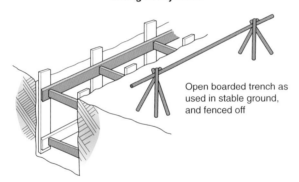

Open boarded trench as
used in stable ground,
and fenced off

Lifting heavy loads

Poor working practice climbing on incomplete scaffold, no boots, hard hat not fitted correctly and no high-visibility jacket. An accident waiting to happen!

Personal Protective Equipment

Sooner or later you will need to use protective clothing and equipment. Overalls and safety footwear may be provided by the employer, or you may have to buy these items yourself. Employers have a legal duty to provide all other protective equipment free of charge and the employee must use it correctly and report any defect or damage. Visitors to the site or other workers are also entitled to the same protection.

The Construction (Head Protection) Regulations 1989

The law requires the use of suitable head protection on all building sites unless there is no risk of head injury other than by the person falling. Safety hats should be adjusted to fit correctly; your failure to make the correct adjustment may mean that you are not providing the necessary level of safety.

The Personal Protective Equipment at Work Regulations 1992

Within these Regulations will be found eye protection and full face protection. Safety glasses, goggles or eye shields must be worn where there is any foreseeable risk of eye injury. Eye injury can result from:

- The use of power tools   (drilling, grinding and threading)
- Hammering and driving tools   (cutting, chipping and chiselling)
- Flying particles   (dust and chemical splashes)
- Welding processes   (sparks and molten splashes)
- Glare from light   (electric arc welding)

The Control of Noise at Work Regulations 2005

Sound levels are measured in decibels (dB). Employers are required, as far as possible, to keep noise levels below 80–85 dB; where above this limit, suitable ear protectors must be provided and the work area designated an ear protection zone.

Protection of skin and hands

Your skin, and particularly your hands, should be protected not only from cuts and abrasions but also from materials and substances which on contact can lead to infection. Skin conditions such as industrial dermatitis and skin cancer can be the result of neglect.

The two main types of gloves available are: (1) hide, leather or similar materials, for damage due to roughness, heat, etc., and (2) PVC, rubber or Neoprene to prevent damage due to contact with chemicals, cement, oils, etc.

Barrier creams can be used to give minimum protection. Hands should be washed regularly to prevent illness due to ingesting toxic substances such as lead and copper, which may be absorbed on the skin.

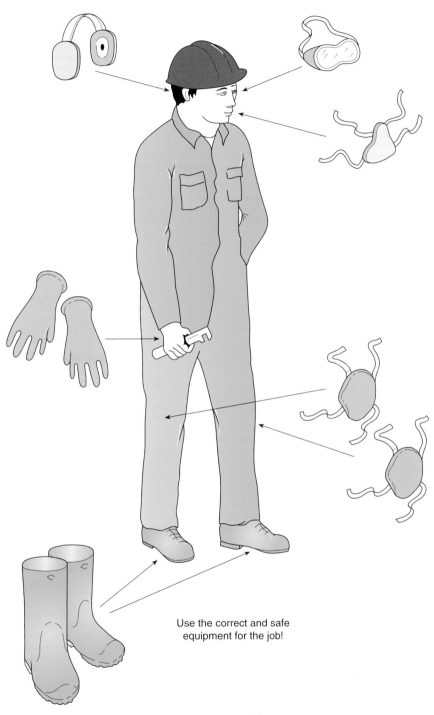

Use the correct and safe
equipment for the job!

Protective clothing

Safety Signs

Relevant British Standard
BS 5499

The Health and Safety (Safety Signs & Signals) Regulations 1996
This Regulation put into practice the European Safety Signs Directive, designed to standardise signs across Europe. There are five main types of safety sign that will be encountered in all places of work, that need to be observed, each identifying specific actions to be taken as appropriate. As you enter a large construction site there is usually a hording displaying an assortment of different safety signs, such as the one shown on page 13. These signs can be categorised as follows:

- **Safe condition** (formerly called an information sign): A green rectangular or square sign with white symbols and possibly some supplementary writing underneath. Note, text-only signs no longer comply. GREEN MEANS GO, therefore showing:
  - The safe way
  - Where to go in an emergency.
- **Fire safety**: A red rectangular or square sign with white symbols and possibly some supplementary writing underneath. RED MEANS FIRE FIGHTING, therefore showing:
  - Location of fire fighting equipment
  - Contents of the fire extinguisher
  - Fire extinguisher to use
  - How the equipment should be used.
- **Hazard** (formerly called a warning sign): A black, yellow-filled, triangle with a black symbol inside the triangle, possibly with some supplementary text underneath. YELLOW MEANS RISK OF DANGER, therefore showing:
  - Caution
  - Risk of danger
  - Hazard ahead.
- **Mandatory**: A solid blue circle with a white symbol, possibly with some supplementary text underneath. BLUE MEANS OBEY, therefore showing:
  - What you must do
  - Carry out the action given by the sign.
- **Prohibition**: A red circle with a red diagonal line. A black symbol is added inside the circle, possibly with some supplementary text underneath. RED MEANS STOP, therefore showing:
  - You must not
  - Do not do
  - Stop.

In addition to the above signs are work traffic signs, however these are basically the same as the signs used for public roads.

wear head
protection

wear eye
protection

Mandatory

first aid

indication
of direction

Safe condition

no smoking

pedestrians
prohibited

Prohibition

risk of
electric shock

general risk
of danger

Hazard

Location of fire
alarm

Location of fire
extinguisher

Fire safety

Typical safety signs in place

Control of Substances Hazardous to Health

There are many materials that we use every day that have the potential to harm. These include dust produced from various work activities, through to chemicals such as fluxes and drain cleaning solutions, which give off vapours and fumes. Materials such as lead, copper or cement are also hazardous as absorption of their toxic particles may occur. All the materials that we come into contact with may bring about injuries, such as burns or a skin irritation, which may go after a short period. However, long term damage to our lungs bringing about breathing problems, or to our skin causing dermatitis, can occur, both of which may be irreversible. It is essential, therefore, that the appropriate precautions are taken, where necessary, and work stopped where health may be at risk.

Control of Substances Hazardous to Health Regulations 2002 (COSHH)
These Regulations, first introduced in 1988, provide the legal framework for the control of all substances used in the workplace. It requires employers to:

- Undertake a risk assessment of the potential hazards to health, identifying what precautions are to be put in place to prevent exposure.
- Put in place control measures to ensure that safe procedures are observed and equipment is properly maintained and adequately provided.
- Inform employees of the risks, and train, where necessary, to ensure the appropriate safety precautions are followed.
- Maintain a list of all the appropriate COSHH safety information sheets (supplied by the manufacturer) for all hazardous substances used.

Working with chemicals: Working with any chemical will have particular dangers which need to be identified in the risk assessment. The manufacturer's packaging will also include essential safety information that needs to be observed. The four groups of chemicals to be found include:

- **Corrosives:** These are irritants that can be dangerous, especially to your eyes and respiratory system. Examples include acids and alkalies.
- **Flammables:** Liquids and gases that burn readily when supported by the correct balance of oxygen and heat. Examples include LPG and petrol.
- **Reactives:** This is the result of two or more substances reacting with one another, forming an explosive or gaseous compound.
- **Toxic agents:** These are poisons, which may cause injury or death.

Working with lead: As a plumber you may spend a good deal of your time handling this material. It is absorbed into the blood stream when ingested via the mouth through inadequate washing before eating, and breathed in via dust and vapour particles in the air, possibly when lead welding. Note, solid lead is not absorbed through the skin. Continued uncontrolled exposure over the long term may cause kidney or brain damage. Female plumbers need to take particular care where they may be pregnant as a developing fetus is at particular risk. When working with lead you should have your blood levels checked by a doctor every three months, especially if you are under 18. Where you may be exposed to lead, your employer has a duty to ensure systems are in place to protect your health.

Dangers of asbestos Currently around 3000 people a year die as a result of an asbestos-related disease. The first signs of the disease may take between 15 and 60 years to develop from the time of exposure. There is no known cure, therefore a high level of caution must be observed when working in the presence of this material. Asbestos materials in position and intact do not pose a risk to health. It is the tiny asbestos fibres that can pass into the lungs when you breath in dust that are dangerous. Asbestos cannot be absorbed through the skin. Whenever you encounter any possible asbestos-containing materials on site you must *stop work* until the material has been suitably identified and, where necessary, removed, usually by an HSE-licensed contractor.

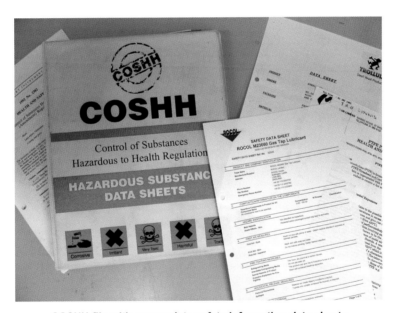

COSHH file with appropriate safety information data sheets

Access Equipment and Working at Heights

Working at any height can be dangerous, either by falling, causing personal injury, or by dropping things from the elevated position onto someone working below. Particular care should, therefore, always be observed.

Work at Height Regulations 2005

This Regulation is designed to reduce the high level of injuries that result every year from operatives falling from work platforms and ladders. It places responsibility on employers and self-employed people to ensure that equipment is safe to work on. Schedule 7 of these regulations requires the person who inspects the equipment to prepare a written report identifying its safe condition and what safety measures are put in place.

Ladder work: When using a ladder, check it over to see that it is in good condition. The stiles (not the rungs) should be secured to the building or platform when in use. A ladder should be stood on a firm, level base and be used at an angle of approximately 75° (ratio: 4 up to 1 out). Above all, when working from a ladder you should never over reach. Where a ladder is used to gain access to a roof or working platform, at least 1 m, or five rungs, should be above the access point. At all times when ascending or descending you should have both hands free to grip the rungs or stiles. Specific points to look for include:

- Split or cracked timber stiles (upright section) or rungs (treads)
- Tie rods missing or bent (located beneath the rungs of wooden ladders)
- Temporary repairs. Note, painted ladders often hide defects
- Mud on the rungs, possibly causing you to slip
- When fully extended there should be three runs overlap at each extension
- Never use a ladder resting against a plastic gutter or fragile surface, as it could suddenly give way, resulting in an accident.

Step ladders should be used on a firm level base, with all four feet in contact with the ground and the ladder opened to the full extent of the rope or hinged bracket. You should not use the top steps unless the steps have been specifically designed with an extended hand rail. Due to their electrical conductivity, aluminum ladders or steps should not be used in close proximity to electrical work.

Working platforms: Any work above 2 m requires the use of a purpose-designed platform, such as a mobile tower. It may be that a full independent or putlog scaffold is erected, which can only be assembled by an appropriately qualified operative. Platforms that have to be erected for some time need to be inspected weekly, and after adverse weather conditions, to ensure they remain in a safe condition. However, on every occasion you should always do a visual inspection to identify any obvious defects. Above all, never alter or work from ineffective scaffolding platforms. Specific defects to look for include:

- Missing components, e.g. toe boards or guard rails
- Poor assembly, e.g. loose, overlapping or protruding boards
- Damaged scaffolding, e.g. split boards and bent and rusty poles
- Unstable scaffolding, e.g. no bracing, no tie-ins and no base plates
- Obstructed or overloaded scaffolding
- Poor ground stability, or wheel breaks not on for mobile towers.

One of the greatest risks of working at height is complacency, especially when working at fairly low heights on a regular basis. One must always be aware of the risk of falling; it only takes one fall to ruin your life.

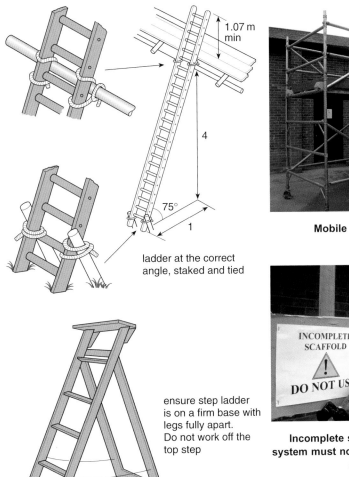

1.07 m min

4

75°

1

ladder at the correct angle, staked and tied

Mobile tower

ensure step ladder is on a firm base with legs fully apart. Do not work off the top step

Incomplete scaffold (this scaffold system must not be used until complete and safe)

Safe Working with Hand Tools and Machinery

Hand tools

Hand tools should always be maintained in a good, clean, safe condition and only used for their intended purpose, for example the metal that a file is made of is very brittle and should not be used as a makeshift chisel as a splinter may break away and cause injury. Toothed wrenches should not be used where a flat face spanner is required and, where used, should not be extended by inserting them into a pipe for increased leverage. Specific things to look for include:

- Handles: These need to be firmly fitted and not show any signs of damage, such as splinters. The unprotected tang of a file or chisel head could cause a serious puncture wound if not suitably protected.
- Cutting tools: These need to be kept in a sharp condition in order to avoid the operative having to use brute force to do the job.
- Mushroomed heads on chisels need to be removed in order to prevent bits flying off as the tool is struck.

Power tools

The modern site utilises an array of different power tools, from drills and grinders to threading machines and saws, all of which have the potential to harm the user. It is therefore essential that you have had adequate training on the operation of such machinery to ensure that you or those around you are not harmed by your actions. These tools may use electrical or battery power. Where electrical power is used this should ideally be 110 volts to ensure that in the event of an electrical fault a reduced voltage is transmitted. Construction sites only permit the use of 110 V tools. However, this is not the case where you may be working in a domestic premises. When using mains voltage power tools you should always operate these via an RCD (residual circuit device), which would break the circuit in the event of a fault. The following identify a few points to look for when using a power tool:

- When using power tools ensure that they are PAT* tested, with no frayed or damaged cables
- Electric tools should be suitably earthed or double insulated
- Use safety goggles when undertaking grinding or drilling operations
- Always use battery-powered or 110 V tools where possible
- Keep trailing extension leads out of the way.

Cartridge-operated tools

These tools use explosives to fire fixings into walls, etc. They are potentially quite dangerous in the wrong hands. Under health and safety law, no one under the age of 18 is permitted to use this type of tool and you must have received specific training and instruction in their use.

---

*PAT testing refers to Portable Appliance Testing. All portable electrical tools and equipment need to be certified safe and labelled up and dated as having been inspected and tested.

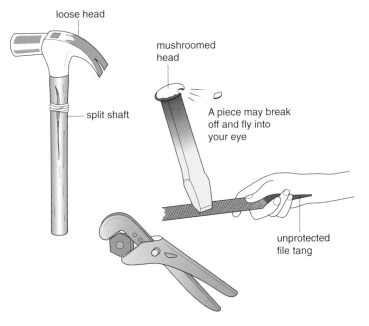

loose head

mushroomed head

split shaft

A piece may break off and fly into your eye

unprotected file tang

Unsafe and inappropriate use of tools

Impulse cartridge-operated nail gun

Typical 110-volt power tools

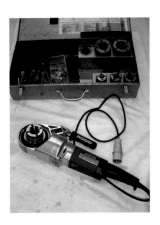

Fire Safety

Relevant British Standards
BS 7937 and BS EN 3

The Fire Precautions (Workplace) Regulations 1999
These Regulations were amended in 1999 to provide a common European method of fire fighting recognition. Prior to these Regulations, extinguishers were colour-coded (e.g. black indicated CO_2) to identify their contents. Today, however, all modern fire extinguishers are painted red, with only a 5% coloured stripe to indicate their contents (see chart opposite). The change was designed to prevent colour blind people selecting the wrong extinguisher, requiring closer examination to identify the contents. The Regulations impose a minimum set of standards and precautionary measures to be put in place in the working environment. Any person who employs staff is responsible for complying with the requirements of these regulations.

Generally, the most important point to consider is that you have made suitable provisions to deal with an outbreak of fire should one occur. In fact, it is likely your insurance cover would be invalid if you had not done so. Always check that nothing is left smoldering on completion of your work. In the event of a fire, gas and electrical supplies should be turned off as quickly as possible and combustible materials moved out of reach of the fire. It is essential that you do not tackle the blaze unless you have been suitably trained in the use of the fire-fighting apparatus. Using the incorrect type of extinguishing agent could make things worse.

Hot Works Certificate
Many construction sites today restrict the use of using a naked flame and sometimes it can only be used with prior notification to the safety co-ordinator, who will issue a 'Hot Works Permit'. Specific requirements would be identified when issuing such a certificate, including adequate fire prevention and remaining on the job at least one hour after the flame has been extinguished to ensure that no fire breaks out.

Extinguishing a fire
There are several methods of extinguishing a fire, and in all cases it involves removing one of the three essential requirements of combustion. (See combustion triangle opposite). Therefore, cool a piece of burning wood below its ignition temperature and it will go out; this may be achieved simply by cooling it with water.

How to effectively use a fire extinguisher
Water: Water cools the burning material, therefore direct the jet of water at the base of the fire.

Foam: Foam forms a blanket over the surface and therefore smothers a fire. Aim the jet at the base of the flames and keep moving it across the area of the fire.

CO_2: This works by smothering the flames by displacing the air. Therefore, the discharge horn should be directed at the base of the flames, moving the jet across the fire. Note, CO_2 can be harmful in a confined space and care needs to be taken not to hold the discharge horn as you could get a frost burn.

Powder: This puts out the flames with some cooling and smothering effect. Point the jet at the base of the fire with a rapid sweeping motion.

Classification of a fire

Fires are classified into a basic group type, according to the fuel and how to extinguish the flame. There used to be a group E, which denoted electrical fires, however this has now been removed due to the fact that the electrical supply should always be switched off, thereby changing the fire to a different classification. The table below identifies the class and correct extinguishing method to use.

Class	Fire risk	Water	Foam	CO_2	Powder	Blanket	Wet chemical
A	Wood, paper and fabrics	✓	✓	✗	✓	✗	✓
B	Flammable liquids: petrol, oil, paint, etc.	✗	✓	✓	✓	✓	✗
C	Flammable gases	✗	✗	✗	✓	✗	✗
D	Flammable metals	✗	✓	✗	✓	✗	✗
F	Cooking oils and fats	✗	✗	✗	✗	✓	✓
Extinguisher colour stripe		Red	Cream	Black	Blue	Red	Gold

There used to be a green 'Halon' gas fire extinguisher, however these are now illegal to use as they destroy the ozone layer. They can be taken into a fire station to be disposed of.

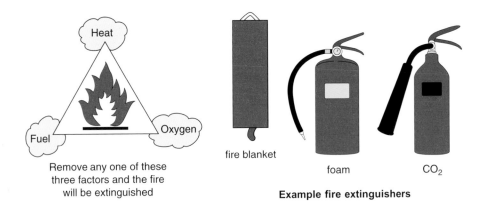

Heat

Fuel

Oxygen

Remove any one of these three factors and the fire will be extinguished

fire blanket

foam

CO_2

Example fire extinguishers

Regulations Governing Plumbing Work

The principal laws concerning plumbing works are the result of various Acts of Parliament, including the Water Acts, Gas Acts, Electricity Acts and Building Acts. These have resulted in the following 'Statutory' Regulations being implemented:

Water Supply (Water Fittings) Regulations 1999 These regulations have replaced the Local Water Bylaws, although there are currently no major changes to the requirements which were laid down in the Water Bylaws. The main purpose of these Regulations is to set down the requirements to be observed to prevent wastage, undue consumption, misuse, contamination of water or erroneous measurement.

Gas Safety (Installation and Use) Regulations 1998 These regulations are concerned with work on gas fittings, including pipework, appliances, ventilation and the extract of flue gases, to ensure that the public is protected from any dangerous situations which may arise. For example, Regulation 3.3 prohibits anyone carrying out gas work unless they are members of a class approved by the Health and Safety Executive (i.e. CORGI registered).

The Electricity Safety Quality and Continuity Regulations 2002 and the Electricity at Work Regulations 1989 These Regulations apply to all electrical equipment and installations. They require that in the event of an electrical fault, the supply must not give rise to danger but be isolated, or cut off. It is now illegal to work on live electrical systems unless there is no other way in which the work can be done. Compliance with the Institution of Engineering and Technology (IET) Wiring Regulations will, in general, satisfy the requirements of these Statutory Regulations although the IET Regulations themselves are not statutory.

The Building Regulations Building requirements differ, depending where in Britain the work is carried out, but in general the Building Regulations apply. These are administered by the local authority via the building control officer. The Building Regulations contain no technical detail; these can be found in a series of approved documents designated A to N (e.g. H1 deals with sanitary pipework and drainage). The purpose of the Building Regulations, among other things, is to conserve fuel and secure the welfare, health, safety and convenience of people in or around buildings.

Reporting of Injuries, Diseases and Dangerous Occurrences Regulations 1995 Usually referred to as 'RIDDOR', these Regulations put a legal responsibility on anyone finding a dangerous occurrence as a result of bad working practices or activities. The incident must be reported to the Local Authority environmental health department or local HSE within a certain limited time. For an example of its use, see page 239.

Control of Substances Hazardous to Health Regulations 2002   These regulations, usually referred to as 'COSHH', identify steps which should be observed when working with substances which may lead to a health risk. It is a duty of everyone working within industry to comply with these requirements.

British Standards   These are not statutory documents although they are constantly referred to by those seeking to meet the requirements of the law. They are methods of design and installation practice which should be followed whenever possible. Throughout this book, the relevant BS number is indicated at the beginning of most topics.

Examples of Regulations and British Standards governing plumbing work

Identification of Pipework

Relevant British Standards
BS EN ISO 11091, BS 1553-1 and BS 1710

In order to understand pipework drawings, one needs to be able to recognise a system of symbols which is standard across the country; this prevents unnecessary labelling and assists in the clarification of various details. The British Standards Institution has produced a series of symbols which is consistently used; however, some different symbols do exist, these usually being ones used prior to current British Standards or because no BS symbol can be found.

Sometimes one is faced with a series of pipes running along the wall of, for example, a boiler room, and the nature and/or contents of each would soon be forgotten if the pipework were not labelled. One could simply write the name of the pipe contents on the pipe or its lagging, but this might be somewhat time consuming and difficult to do neatly. Therefore, a system of colour coding has been designed, again conforming to a British Standard, to enable pipe contents identification.

Pipe contents identification chart

Pipe contents	Basic colour*	Specific colour†			Basic colour*
Untreated water	Green		Green		Green
Reuse water (e.g. rainwater)	Green		Black		Green
Drinking water	Green		Auxiliary blue		Green
Cold down service	Green	White	Blue	White	Green
Hot water supply	Green	White	Crimson	White	Green
Central heating <100°C	Green	Blue	Crimson	Blue	Green
Boiler feed	Green	Crimson	White	Crimson	Green
Chilled	Green	White	Em. green	White	Green
Fire extinguishing	Green		Red		Green
Condensate	Green	Crimson	Em. green	Crimson	Green
Steam	Silver grey		Silver grey		Silver grey
Natural gas	Yellow ochre		Yellow		Yellow ochre
Diesel fuel oil	Brown		White		Brown
Furnace fuel oil	Brown		Brown		Brown
Compressed air	Light blue		Light blue		Light blue
Drainage	Black		Black		Black
Electrical conduits	Orange		Orange		Orange

*150 mm (see bottom of figure opposite); †100 mm (see bottom of figure opposite)

Occasionally it will also be necessary to indicate the direction of flow; this is shown by an arrow situated in the proximity of the pipe contents identification code. With central heating pipework the word flow or the letter F is shown on one pipe and return or R on the other.

30

1 Key Principles

Pipework symbols

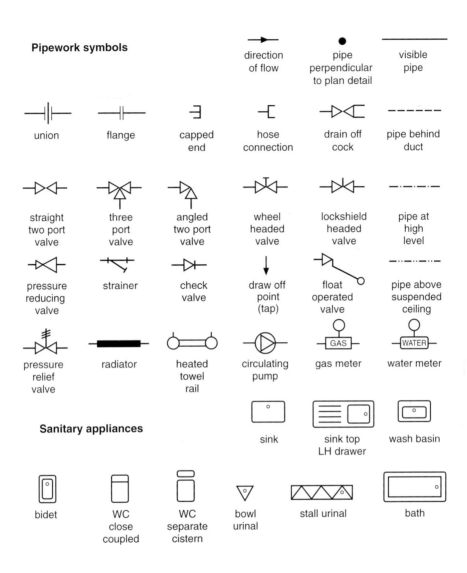

direction of flow — pipe perpendicular to plan detail — visible pipe

union — flange — capped end — hose connection — drain off cock — pipe behind duct

straight two port valve — three port valve — angled two port valve — wheel headed valve — lockshield headed valve — pipe at high level

pressure reducing valve — strainer — check valve — draw off point (tap) — float operated valve — pipe above suspended ceiling

pressure relief valve — radiator — heated towel rail — circulating pump — gas meter — water meter

Sanitary appliances

sink — sink top LH drawer — wash basin

bidet — WC close coupled — WC separate cistern — bowl urinal — stall urinal — bath

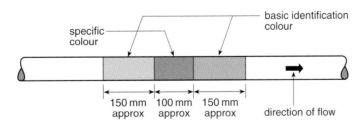

basic identification colour

specific colour

150 mm approx — 100 mm approx — 150 mm approx — direction of flow

Application of pipe contents identification colours

Identification of pipework

The International Metric System

Relevant British Standard
BS ISO 1000

The metric system was first introduced by the French National Assembly late in the sixteenth century; it was adopted by the British early in the 1970s. One of the main characteristics of the system is its decimal nature; the conversion from smaller to larger units and vice versa is thus made by moving the decimal point to the left or right.

The SI system of units (Système International d'Unités), developed from the metric system, has been defined and recommended as the system of choice for scientific use worldwide, used internationally by most nations. It should be noted that not all metric units are SI units, for example hectare and litre.

Examples of SI-derived units and British imperial equivalents

| Quantity | Metric | | Imperial | | |
	Unit	Symbol	Unit	Symbol	Conversion factor
Length	metre	m	yard	yd	0.9144
Area	square metre	m^2	square yard	yd^2	0.8361
Area	hectare	ha	acre	–	0.404686
Volume	cubic metre	m^3	cubic yard	yd^3	0.7646
Capacity	litre	l	gallon	gal	4.546
Mass	kilogram	kg	pound	lb	0.4536
Force	newton	N	pound-force	lbf	4.448
Pressure	newtons per square metre	N/m^2	pound-force per square inch	lbf/in^2	6894
Pressure	pascal	Pa		lbf/in^2	6894
Pressure	bar	bar		lbf/in^2	0.06894
Velocity	metres per second	m/s	foot per second	ft/s	0.3048
Temperature	kelvin	K	centigrade	°C	see opposite
Temperature	celsius	°C	centigrade	°C	see opposite
Energy	joule	J	British thermal unit (Btu)	Btu	1055
Power	kilowatt	kW	Btu per hour	Btu/h	0.0002931
Power	kilowatt	kW	horse power	hp	0.7457

To convert metric to imperial divide by conversion factor
To convert imperial to metric multiply by conversion factor
Note: $1\,Pa = 1N/m^2$; $1\,bar = 100\,kN/m^2$.

Examples of converting metric to imperial units:

(1) Express $15\,m^2$ in yd^2: $15 \div 0.8361 = 18\,yd^2$
(2) Express $17\,kW$ in Btu/h: $17 \div 0.0002931 = 58\,000\,Btu/h$

Examples of converting imperial to metric units:

(3) Express 50 gal in litres: $50 \times 4.536 = 227$ litres
(4) Express $14.7 \, \text{lbf/in}^2$ in N/m^2: $14.7 \times 6894 = 101\,342 \, \text{N/m}^2$

The metric system is designed in such a way that, by prefixing the required unit with one of the symbols identified in the following table, its value can be increased or decreased in multiples of ten, e.g. kilogram, centimetre.

Prefix	Symbol	Multiplication factor
atto-	a	10^{-18}
femto-	f	10^{-15}
pico-	p	10^{-12}
nano-	n	10^{-9}
micro-	mu	$0.000001 \ (10^{-6})$
milli-	m	0.001
centi-	c	0.01
deci-	d	0.1
No prefix (SI unit only, e.g. metre)		1
deca-	da	10
hecto-	h	100
kilo-	k	1000
mega-	M	$1\,000\,000 \ (10^6)$
giga-	G	10^9
tera-	T	10^{12}
peta-	P	10^{15}
exa-	E	10^{18}

If we put the prefix system to use we find examples such as the following expressions of derived units:

(1) $2 \text{ mm} = (2 \times 0.001) = 0.002 \text{ m}$ (2) $4 \text{ cm} = (4 \times 0.01) = 0.04 \text{ m}$

(3) $7 \text{ km} = (7 \times 1000) = 7000 \text{ m}$ (4) $4 \text{ hm} = (4 \times 100) = 400 \text{ m}$

Converting temperature scales   Three temperature scales are found in general use: the Fahrenheit, the Celsius (centigrade) and the Kelvin (absolute) scales. The Kelvin scale is measured in degrees above absolute zero (a hypothetical temperature, the lowest achievable, characterised by the complete absence of heat energy). $0 \, \text{K} = -273.15°\text{C}$; for every 1 K rise in temperature a 1°C rise in temperature is also experienced; for example, $0°\text{C} = 273 \, \text{K}$ and $1°\text{C} = 274 \, \text{K}$. Converting to and from the old Fahrenheit scale is achieved using the following calculations:

$$\text{Celsius to Fahrenheit} = (°\text{C} \times 1.8) + 32 = °\text{F}$$

$$\text{Fahrenheit to Celsius} = (°\text{F} - 32) \times 0.56 = °\text{C}$$

$$\text{Examples: } 21°\text{C} = (21 \times 1.8) + 32 = 69.8 \text{ F}$$
$$100°\text{F} = (100 - 32) \times 0.56 = 38.1°\text{C}$$

Areas, Volumes and Capacities

Plumbers frequently need to calculate areas, volumes and capacities. Given here are some standard examples.

Areas
Areas are the product of two linear dimensions; therefore, the answer is expressed in the unit squared, e.g. m^2 or mm^2.

Area	Formula	Example no.	Answer
Rectangle or square	L × B	1	$1.2 \times 0.7 = 0.84\,m^2$
Triangle	L × (D ÷ 2)	2	$1.2 \times (0.7 \div 2) = 0.42\,m^2$
Parallelogram	L × D	3	$1.2 \times 0.7 = 0.84\,m^2$
Trapezium	Average L × D	4	$[(3 + 1) \div 2] \times 0.7 = 1.4\,m^2$
Circle	$^*\pi R^2$	5	$3.142 \times 0.6 \times 0.6 = 1.13\,m^2$

$^*\pi = 3.142$, which is the number of times the diameter will go around the circumference of a circle.

Volumes
Volumes are calculated by multiplying the area of the vessel or space by the height of the void. Therefore the answer is expressed in the unit cubed, e.g. m^3 or mm^3.

Volume	Formula	Example no.	Answer
Room or tank	L × B × H	6	$1.2 \times 0.7 \times 0.6 = 0.504\,m^3$
Cylinder/pipe	$\pi R^2 \times H$	7	$3.142 \times 0.6 \times 0.6 \times 2.3 = 2.6\,m^3$

Sometimes volume is expressed as *capacity*: to convert a volume in units of m^3 to litres the value is simply multiplied by 1000. Thus in the last example, the cylinder with a volume of $2.6\,m^3$ would hold $(2.6 \times 1000) = 2600$ litres.

Because 1 litre of water has a mass of 1 kg we can further see that the cylinder contents would weigh 2600 kg or 2.6 tonnes.

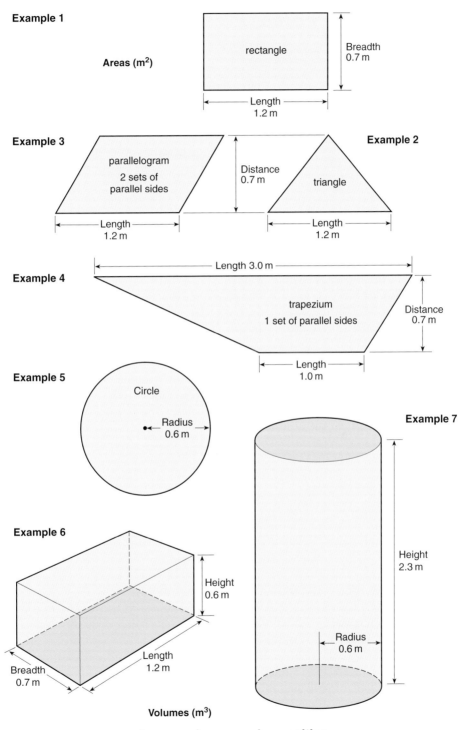

Example 1

Areas (m²)

rectangle

Breadth
0.7 m

Length
1.2 m

Example 3

parallelogram
2 sets of
parallel sides

Distance
0.7 m

Example 2

triangle

Length
1.2 m

Length
1.2 m

Example 4

Length 3.0 m

trapezium
1 set of parallel sides

Distance
0.7 m

Length
1.0 m

Example 5

Circle

Radius
0.6 m

Example 7

Example 6

Height
0.6 m

Length
1.2 m

Breadth
0.7 m

Height
2.3 m

Radius
0.6 m

Volumes (m³)

Areas, volumes and capacities

Mass, Weight, Force, Pressure and Density

Mass and weight

Mass and weight are often used interchangeably when we speak, but this is not correct. *Mass* is the reluctance of a body to change its motion when a force acts on it; it is measured in kilograms or tonnes. *Weight* is the force with which a body is attracted to another body, and is equal to the product of the object's mass and the acceleration due to gravity. So, weight varies as gravity varies and is measured in newtons.

Force and pressure

Force can be defined as any action that alters the position of a body. One newton is the force that gives a mass of 1 kg an acceleration of 1 meter per second per second (i.e. $1\,N = 1\,kg \times 1\,m/s^2$). The force produced by a 1 kg mass on Earth is 9.81 N.

Pressure is the force acting upon a given area: a force of one newton acting upon an area of $1\,m^2$ produces a force of one pascal ($1\,N/m^2 = 1\,Pa$). Sometimes pressure is expressed in other units, such as the bar ($1\,bar = 100\,000\,N/m^2 = 100\,kN/m^2$) or the pound-force per square inch ($1\,lbf/in^2 = 6894\,N/m^2 = 6.894\,kN/m^2$).

Atmospheric pressure

Atmospheric pressure at sea level is the weight of a column of air of unit cross-sectional area extending upwards indefinitely; the standard value is $101\,300\,N/m^2$. Because a gas is easily compressible, the air nearest the Earth's surface is denser than that higher up, so the atmospheric pressure at the top of a mountain is less than at sea level.

Density

Density = mass ÷ volume. It is measured in kilograms per metre cubed (kg/m^3). $1\,m^3$ of water has a mass of 1000 kg at 4°C, but only 967 kg at 82°C because the molecules are packed more closely together at the lower temperature and the liquid is therefore denser. Peculiar to water, when cooled below 4°C water begins to expand again, so it is at its maximum density at 4°C.

Relative density (formerly known as specific gravity)

Relative density is the ratio of the density of a substance to the density of a reference substance. For solids and liquids, the reference substance is usually water. For example, the mass of $1\,m^3$ of water at 4°C is 1000 kg, whereas that of $1\,m^3$ of lead is 11 300 kg; therefore the relative density of lead is 11.3 – lead is 11.3 times more dense than water. Water has a relative density of 1.0: any material with a relative density greater than 1.0 will sink in water and any with a relative density less than 1.0 will float. For gases, the reference substance is usually air, at standard temperature and pressure. By calculating relative densities we can compare materials with each other: from the following table it can be seen that cast iron is less dense than tin, and that natural gas is less dense than air, so will rise upwards should there be a leak.

Relative density at atmospheric pressure

Solids and liquids	Relative density	Gases	Relative density
Water	1.0	Air	1.0
Class C fuel oil	0.79	Methane (natural gas)	0.6
Linseed oil	0.95	Propane	1.5
Aluminium	2.7	Butane	2.0
Zinc	7.1		
Cast iron	7.2		
Tin	7.3		
Mild steel	7.7		
Copper	8.9		
Lead (milled)	11.3		
Mercury	13.6		
Gold	19.3		

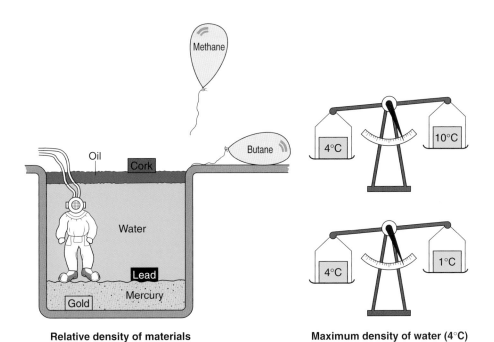

Relative density of materials

Maximum density of water (4°C)

Mass/weight and pressure

Water Pressure

There are two basic means of creating pressure in any plumbing system: (1) by means of a pump connected to a pipeline, or (2) by means of the weight of the water itself. For water in a container the pressure at a point below the surface is proportional to the depth, i.e. the greater the head the greater the pressure (see diagram). A container of cross-sectional area 1 m² and 1 m high would contain 1000 litres of water of mass 1000 kg; therefore the 1000 kg of water would weigh 1000 × 9.81 = 9810 newtons per square meter (9.81 kN/m²).

Example: The pressure at the base of a 4 m high container of water with a cross-sectional area of 1 m² is 4 × 9.81 = 39.24 kN.

The higher a column or head of water, the greater the pressure at its lowest point; therefore it is essential to install feed and storage cisterns as high as possible, giving a good pressure at the draw-off points. As can be seen from the diagram, the pressure at tap A is much greater than at top B; the volume of the feeder tank has no effect on the pressure at the tap.

The pressure found so far is known as the intensity of pressure, i.e. the pressure over 1 m². If the total pressure needs to be found acting upon an area larger or smaller than 1 m² the intensity of pressure simply needs to be multiplied by the area acted upon.

Example: The intensity of pressure at the base of a boiler which measures 500 mm long by 300 mm wide, fitted 2.5 m below the water level in a storage cistern (see diagram) is: 2.5 × 9.81 = 24.525 kN/m² (24.525 kPa). The total pressure = 24.525 kN/m² × 0.5 × 0.3 = 3.6788 kN.

To find the height to which water would rise in a water main, the equation can be rearranged:

$$\text{Intensity of pressure} = \text{head} \times 9.81$$

so when rearranged,

$$\text{Head} = \text{intensity of pressure} \div 9.81$$

Example: The mains water pressure is 400 kN/m² (4 bar). Ignoring any frictional resistances, the height to which the water will rise in a vertical pipe = 400 kN/m2 ÷ 9.81 = 40.77 m.

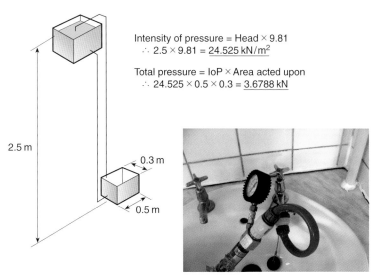

Intensity of pressure = Head × 9.81
∴ 2.5 × 9.81 = 24.525 kN/m²

Total pressure = IoP × Area acted upon
∴ 24.525 × 0.5 × 0.3 = 3.6788 kN

2.5 m

0.3 m

0.5 m

Using a pressure gauge to find the supply pressure

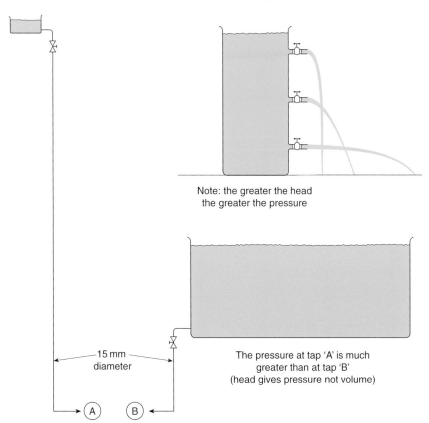

Note: the greater the head
the greater the pressure

15 mm
diameter

The pressure at tap 'A' is much
greater than at tap 'B'
(head gives pressure not volume)

A   B

Water pressure

Physical Properties of Materials

All materials have unique physical properties.

Boiling point Temperature at which the vapour pressure of a liquid is equal to standard atmospheric pressure.

Density The mass of molecules per unit volume; units are kg/m^3.

Ductility (plastic deformation under tension) The ability of a material to withstand distortion without fracture, e.g. a metal drawn out into a fine wire.

Durability The ability to resist wear and tear, i.e. long lasting.

Elasticity The ability of a material to return to its original shape after being distorted.

Fusibility The ability of a material to be melted.

Hardness The ability of a material to resist penetration or wear. The hardest natural substance is diamond.

Malleability (plastic deformation under compression) The ability of a material to be worked without fracture.

Melting point Temperature at which a solid changes to a liquid.

Plasticity The opposite of elasticity: a material which does not return to its original shape after being deformed.

Relative density (formerly called specific gravity) For solids and liquids this is the density of a substance compared to the density of water. Water has a relative density of 1; materials greater than 1 will sink in water.

Specific heat capacity The amount of heat required to raise the temperature of 1 kg of material by 1 K.

Tensile strength The force required to break a uniform rod of unit cross-sectional area by a straight pull; it is measured in N/m^2.

Thermal expansion The amount a material expands when heated.

Metals

Pure metals, examples of which are given in the table, are too soft and not strong enough for most engineering purposes. An alloy is a mixture of two or more metals or a metal and a non-metal; alloying is often used to increase the useful properties, e.g. strength and hardness.

Ferrous metals contain iron and, except for stainless steel, will rust on exposure to water, carbon dioxide and oxygen; *non-ferrous metals* do not contain iron. Ferromagnetic substances, e.g. iron, cobalt, nickel and their compounds, can be magnetised.

Physical properties of common plumbing metals

Metal	Chemical symbol	Melting point (°C)	Boiling point (°C)	Density (kg/m³)	Relative density	Specific heat capacity (kJ/kg K)	Tensile strength (MN/m²)
Aluminium	Al	660	2467	2705	2.7	0.887	81
Copper	Cu	1083	2567	8900	8.9	0.385	308
Iron	Fe	1535	2750	7860	7.8	0.554	278
Lead	Pb	327	1740	11300	11.3	0.125	16
Tin	Sn	232	2260	7310	7.3	0.226	23
Zinc	Zn	419	907	7130	7.1	0.397	170

Some common alloys

Alloy	Main elements
Brass	Zinc and copper
Bronze	Copper and tin
Cast iron	Iron and 2–4% carbon
Gunmetal	Copper, tin and zinc
Mild steel	Iron and approximately 0.2% carbon
Stainless steel	Iron, chromium and nickel
Solder	Lead and tin
Solder (lead free)	Tin and copper

Plastics

Plastics are man-made materials and can be divided into two broad groups: thermoplastics and thermosetting plastics. A thermoplastic, when heated, will soften; in this state it can be formed to the shape required and on cooling will harden. If required it can be reheated to form a new shape, without any significant changes in properties. A thermosetting plastic, however, can be heated initially to soften and form the shape, but when it cools, the shape is permanent and no reheating will soften it again.

- **Thermoplastics**   These include acrylics (Perspex), nylon, polythene, polypropylene and polyvinyl chloride (PVC).
- **Thermosetting plastics**   These include Bakelite, Formica, melamine and polyester.

Heat and its Effects

Temperature and heat

The temperature of a body and the heat contained in that body are two different things. *Temperature* can be measured (using a thermometer or thermocouple) in degrees celcius (°C) or in kelvin (K), thus giving a value for the *kinetic energy* of the molecules in the body. The *heat* contained in a body is the sum of the kinetic energies of all its molecules, i.e. its *internal energy*, and is measured in joules (J).

1 joule (J) = power supplied by 1 watt (W) in 1 second

Internal energy depends upon the mass of the material: 500 litres of water at a temperature of 65°C will have 100 times the internal energy of 5 litres of water at this temperature, and 100 times more heat energy can be obtained from it.

Specific heat capacity

Materials vary in the quantity of heat needed to produce a given temperature rise in a given mass. The specific heat capacity is the quantity of heat required to produce a temperature rise of 1 K (i.e. 1°C) in 1 kg of material. For example, it would require 4.186 kJ of energy to raise the temperature of 1 kg of water by 1°C, but only 0.385 kJ to raise the temperature of 1 kg of copper by 1°C.

Specific heat capacities

Substance	kJ/(kg K)	Substance	kJ/(kg K)
Water	4.186	Lead	0.397
Aluminium	0.887	Copper	0.125
Cast iron	0.554	Zinc	0.385

Physical change

When a substance converts from solid to liquid (fusion) or from liquid to gas (vaporisation) it is said to change state. For example, water (liquid) changes to ice (solid) upon cooling to 0°C, or to steam (gas) by heating to 100°C. In all three states (solid, liquid and gas) it remains H_2O, and the changes are reversible. In the same way, lead melts if heated to 327°C and vaporises if heated further to 1740°C.

Heat transfer

Heat energy can be transferred in the following ways:

- **Conduction:** The transfer through a material of heat energy down a temperature gradient.
- **Convection:** In liquids and gases, heated molecules move more vigorously and thus take up more space than cooler molecules, so hot fluids are less dense than cold, and are therefore pushed upwards by denser fluids.
- **Radiation:** The energy produced by a source, e.g. the sun, because of its temperature, which travels as electromagnetic waves.

Calorific value

Calorific value is the amount of heat produced by a specific amount of fuel; for example, the amount of heat produced by burning 1 kg of wood is less than that produced by burning 1 kg of coal. Natural gas has an even higher calorific value.

The calorific value of gases is usually expressed as the amount of heat energy in joules they contain per cubic metre (m³), for solid and liquid fuels it is given in joules per kilogram, while for electricity it is expressed as joules per kilowatt hour.

Calorific values

Fuel	Calorific value
Anthracite (coal)	32 MJ/kg
Wood	19 MJ/kg
Domestic grade oil	45 MJ/kg
Liquefied petroleum gas	30 MJ/kg
Natural gas	38 MJ/m³
Electricity	3.6 MJ(kWh)

Thermal expansion

Most materials expand when heated. All substances are made up of molecules which, when heated, move about more vigorously and thus move further apart, which results in the materials becoming larger. When the material cools the molecules slow down and move closer together; thus the material gets smaller or contracts. The amount the material expands in length when heated can be simply calculated using the following formula:

Expansion = original length × temperature rise × coefficient of thermal expansion

The coefficients of thermal expansion for typical plumbing materials are given in the following table:

Material	Coefficient K
Plastic	0.00018
Zinc	0.000029
Lead	0.000029
Aluminium	0.000026
Tin	0.000021
Copper	0.000016
Cast iron	0.000011
Mild steel	0.000011
Invar	0.0000009

Example: Find the amount a 9 m long plastic discharge stack will expand due to a temperature rise of 24°C.

$$\text{Expansion} = 9 \times 24 \times 0.00018 = 0.039 \text{ m or } 39 \text{ mm}$$

Corrosion

Relevant British Standard
BS 7593

Corrosion is the change of a metal into a metallic compound resulting from a chemical reaction on its surface. Several types of corrosion can be identified, the most common being atmospheric and electrolytic corrosion and dezincification of brass.

Atmospheric corrosion (oxidation)
Corrosion caused by moisture and gases in the air. When oxygen mixes with the surface of iron or some steels it forms an iron oxide, commonly called rust. This rust falls away, exposing fresh metal underneath, and the process continues until the metal rusts away completely. Non-ferrous metals are attacked by gases such as carbon dioxide and sulphur dioxide, but the surface corrosion which occurs does not flake off like rust, and therefore acts as a skin on the surface (called patina) and protects the metal from further corrosion. The colour of patina is different from that of the metal itself, the most striking example being copper which turns green. In coastal areas there is a lot of salt in the air; the rain is therefore alkaline in these areas, and tends to destroy aluminium.

Electrolytic corrosion (galvanic action or electrolysis)
The dissolution of one metal (the anode) in the presence of another (the cathode) when connected together via an electrolyte (any liquid or moisture which carries electrically charged particles from an anode to a cathode), such as impure water. The rate of corrosion depends upon the electrolyte and the position of the metals in the electrochemical series. If the water is acidic or hot, the corrosion will be increased. The following elements are to be found in the electrochemical series:

- Copper
- Tin
- Lead
- Nickel
- Iron
- Zinc
- Aluminium
- Magnesium.

Note: Those high on the list will cause those lower on the list to dissolve, e.g. copper will make zinc dissolve.

Dezincification of brass
Brass comprises copper and zinc; dezincification is a form of accelerated electrolytic corrosion in which the zinc content of the alloy is converted into a basic carbonate, which may cause a blockage in the pipeline, leaving the fitting porous and brittle.

Occurrence of corrosion
To prevent electrolytic corrosion on sheet roofwork, one must avoid the use of mixed metals and never let metal roofs higher on the electrochemical series discharge

onto roofs made of metal lower on the list, e.g. run-off from lead onto aluminium will cause excessive corrosion.

Corrosion within plumbing systems occurs due to several design faults. To prevent atmospheric corrosion, never use 'black iron' pipes (i.e. not galvanised) if fresh water is to pass through the system, such as on a mains supply or hot and cold distribution pipework, because air will have been absorbed into the water when it fell as rain.

Black iron pipes and steel radiators can be used in heating systems without fear of atmospheric corrosion because, providing the system is designed correctly, once the water has filled the heating pipework, it simply circulates round the system and is never replaced. Thus the air is expelled from the water via its circulation, so corrosion cannot occur.

Corrosion of metals can be caused by materials such as new timbers made of red cedar, chestnut, fir, oak and teak, which can leach harmful acids, as do moss and lichen growing on roofs. Alkaline attack can result from the presence of fresh cement, lime and plaster.

Prevention of corrosion

Protective coatings The simplest way of protecting metals is to paint them or cover them with plastic. Iron is often dipped in zinc (a process known as galvanising) to give it a zinc coating which prevents the air coming into contact with the surface of the metal and attacking it.

Cathodic protection To prevent electrolytic corrosion, a metal is often incorporated within the system which will be dissolved before all others; this is known as the sacrificial anode. Examples include galvanising, or the rod of magnesium which is fixed into many hot storage vessels. Sometimes a piece of aluminium is placed into the cold storage cistern to give the same effect.

System design To reduce corrosion problems, systems need to be designed to prevent air entering the pipework. The most important consideration in vented hot water heating system design is to ensure that the water is filled via a feed and expansion (f & e) cistern separate from that for the domestic water distribution pipework to the sanitary appliances; otherwise the waters will mix in the storage cistern upon expansion.

Corrosion inhibitors These are used as a method of preventing corrosion in heating systems. A chemical is added to the primary heating circuit via the f & e cistern. This neutralises any flux residues, etc., which make the water acidic; it also coats the system with a fine film.

Cathode, as taken from a hot storage vessel, showing signs of corrosion

Fitting showing evidence of dezincification of brass

Plumbers' Tool Kit

The tools that a plumber would use throughout their career consist of a vast number of different implements, from screwdrivers and hammers through to test meters, flue gas analysers and welding equipment. Much of the specialist equipment is supplied by the employer, or hired as and when necessary. However, for the plumber travelling to a building site, travelling around undertaking small contracts, or for the operative just starting out in their new career, a complement of tools, such as those listed below, would suffice. The specialist lead working tools and extra large pipe cutters and wrenches, etc., identified over the following two pages would not be carried on a day-to-day basis.

General tool list
- Tool box or bag
- Handy size copper tube bender
- Various sized screwdrivers, both cross and straight head
- Claw hammer, or similar, and possibly a club hammer for heavy work
- Chisels, both for wood and brickwork
- Basin spanner
- 3 Various sized adjustable spanners
- 1–2 adjustable wrenches, e.g. 'stillson' wrench
- Large-framed hacksaw plus spare blades
- Junior hacksaw plus spare blades
- Wood saws (panel and pad and floorboard saws)
- Copper tube pipe cutters
- 2–3 various files and a rasp
- Side cutters and pliers
- Tape measure
- Plumb bob
- Large and small sprit levels
- Temporary bonding wires
- Stopcock key
- Blowlamp and spark igniter
- Cutting knife and tin snips small and large.

Additional extras
- Battery or electric drill and drill bits
- Goggles, ear defenders, gauntlets, high visibility jacket and hard hat
- Multi-meter
- Tap reseating tools and hole cutters
- Pressure and flow test kit
- Pointing trowel
- Dust sheet and dust brush.

The tools used by anyone should always be kept in a clean serviceable condition and metal tools should be oiled occasionally. Tools which are blunt invariably cause more damage than necessary and will clearly make it harder to work. It is essential to take tools out of service when they are damaged in any way and, to ensure good long service, the right tool should always be selected for the job. A good, well-stocked tool kit is a sign of a professional. A bucket with a few cheap, damaged, blunt tools thrown in suggests a 'don't care' attitude. Start as you mean to go on and you will go a long way.

Poor stock of tools, creating the wrong impression

Typical extra tools the professional may have

Well-stocked plumbers' tool kit

Specialist Hand Tools for Pipework

Pipe cutters

Apart from hacksaws, which obviously could be used, pipe cutters are of three basic types: roller, wheel and link cutters.

The **roller cutter** is the most commonly used, especially for smaller pipes such as copper tube; in fact the 'pipe slice' is now a very familiar tool, enabling easy access in awkward locations. The roller cutter consists of two rollers and one cutting wheel. The pipe cut is achieved by rotating the cutter fully around the pipe. The rollers prevent a burr from forming on the external pipe wall. An internal burr will, however, be created by using these cutters, and this must be removed, using either a round file or a deburring reamer.

The **wheel cutter** consists of three or four cutting wheels and is suitable for cutting pipe where a full 360° turning circle cannot be achieved. It does, however, have the drawback of leaving an additional external burr on the pipe.

Link cutters are used on large diameter cast iron pipes. When cutting cast iron, no burr is produced because the pipe actually breaks as a result of the even pressure applied and the brittle nature of the pipe wall.

Pipe wrenches and spanners

Among the most important of these tools are straight pipe wrenches, pipe grips, chain wrenches, basin spanners and adjustable spanners. The wrench has teeth which are used to bite into the pipe. The spanner has smooth-faced jaws which are designed to be used across the flats of nuts, thus preventing the damage and slipping which may occur using a wrench.

Pipe threading equipment

Sometimes a thread needs to be cut on steel pipework; this is achieved using dies. The dies are held in a die stock which can be of several designs and be either hand held or machine operated. The hand held design is commonly a drop head stock, in which a separate die head is used for each size of pipe; this is not so bulky as those requiring adjustment of the thread size.

Pipe bending equipment

Many designs of pipe bender will be found. In general, those used for copper pipe are of the lever design, the type used for smaller pipe sizes having fixed rollers. Where roller adjustment is required, incorrect pressure will have one of two effects: if the adjustment is too low, the distance between the rollers will be too small and cause excessive throating (flattening) to the finished bend; if it is too high the rollers will be too far apart and ripples on the pipe will be produced. When bending copper tube, full support of the pipe is required, which is achieved using a backguide. For mild steel pipes, a hydraulic bending machine will be used, with no backguide support necessary.

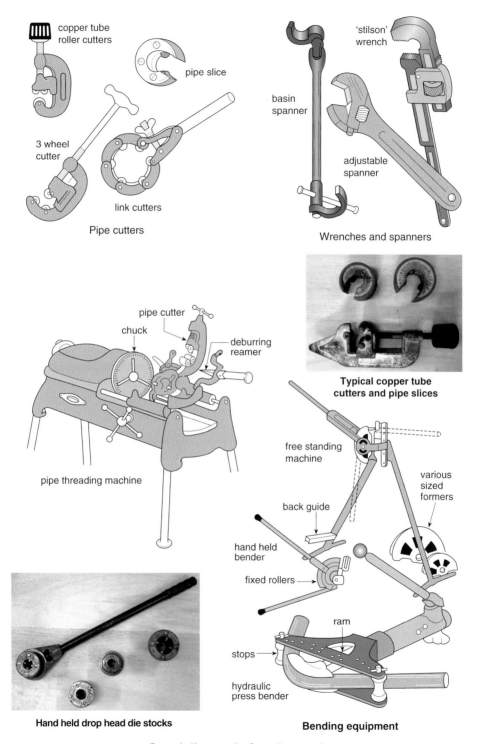

copper tube roller cutters

pipe slice

3 wheel cutter

link cutters

Pipe cutters

'stilson' wrench

basin spanner

adjustable spanner

Wrenches and spanners

pipe cutter

chuck

deburring reamer

pipe threading machine

Typical copper tube cutters and pipe slices

free standing machine

back guide

hand held bender

fixed rollers

various sized formers

ram

stops

hydraulic press bender

Bending equipment

Hand held drop head die stocks

Specialist tools for pipework

Specialist Hand Tools for Sheetwork

Many tools are used when carrying out the weathering to a roof using lead and copper sheet, etc. Among these are tin snips, mallets and hammers. The plumber also uses specialist tools which are made from suitable hardwoods such as box, beech or hornbeam, or, to avoid high cost, materials such as high density polythene. Typical tools include the following:

Flat dressers: used to dress the metal to a flat surface, making it lie flat without any undue humps.

Bossing tools: used to boss (form) sheet lead into the required shape. The tool used to boss the lead will be at the preference of the user and include the bossing mallet, bossing stick or bending stick. Today many plumbers use a rubber mallet.

Setting-in stick: a tool used to set in a crease in sheet lead or to reinforce and square up the metal on completion. The tool is struck with a mallet.

Chase wedge: used as a setting-in stick or to chase in angles and drips when used in conjunction with a drip plate.

Step turner: used to turn in the individual steps on step flashings to allow fixing into the brickwork.

Drip plate: a piece of steel plate, 100 mm × 150 mm × 1 mm thick, which is positioned between two pieces of lead being bossed close together; it allows one sheet to slide freely over the other.

Dummy: a home-made tool designed to assist in starting the corner when lead bossing; it is also used to assist in the bossing of difficult details and angles.

Lead knife: a special knife which is used to cut sheet lead. The lead is scored with the knife several times and pulled apart to give a neat, straight cut.

Shavehook: used to remove the thin layer of oxide on sheet lead before lead welding.

Seaming pliers: these are generally only used for the harder sheet materials such as aluminum, copper and zinc, where they are used to assist in folding and forming welts.

To give a long life to wooden tools, they should regularly be soaked in linseed oil and never struck or stored with steel tools. The edges should be kept slightly rounded to prevent damage to the sheet metal being worked. Tools used for lead work should be kept separate from those used for copper roofs.

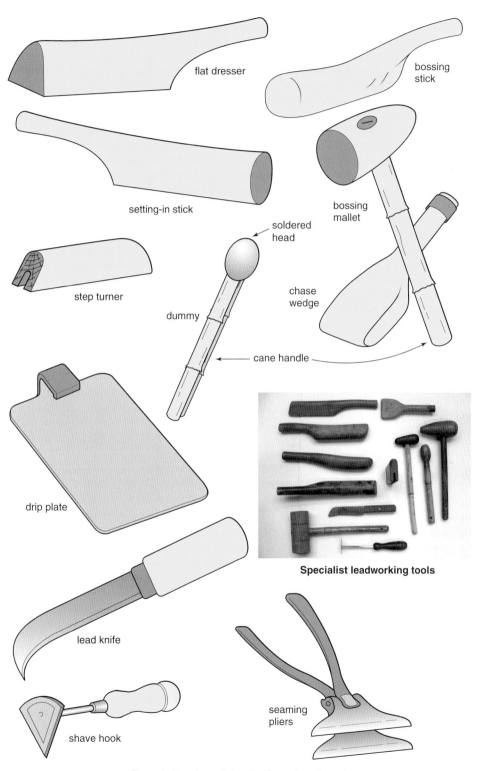

flat dresser

bossing stick

setting-in stick

bossing mallet

soldered head

step turner

dummy

chase wedge

cane handle

drip plate

Specialist leadworking tools

lead knife

seaming pliers

shave hook

Specialist hand tools for sheetwork

Plastic Pipe and Fittings

Relevant British Standards
BS EN 1452, BS 4514, BS 5255 and BS 4346

Two common types of plastic will be found in the construction industry. Firstly there are those produced by the polymerisation of ethylene (polythene, polyethylene) and propylene (polypropylene), which are used primarily for mains supply pipework, hot water and heating systems and certain waste pipe applications. The second type include uPVC (unplasticised polyvinyl chloride) and ABS (acrylonitrile butadiene styrene) which are mainly used for waste distribution pipework and cold water installations.

Plastic can be joined by compression fittings, push fit connections, fusion welded joints or solvent welded joints; the method chosen will depend on the type of plastic and its use. For example, it is possible to form a solvent welded joint to uPVC, but it cannot be fusion welded; conversely, polythene cannot be solvent welded.

Compression fittings

These are of two types: (1) those used for polythene, made in gunmetal or similar, and used on supply pipework to withstand high pressures. These usually use a copper compression ring with a copper liner, which is inserted into the pipe to prevent the wall collapsing when the nut is tightened; (2) those (as used on uPVC) usually made of a plastic material and used for waste discharge pipework. This type utilises a rubber compression ring, the fitting only being made hand-tight.

Push-fit connections

Many types of push-fit joint are now being marketed both for high pressure pipework and for waste systems. The main difference between them is that those used on high pressures incorporate a grip ring which prevents the pipe from pulling out. Those used for waste pipework simply have a rubber 'O' ring or the equivalent. Sometimes a push-fit connection is used on waste pipework to allow for expansion and contraction; if this is the case when first assembling the pipe fitting, push the pipe in fully and then withdraw it by about 10–15 mm.

Fusion welded joints

This is a joint in which the plastic is melted onto the fitting. Materials such as polythene and polypropylene are joined by this method. The join can be achieved by the use of a specially heated tool which melts the pipe and fitting, or by an electric charge applied to a wire which is located just below the surface plastic of special fittings. This wire heats up and melts the plastic.

Solvent welded joint

This joint, made from a special solvent cement, is used to join materials such as uPVC and ABS. Solvent cement is not a glue; it does not simply stick the pipes together but when the liquid is applied to the plastic, it temporarily dissolves it. A solvent welded joint will set in 5–10 minutes, but will require at least 12–24 hours to become fully hardened.

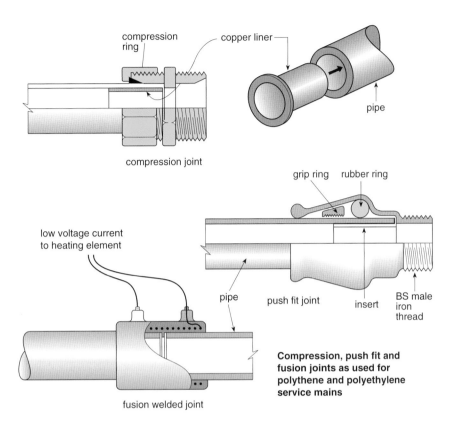

compression ring — copper liner —

pipe

compression joint

grip ring   rubber ring

low voltage current to heating element

pipe   push fit joint   insert   BS male iron thread

Compression, push fit and fusion joints as used for polythene and polyethylene service mains

fusion welded joint

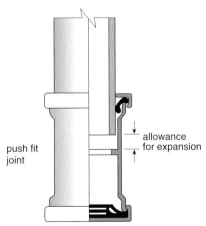

push fit joint

allowance for expansion

Push fit joint as used for low pressure waste pipework

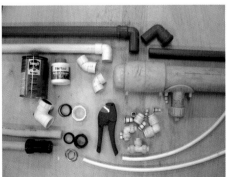

Various types of plastic fittings used on high and low pressure pipework

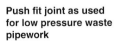

Plastic pipe and fittings

Low Carbon Steel Pipe and Fittings

Relevant British Standard
BS 1387

Low carbon steel pipe, often referred to as mild steel, is available either painted black or galvanised (coated inside and out with zinc). Black iron pipes should only be used for wet heating systems or oil and gas supply pipework. If black iron pipes were used where fresh water is continuously being drawn off through the pipeline, they would soon corrode. Steel pipes are manufactured in three grades and colour-coded accordingly: heavy gauge, red; medium gauge, blue; and light gauge, brown. The outside diameter in each case is the same. Only heavy gauge is permitted to be used below ground. All grades are used above ground although light gauge tube is rarely employed, being restricted, possibly, to dry pipe sprinkler systems.

Threaded joints Threads are cut into the pipe to give a British Standard pipe (BSP) thread to BS 21 using stocks and dies. The thread is cut on site using hand dies or powered threading machines. An assortment of fittings are used to join the pipe, which may have either a parallel or tapered thread, and may be made of either malleable cast iron or steel. Steel fittings tend to be stronger, although more expensive.

The length of thread on the pipe should be such that one and a half or two threads are showing when the fitting is assembled. This is because these first two threads have not been cut to the correct depth, owing to the cutting process. A few strands of hemp may be applied to the thread in a clockwise direction followed by some jointing paste to give a sound joint. An alternative jointing material is poly-tetrafluoroethylene (PTFE) tape which can be used, but should be avoided on larger threads or where the thread is slightly distorted. The jointing medium used will depend on the contents of the pipe. The use of hemp as suggested may be prohibited, such as for gas pipework and potable drinking supplies.

Compression joints It is possible to purchase several designs of compression joint for use on steel pipe; these incorporate a rubber compression ring. They tend to be rather expensive, but a saving can be made on installation time. They prove particularly useful when jointing to different materials and are often referred to as a transitional fitting.

Press-fit joints It is now possible to use a press-fit joint as used for copper tube to join pipework in certain situations. This method of jointing is explained on p. 56.

Disconnection joints Because of the way in which screwed fittings are made, i.e. rotated on the pipe, it is impossible to remove or assemble pipework where this rotation is not possible. Therefore, a special fitting will be required such as that provided by a union, a long screw or a flange, which allows the joint to be made and disconnected without any turning of the pipe or fitting. Alternatively a compression joint may be used.

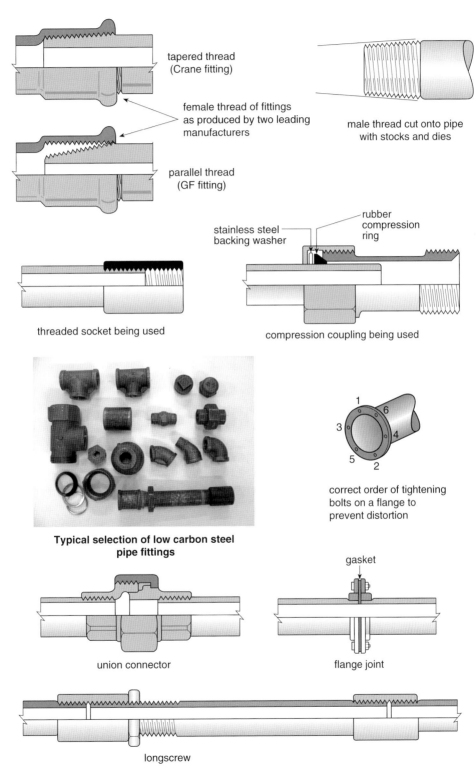

tapered thread
(Crane fitting)

female thread of fittings
as produced by two leading
manufacturers

male thread cut onto pipe
with stocks and dies

parallel thread
(GF fitting)

rubber
compression
ring

stainless steel
backing washer

threaded socket being used

compression coupling being used

**Typical selection of low carbon steel
pipe fittings**

correct order of tightening
bolts on a flange to
prevent distortion

union connector

gasket

flange joint

longscrew

Low carbon steel pipe and fittings

Copper Pipe and Fittings

Relevant British Standards
EN 1057 and BS 2051

Copper tube is available in a wide range of sizes, from 6 mm to 159 mm, and in various grades of temper and thickness. The material strength is expressed as an 'R' number. Basically the higher the number the harder the material.

- R220 is annealed (soft) and supplied in coils (formerly grades W and Y)
- R250 is a half hard condition supplied in straight lengths and suitable for bending (formerly grade X)
- R290 is a hard condition and unsuitable for bending (formerly grade Z).

Jointing methods

Compression joints   There are two main types, the manipulative and non-manipulative. The difference is that with the manipulative joint the end on the tube needs to be worked to form a bell mouth, i.e. it is manipulated. With the non-manipulative fitting the compression ring (olive) is simply slipped onto the tube.

Press-fit joints   A heat-free joint achieved by the use of an electric pressure tool, which exerts an equal pressure around the fitting, positioned onto the pipe. It compresses onto the collar of the joint that houses an 'O' ring to form a tight fit onto the tube creating a sound joint. This system is suitable for both water-filled and gas installations, however a specific 'O' ring needs to be incorporated within the fitting where gas is supplied.

Soldered joints   A soldered joint is one in which an alloy with a lower melting temperature than copper has been used to join the pipe. There are two basic classifications of soldered joint: the soft soldered and hard soldered joint.

Soft soldered joints are made with a propane blowtorch, using a tin composite solder with a melting range of 180–230°C. The solder is drawn into the tube by capillary action. Hard soldering requires a much higher temperature, in excess of 600°C, achieved by using oxyacetylene equipment. A hard soldered joint can be formed in the same way as a soft soldered joint, alternatively the pipe can be manipulated to form a bell mouth opening into which large deposits of solder are melted, to form what is referred to as a bronze-welded joint (see p. 70). It is possible to purchase soft soldered joints with a ring of solder already in, thus ensuring the correct amount of solder is used alternatively the solder may be added, referred to as an end-feed fitting. Note, solders containing lead are not permitted to be used on hot and cold drinking water supplies.

Push-fit joints   There are several types of push-fit joint designed for use on hot and cold water supplies. They are made in both plastic and brass and incorporate an 'O' ring seal. The advantages include ease of use and flexibility in rotating the pipe in any direction, however they can be more expensive and bulky in size, with an unsightly appearance. They also have difficulty in maintaining the path of electricity for bonding purposes. Pulling the fitting apart generally requires a specific technique identified by the supplier.

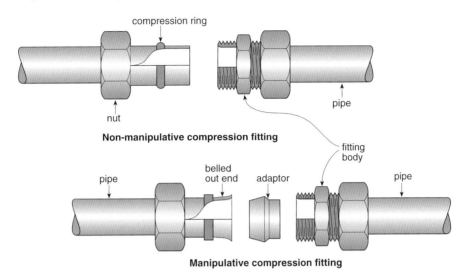

compression ring

nut

Non-manipulative compression fitting

pipe

fitting body

pipe

belled out end

adaptor

pipe

Manipulative compression fitting

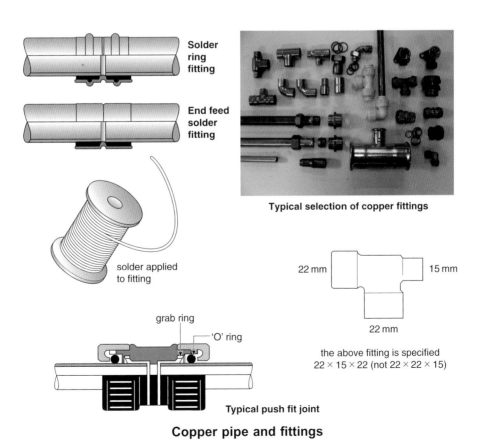

Solder ring fitting

End feed solder fitting

Typical selection of copper fittings

solder applied to fitting

grab ring

'O' ring

Typical push fit joint

22 mm

15 mm

22 mm

the above fitting is specified
22 × 15 × 22 (not 22 × 22 × 15)

Copper pipe and fittings

Low Carbon Steel Pipe Bending

Steel pipes are bent either by using a hydraulic bending machine, or by applying heat to the section to be bent. Due to the thickness of the pipe walls, it is not necessary to support the pipe fully, as with bending copper tube.

Hydraulic press bender

To use a hydraulic press bending machine, a former of the correct size is selected and fitted onto the end of the ram. The stops are positioned into the pin holes as indicated on the machine. When the pipe is inserted into the former and the handle operated, the ram moves forward to force the pipe against the stops; as the pumping continues, the pipe is forced to bend. To release the pumping action, a by-pass valve is turned which allows the hydraulic fluid to escape to another chamber in the machine.

Bending using the hydraulic press bender It is very simple to form a 90° bend from a fixed point on these machines. First, a line is marked on the pipe at a distance from the fixed point where the centre line of the finished bend is required (distance 'X', see figure). Then, from this measurement one can deduct the internal bore of the pipe. The pipe is now placed in the bending machine at this new point and lined up with the centre line of the correct sized former. The machine can then be pumped to apply pressure and bend the pipe to the required angle. Because there is a certain amount of elasticity, the bend is over-pulled by about 5°.

To make an offset using these machines, first measure along the pipe from the fixed point (this time making no deductions), and pull this round to the required angle, as shown opposite. Then, using a straight edge or positioning the pipe against two parallel lines the distance of the offset apart, the second bending point is marked on the tube. The pipe is replaced in the machine and this mark is positioned at the centre line of the former. When pulling this second bend, it is essential to ensure that it is pulled running true, and on the same plane as the first bend.

Bending using heat

This is described on page 62, Pipe bending using heat. Some of the main points to observe when using heat to bend steel tube include:

- Mark out the heat length correctly and apply heat, until cherry red, along the whole heat length
- Pull as much as possible in one go, pulling from each end alternately
- Should a second heat-up be required, reheat the whole heat length
- For safety, always cool completed work.

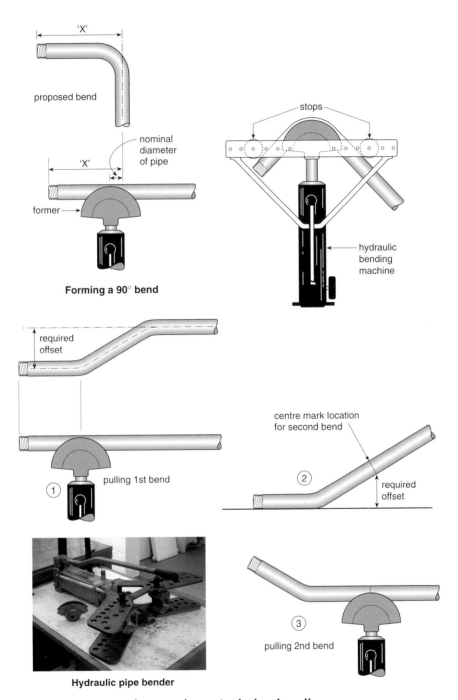

'X'

proposed bend

nominal diameter of pipe

'X'

former

Forming a 90° bend

stops

hydraulic bending machine

required offset

pulling 1st bend

①

centre mark location for second bend

②

required offset

Hydraulic pipe bender

pulling 2nd bend

③

Low carbon steel pipe bending

Copper Pipe Bending

Hand bending

This is a method of preventing the pipe from collapsing, either by using a spring or by packing the pipe with dry silver sand. The spring could be inserted internally or fitted externally to the pipe. When using an internal spring, the radius of the bend must not be too small or it will be impossible to withdraw the spring. To assist spring removal, slightly overpull the bend, then pull back to the required position. Larger pipes will need the application of heat (see page 62).

Machine bending

Bending machines are used for copperwork on the principle of leverage and are either hand held or free standing. The larger machines have different sized formers and backguides for the various pipe sizes. Once the machine is set up, with the rollers correctly adjusted to prevent throating or rippling, the various bends can be pulled as follows:

90° bend   A line is marked on the pipe at a distance from the end of the pipe equal to that between the fixed point and the back of the finished bend (distance 'X', see figure). The pipe is then inserted into the bending machine as shown. Make sure that the line marked on the pipe is square with the back of the former. The bend is simply pulled round by the lever arm.

An offset   Initially, the first bend is pulled to any angle, unless one has been specified. The pipe is then pushed further along the bending machine and turned around to make that first bend look upwards away from the former. A straight edge and rule are now used, as shown, to measure the required offset. When marking this out, ensure that the straight edge is running parallel with the piece of pipe looking upwards away from the former. Once the correct measurement is obtained, the bend can be pulled round until it is at the same angle as the first bend and can be seen to run parallel; this is best checked using a straight edge.

The passover   There are several methods of forming a passover. One of the simplest is the following: a bend is first pulled, the angle of which depends upon the size of the obstruction; the bend should not be too sharp, otherwise difficulty will be experienced when pulling the offset bends. A straight edge is positioned over the bend centrally at a distance equal to the required passover. The pipe is marked with two lines (see figure), which will be the back of the finished offsets. The pipe is then inserted into the bending machine, and when the first mark is in line with the back of the former, the first set is pulled. Once this bend has been completed, the pipe is simply reversed in the machine, and the second mark lined up in the former and pulled to complete the passover.

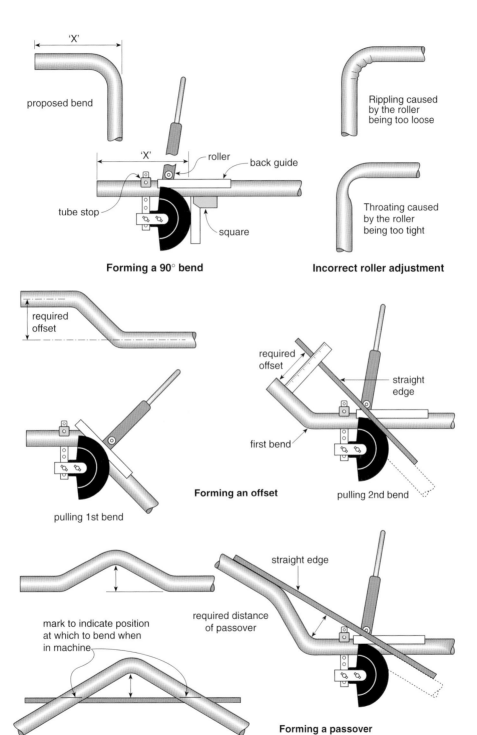

Forming a 90° bend

Incorrect roller adjustment

proposed bend

'X'

'X'

roller

back guide

tube stop

square

Rippling caused by the roller being too loose

Throating caused by the roller being too tight

required offset

pulling 1st bend

Forming an offset

required offset

straight edge

first bend

pulling 2nd bend

straight edge

required distance of passover

mark to indicate position at which to bend when in machine

Forming a passover

Copper pipe bending

Pipe Bending Using Heat

By the application of heat, it is possible to pull a bend on plastic, copper and low carbon steel pipe without the use of a bending machine. Two principles need to be understood in order to achieve a successful bend. Firstly, the internal diameter of soft- or thin-walled pipes must be protected from flattening. This is achieved by inserting a bending spring, or packing the pipe with dry silver sand. Steel is bent without any internal support. Secondly, when heating the pipe, the bend must be heated to the correct length only. The heat length is found either by calculation, or by producing a drawing from which to take a measurement.

The heat length (by calculation)
The amount of pipe to be heated is simply found using the following formula:

$$\text{Heat length} = 2R\pi \div 4$$

where R = centre line radius of proposed bend; $\pi = 3.142$
   In the absence of a specified radius, one generally uses four times the nominal diameter of the pipe.

Example: Find the heat length for a 15 mm ($\frac{1}{2}$ in BSP) diameter low carbon steel pipe which is to be pulled to an angle of 90°.

$$\text{Radius of bend R} = 15 \times 4 = 60 \text{ mm}$$

$$\text{Therefore } 2R\pi \div 4 = (2 \times 60 \times 3.142) \div 4 = 94 \text{ mm}$$

Sometimes, to save time on site and to avoid using a calculator, the plumber simply breaks down the formula to the simplified version of one and a half times the centre line radius. Therefore, if we again find the heat length for the previous example, using this method we find that:

$$R = 15 \times 4 = 60 \text{ mm}$$

$$\text{Therefore } 60 \times 1\frac{1}{2} = 90 \text{ mm}$$

which is not too far from the previous example, and hence good enough.
   To bend the pipe, the heat length is chalked onto the pipe, the pipe is heated as necessary and pulled 45°, quickly reversed and the second 45° pulled, if possible, to achieve the 90° bend. It is not likely to be possible to pull bends of this angle in pipes greater than 20 mm diameter in one go.
   When bending angles of less than 90°, the heat length is generally taken to be half the total heat length for a 90° bend. When bending a metallic pipe, it should be heated to an annealed temperature (red hot). With thermoplastics, on the other hand, the material is heated to a temperature at which it becomes soft and pliable. The softening temperature of plastic is very near the temperature at which it chars; therefore, care must be taken not to damage the tube when heating, which is best carried out using only the hot air of the flame.

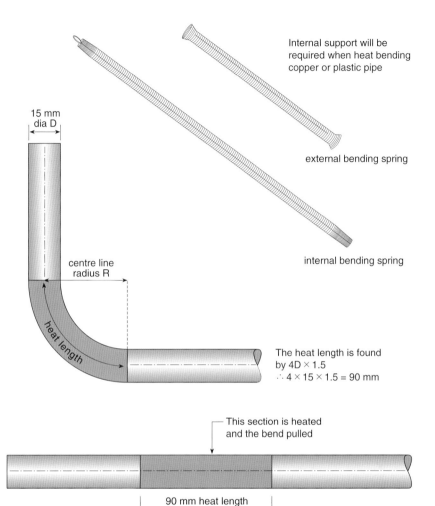

Internal support will be required when heat bending copper or plastic pipe

external bending spring

internal bending spring

15 mm dia D

centre line radius R

heat length

The heat length is found by 4D × 1.5
∴ 4 × 15 × 1.5 = 90 mm

This section is heated and the bend pulled

90 mm heat length

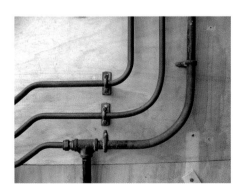

**Bends pulled with increased radius
(made using bending springs and heat)**

**Bend pulled with bending machine
(all radii remain the same)**

Pipe bending using heat

Soldering

Unlike welding, in which the joint is made by fusing with the parent metal, soldering is a process in which alloys are used to join metals below their melting temperature. In order to solder successfully, proceed as follows:

1. Thoroughly clean the joint (some fluxes have self-cleansing capabilities)
2. Apply flux (it should be noted that some soldering processes using an oxyacetylene flame do not require a flux)
3. Apply heat (either directly or via a soldering iron)
4. Feed solder to the joint
5. Allow the joint to cool, without movement
6. Remove excess flux (failure to do so could result in corrosion problems).

Many soldered joints, be they soft or hard, are formed by allowing the solder to flow between two close surfaces. These joints are called capillary or brazed joints. A brazed joint is one which remains strong at high temperatures (i.e. hard soldered). Capillary joints can be made at any angle because the solder does not fill the joint like filling a cup; it is drawn between the two close-fitting surfaces.

The correct flux must always be used, as specified by the manufacturer. It is applied in order to prevent the oxygen in the atmosphere coming into contact with the cleaned metal, thus preventing oxidation.

Soft soldering
The term soft soldering refers to any soldering process in which the solder used makes a joint which will not withstand too much stress. Generally, soft soldered joints are made with a flame temperature no higher than about 450°C.

Soft solders include lead/tin and lead-free solders which are composed mainly of tin and small amounts of copper, silver or antimony to give strength.

Hard soldering
There are several types of hard solder including silver, silver alloys with varying percentages of copper, and the cheaper-to-purchase copper/phosphorus alloys (cupro-tected). Hard soldered capillary joints are made in much the same way as soft soldered joints, the main difference being the temperatures required, which range from 600°C to 850°C. These temperatures ensure that a very secure job is made.

Bronze welding is another form of hard soldered joint (see page 70).

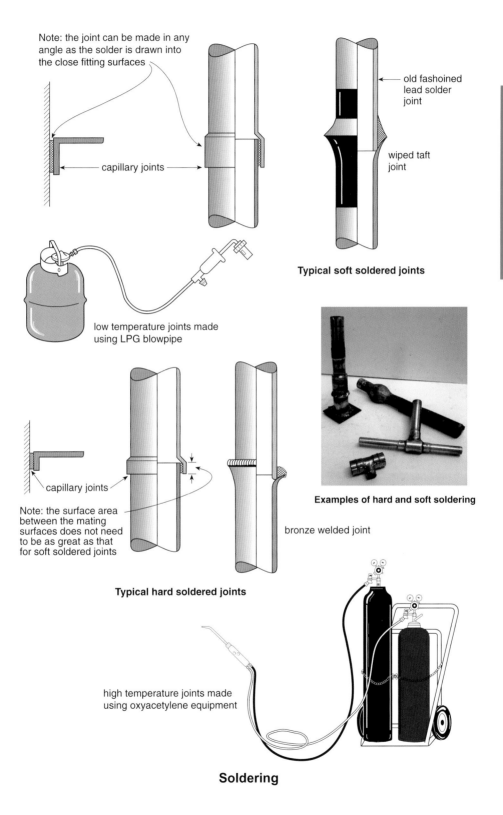

Note: the joint can be made in any angle as the solder is drawn into the close fitting surfaces

capillary joints

old fashoined lead solder joint

wiped taft joint

Typical soft soldered joints

low temperature joints made using LPG blowpipe

capillary joints

Note: the surface area between the mating surfaces does not need to be as great as that for soft soldered joints

Typical hard soldered joints

bronze welded joint

Examples of hard and soft soldering

high temperature joints made using oxyacetylene equipment

Soldering

Welding Equipment and Safety

Relevant British Standard
BS EN 730

The gases most commonly used in welding processes include oxygen, which is supplied in a black cylinder with a right-handed thread for the connection of the regulator; and acetylene, which is supplied in a maroon-coloured cylinder with a left-handed thread. The oxygen hose is blue or black and the acetylene hose is red. To the other end of the hoses is fitted the required blowpipe with the correct welding nozzle.

When carrying out any welding processes, you should take several precautions to ensure your own safety and that of those around you; the following checklist should be observed:

- Always wear protective clothing, especially the correct eye goggles or shields.
- At all times ensure good ventilation when welding.
- Erect any necessary signs or shields to give protection to people and to warn them of the welding process taking place.
- Always have fire-fighting apparatus to hand.
- Always repair or replace perished or leaking hoses with the correct fittings; do not use odd bits of tubing to join the hoses. On no account should piping or fittings made of copper be used with acetylene, as an explosive compound would be produced. Acetylene should never be allowed to come into contact with an alloy containing more than 70% copper.
- Store the gas cylinders in a fireproof room. If it can be avoided, oxygen should not be stored with combustible gases such as acetylene. Full and empty cylinders should be stored separately.
- Acetylene cylinders should always be stored and used upright to prevent any leakage of the acetone.
- Oxygen cylinders must not be allowed to fall over because, should the valve be broken off, the gas escaping at high pressure would cause the cylinder to shoot off like a torpedo, causing extensive damage; for this reason, oxygen cylinders should always be secured.
- Oil or grease will ignite violently in the presence of oxygen; therefore, cylinders should be kept clear of such materials.
- Allow an adequate flow of fuel gas to discharge from the blowpipe before lighting up.
- In the event of a serious flashback or backfire, plunge the blowpipe in a pail of water to cool it, leaving the oxygen running to prevent water entering the blowpipe.
- Hose check valves should be fitted on the blowpipe to prevent any flashback into the hoses. Flashback arrestors should also be fitted to the regulators to prevent a flashback occurring within the cylinder itself.

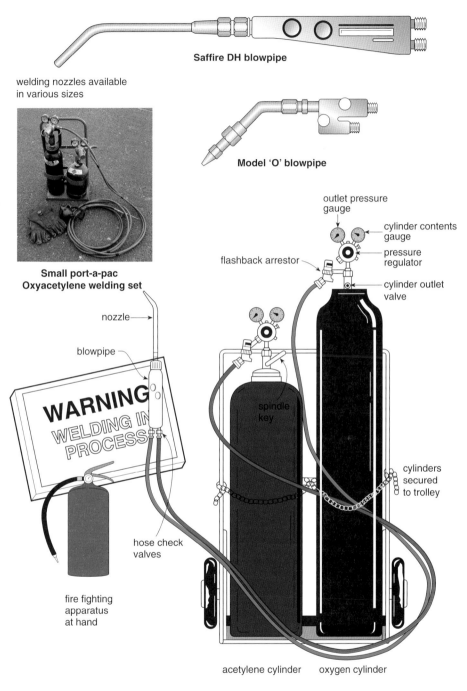

Saffire DH blowpipe

welding nozzles available
in various sizes

Model 'O' blowpipe

**Small port-a-pac
Oxyacetylene welding set**

outlet pressure
gauge

cylinder contents
gauge

flashback arrestor

pressure
regulator

cylinder outlet
valve

nozzle

blowpipe

**WARNING
WELDING IN
PROCESS**

spindle
key

cylinders
secured
to trolley

hose check
valves

fire fighting
apparatus
at hand

acetylene cylinder   oxygen cylinder

Standard welding equipment

Welding equipment and safety

Welding Processes

Relevant British Standard
BS 499

Welding can be defined as the coalescence, or joining, of metals using heat, with or without the application of pressure, and with or without the use of a filler rod. The heat required to weld successfully may be supplied either by electricity or by a gas flame. Before modern methods of welding, metals were joined by heating in a forge to welding temperature, then hammering or pressing the two metals together.

With oxyacetylene welding, in which oxygen and acetylene gases are used in approximately equal volumes, a flame temperature of about 3200°C is produced at the tip of the inner cone. The flame can be adjusted by increasing the amount of oxygen or acetylene to alter the flame's characteristics (see figure). Most welding processes, including lead welding, require the use of a neutral flame. Bronze welding (see page 70) requires a slightly oxidising flame.

Successful welding may be carried out by employing either of two basic welding techniques: the leftward or rightward methods. With the leftward welding technique, the filler rod precedes the blowpipe and the weld progresses from right to left with the blowpipe nozzle pointing in the direction of the unwelded surfaces. Conversely, with the rightward welding technique, the welding progresses in the opposite direction, i.e. from left to right. In order to achieve the rightward weld successfully, the blowpipe nozzle is directed into the completed weld at a much lower angle; also, with this technique the blowpipe flame precedes the filler rod along the weld.

A 'true' welded joint uses a filler rod of the same metal as that of the metals being welded, and is referred to as an autogenous weld. Welding processes such as bronze welding are not, in fact, welded joints, but are a process of hard soldering.

Welding is carried out by first ensuring that the edges of the metal are free from oxides, dirt or grease, etc., and are butted together. The correct blowpipe is chosen, with the correct nozzle, and the flame adjusted to a pressure of approximately 0.14–0.21 bar (2–4 lbf/in^2) depending on the metal thickness. The metal is heated to its melting temperature, and the filler rod applied as necessary. Thick steel will require higher gas pressures than those suggested above.

When producing an autogenous weld on metals such as lead or steel using a neutral flame, no flux is required because the oxyacetylene flame prevents the gases in the air coming into contact with the surface metal, and oxidation will therefore not result. This is not to be taken as the rule however, as some welding processes, e.g. those employed when welding aluminium, will require a flux.

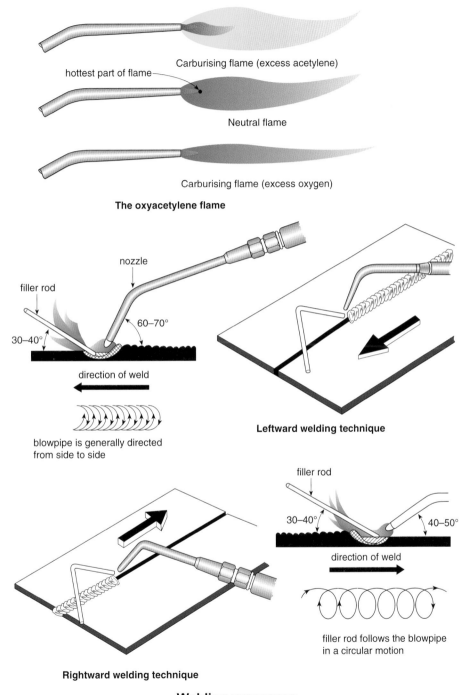

Carburising flame (excess acetylene)

hottest part of flame

Neutral flame

Carburising flame (excess oxygen)

The oxyacetylene flame

nozzle

filler rod

60–70°

30–40°

direction of weld

blowpipe is generally directed
from side to side

Leftward welding technique

filler rod

30–40°

40–50°

direction of weld

filler rod follows the blowpipe
in a circular motion

Rightward welding technique

Welding processes

Bronze Welding

Relevant British Standard
BS 1724

Bronze welding is used to join various metals. The resulting join is referred to as a welded joint but, in fact, no melting of the parent metal takes place. Therefore it is really a type of hard soldered joint. To make a bronze welded joint, a space is required between the two surfaces to be joined. This can be achieved by grinding thick material back to form a bevelled edge, or, in the case of pipe, a space could be left or a bell joint formed. The joint should be thoroughly cleaned to remove any oxide coating, then heat applied to the joint with a slightly oxidising flame, which is worked from side to side, or in small circles over the surface.

A slightly oxidising flame is chosen because the excess oxygen in the flame is required to mix with the zinc in the filler rod, which boils and changes to a vapour below the melting temperature of copper. When the oxygen mixes with the zinc, a zinc oxide is formed which melts at a much higher temperature (pure zinc oxide melts at 1975°C). Failure to oxidise the zinc vapour will cause it to bubble up through the weld, leaving a series of blowholes.

The fluxed filler rod is introduced to the joint and the flame is lifted to stroke the rod, which, upon melting, should adhere to the metal being joined. The process is continued slowly, progressing along the joint, adopting the leftward welding technique (see page 68), forming a series of characteristic weld ripples. The filler rods usually used for bronze welding consist of approximately 60% copper and 40% zinc, with a small amount of tin and silicon to act as deoxidisers and assist in its flowing characteristics.

One of the advantages of bronze welding is the temperature at which the filler rod melts (around 850–950°C). This minimises the distortion which would otherwise take place should the joint be fusion-welded, since this requires a temperature of at least the melting point of the parent metal. Also, bronze welding can be used to join dissimilar metals, such as copper and iron.

A special flux is used to bronze weld: borax and silicon. This is either made into a paste by mixing with water and applied directly to the joint, or a heated filler rod is dipped into the powdered flux which, in turn, sticks to the rod. Special flux-impregnated filler rods can be purchased at a little extra cost. As with all joints requiring a flux, any excess flux must be removed on completion of the weld. It is sometimes possible to loosen the flux residue by quenching in water immediately after bronze welding the joint.

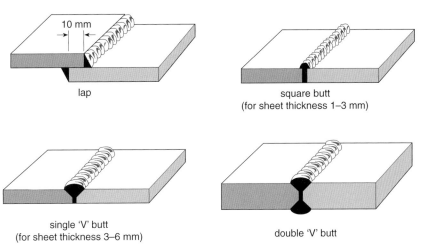

10 mm

lap

square butt
(for sheet thickness 1–3 mm)

single 'V' butt
(for sheet thickness 3–6 mm)

double 'V' butt

Bronze welded joints to sheet material

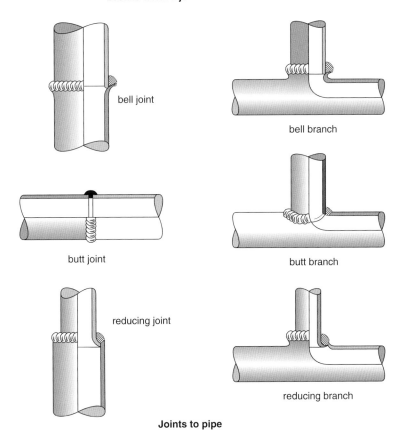

bell joint

bell branch

butt joint

butt branch

reducing joint

reducing branch

Joints to pipe

Bronze welding

Lead Welding (Lead Burning)

Lead can easily be welded together either with oxyacetylene welding equipment, or with propane gas, although propane gas welding is limited to simple butt and lap joints. Filler rods are generally made from strips of sheet lead 6–8 mm wide, cut from waste. One of the prerequisites of successful welding is cleanliness of the filler rod and of the surfaces to be welded. The correct nozzle must be chosen (model 'O' blowpipe, nos 2–3) and adjusted to a neutral flame (see page 68). The pressure at both oxygen and acetylene regulators should be adjusted to 14 kN or 0.14 bar (2 lbf/in^2). A finished weld should look uniform in size and shape, and it is essential that there is sufficient penetration. Equally, there should be no undercutting or overlapping to the edge of the weld.

Butt welds When a butt weld is to be carried out, the meeting surfaces should first be shaved clean, approximately 6 mm wide on each sheet. Then, a tack weld is applied at several intervals along the joint, which prevents the joint opening because of expansion. The welding nozzle is held close to the joint at an angle of about 60° and a molten pool is allowed to become established. The filler rod is introduced close to the nozzle and the blowpipe is slightly raised to melt off a piece of the filler rod which drops and fuses with the molten pool. The flame is returned to the pool in a stroking action (see figure). The blowpipe is then moved forwards to a new position where the process is repeated.

Lap welds The lap joint is formed by carring out a welding process similar to that for the butt joint, except that one sheet (the overcloak) is lapped on top of the sheet to be joined by 25 mm. To prepare this joint, the overcloak is cleaned on both sides, placed on the cleaned undercloak and tacked in position. The first weld is made joining the two sheets. This causes a certain amount of undercutting to the overcloak, so that a second weld is required as reinforcement.

Vertical welds With vertical welding, no filler rod is used and the joint is prepared as for lap welding. The nozzle is held close to the overcloak and as the lead begins to melt the nozzle is circled around and around, in the shape of a no. 6, finishing off at the undercloak. A molten lead bead will follow the flame; when it becomes fused with the undercloak, the flame is removed.

Inclined welds Two different techniques can be adopted to create lead welding up an incline. One produces a true incline joint, in which the seams to be welded are either butted or lapped together, and welding of the sheet is carried out by depositing the lead up the incline in the form of overlapping semicircular beads. The second technique produces a vertical welded joint in which the two adjoining sheets overlap each other at an angle; lead welding vertically in this fashion tends to be easier and stronger than a true vertical lead weld.

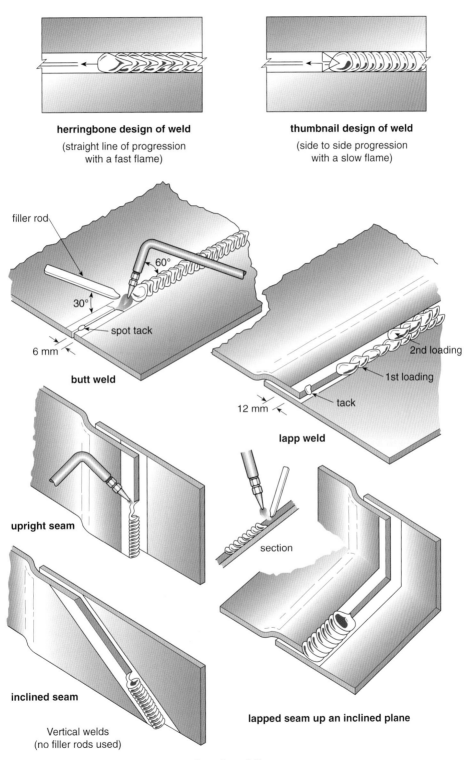

herringbone design of weld
(straight line of progression
with a fast flame)

thumbnail design of weld
(side to side progression
with a slow flame)

filler rod

60°

30°

spot tack

6 mm

butt weld

2nd loading

1st loading

tack

12 mm

lapp weld

upright seam

section

inclined seam

Vertical welds
(no filler rods used)

lapped seam up an inclined plane

Lead welding

In-Line Valves

Relevant British Standards
BS 1010, BS 1552 and BS 5433

Cocks, taps and valves

The definition of a cock, tap or valve can give rise to many an argument. I would suggest that the differences are just matters of local terminology: e.g. bib-tap and bibcock or stop-tap and stopcock. The purpose of a valve is to adjust and regulate the velocity of flow through a pipeline, either in line or at the point of termination.

Screwdown valves   There are several designs of in-line screwdown valve, including the stopcock and disc globe cock. Screwdown valves are characterised by a plate or a disc, shutting slowly, at right angles to the valve seating. Stopcocks are used on high pressure pipework and have a rubber washer fitted to give a sound seal. Make sure that the valve is fitted in the correct direction so that it will lift the washer off the seating should it become detached from the jumper. Failure to do so may result in a no-flow situation.

The disc globe cock is similar in construction to the stopcock, although more robust in its design, and is used in high pressure or high temperature pipework.

Gatevalves   This valve is sometimes referred to as a fullway gatevalve because when it is fully open there is no restriction of flow through the valve, unlike the screwdown valve where the liquid passes up and through a seating aperture. For this reason, gatevalves are recommended for use where there is low pressure in the pipeline, such as when fed from a storage cistern. It is recommended that these valves be fitted in vertical pipework to prevent sludge from settling in the base of the seating and thus stopping the gate from closing fully.

Quarter-turn valves   Two main types of quarter-turn valve will be found: the plugcock and the ballvalve. These valves work by the quarter-turn operation of a square head located at the top of the valve, which aligns the hole in the valve with the hole in the pipe. There is a slot in the head and when it is in line with the direction of the pipe it indicates that the valve is open. The plug cock has traditionally been used for gas pipelines and is widely known as the 'gas cock'. The plug cock has a tapered plug whereas the ballvalve consists of a circular ball through which the liquid or gas can flow when the valve is open. The term 'ballvalve' is rarely used (thus it is seldom confused with float-operated valves), and trade names such as Ballafix or Minuet are used instead. These valves are very common today and are used for a number of operations such as the service valve to a cistern or to a washing machine, some being fitted with a turning handle whilst others need the use of a screwdriver. Large ballvalves are sometimes installed on cold feed and distribution pipework, replacing the more traditional, but less effective, gatevalve.

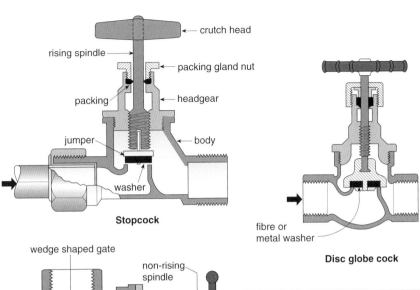

Stopcock

crutch head

rising spindle

packing gland nut

packing

headgear

jumper

body

washer

fibre or
metal washer

Disc globe cock

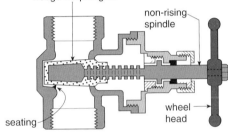

wedge shaped gate

non-rising
spindle

seating

wheel
head

Fullway gatevalve

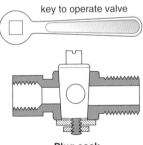

key to operate valve

Plug cock
(in closed position)

Examples of in-line valves

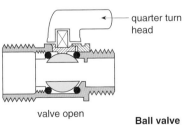

quarter turn
head

'O' rings

valve open

valve closed

Ball valve
(typical design used for washing machines etc.)

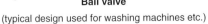

In-line valves

Terminal Valves

Relevant British Standard
BS 1010

A valve fitted at the point of use is sometimes called a terminal fitting, a reference to the point where it 'terminates'. Several terminal valves are available, including draw-off taps, drain-off cocks and float-operated valves.

Draw-off taps   These generally work in a similar fashion to screwdown valves (previously mentioned). Many modern valves make use of ceramic discs. These do not work as screwdown valves but allow two polished ceramic discs to turn and align two portholes through which the water can pass. These are very popular at present, allowing for a quarter-turn design of tap to be used, but consideration must be given to water hammer problems which may arise. The cold outlet tap should be located to the right-hand side when fitting the tap into an appliance.

The **bibtap** is characterised by a horizontal male iron thread which is screwed directly into a fitting. A typical location would be above a butler sink or fitted outside for use with a hosepipe connection (referred to as a hose union bibtap).

The **pillar tap** differs from the bibtap in that it has a long vertical male iron thread which passes through the appliance and is held in position with a back-nut, the supply being fed to this point below the appliance.

The **globe-tap** is now obsolete, but will occasionally be found. It had a female thread as its point of entry and was used for old baths with a side entry. Installation of this valve is no longer permitted because of back-siphonage problems.

Mixer taps are designed to allow the flow of water from hot and cold supplies to be delivered via one outlet spout. Depending on the design of mixer, the water either mixes in the body of the tap or as it comes out of the spout (see figure). It is a requirement that waters under different pressures be mixed as they leave the spout, otherwise it would not be possible to get water from the supply under the least pressure. A monoblock is simply a design of mixer tap where only one hole is required in the appliance, designed for cosmetic reasons.

The **supatap**, available in both 'bib' or 'pillar' design, offers a facility whereby it is possible to re-washer the tap without having to turn off the water supply. This is achieved by an automatic closing valve which drops to stop the flow of water should the nozzle be removed.

A drain off cock is a valve located at the lowest point of any system and has a serrated hose connection outlet. Drain taps fitted to supply pipes must conform to BS 2879 and be of the screwdown pattern.

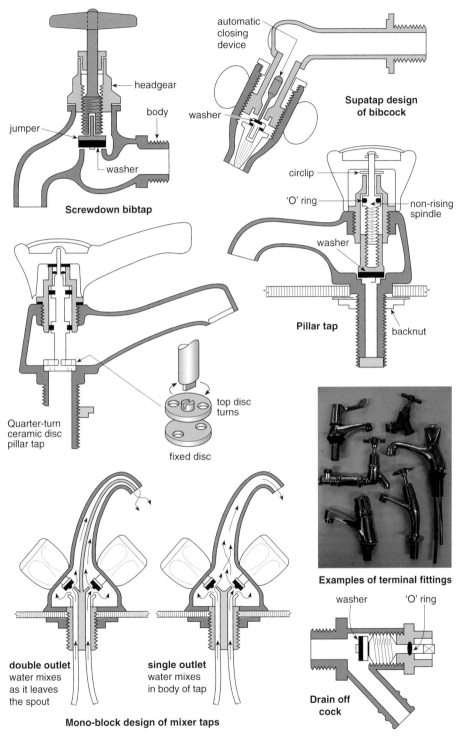

Screwdown bibtap

headgear

body

jumper

washer

Supatap design of bibcock

automatic closing device

washer

Pillar tap

circlip

'O' ring

non-rising spindle

washer

backnut

Quarter-turn ceramic disc pillar tap

top disc turns

fixed disc

Examples of terminal fittings

double outlet
water mixes
as it leaves
the spout

single outlet
water mixes
in body of tap

Mono-block design of mixer taps

washer

'O' ring

Drain off cock

Terminal valves

Float-operated Valves

Relevant British Standard
BS 1212

Float-operated valves are commonly called 'ballvalves', owing to the use of a ball float, although the name is somewhat inappropriate for some modern designs. These valves are fitted into cisterns to regulate the water supply automatically. The valve closes as the water reaches a predetermined level. It works on the principle of leverage: as the water rises, it causes the ball to float up, exerting an upward thrust which is transferred, via the lever and its fulcrum, to the plunger or piston; this gradually closes the washer onto the seating. The valve is available in several designs: the diaphragm, Portsmouth or Croydon.

The **diaphragm ballvalve** consists of a large diaphragm washer which is forced onto a seating by a small plunger. This design conforms to BS 1212, parts 2 or 3, and is the only design permitted in domestic premises. This is because it is able to maintain an air gap and therefore prevent back siphonage.

The **Portsmouth ballvalve** is commonly seen, although it is currently being phased out for domestic use in favour of the diaphragm valve. The Portsmouth valve is more sluggish in use than the diaphragm valve and no longer conforms to current water regulations when installed directly into the side of a cistern.

The **Croydon ballvalve** is now obsolete. It worked in a similar fashion to the Portsmouth, although the piston moved up and down instead of from side to side.

Some designs, referred to as equilibrium valves, allow water to pass through the washer to its other side. This means that the water pressure acts on the valve from both sides keeping the valve in a state of equilibrium (pressure pushing on both sides of the washer). The float of this type of valve only has to lift the weight of the arm, whereas in simple lever-type valves, the effort provided by the float not only has to overcome the weight of the arm, but also the pressure of the incoming water.

Equilibrium valves are useful in areas where the water supply pressure is very high or where persistent water hammer may be encountered.

It is possible to purchase float-operated valves which use ceramic discs, as it is with draw-off taps, mentioned on page 76.

It is a regulation that whenever a float-operated valve is fitted, a means of servicing the valve must be provided. This is achieved by the installation of a service valve as close as practicably possible to the cistern in which the float-operated valve is fitted.

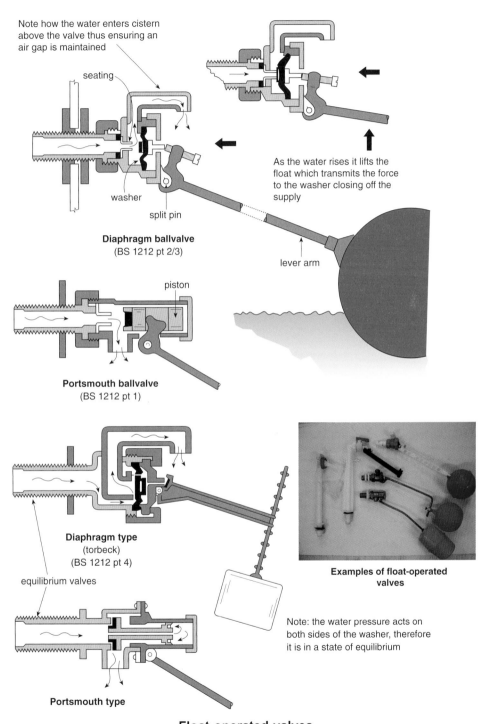

Note how the water enters cistern above the valve thus ensuring an air gap is maintained

seating

washer

split pin

Diaphragm ballvalve
(BS 1212 pt 2/3)

As the water rises it lifts the float which transmits the force to the washer closing off the supply

lever arm

piston

Portsmouth ballvalve
(BS 1212 pt 1)

Diaphragm type
(torbeck)
(BS 1212 pt 4)

equilibrium valves

Examples of float-operated valves

Note: the water pressure acts on both sides of the washer, therefore it is in a state of equilibrium

Portsmouth type

Float-operated valves

Part 2
Hot and Cold Water Supplies

Plumbing, 4th Edition. R. D. Treloar.
© 2012 Blackwell Publishing Ltd. Published 2012 by Blackwell Publishing Ltd.

Classification of Water

Water is a compound of hydrogen and oxygen. When we burn natural gas (a hydrocarbon, CH_4) dihydrogen monoxide (H_2O, i.e. water) and carbon dioxide (CO_2) are obtained as combustion products.

Pure water is a transparent, tasteless liquid which can be found in three physical states: solid (ice), liquid (water) or gas (steam or vapour). At atmospheric pressure, between 0 to 100°C, water is a liquid. At 0°C, water changes to ice with an immediate expansion in volume of 10%. At 100°C, it changes to steam, its volume expanding some 1600 times.

To convert water back to its constituent elements, an electric current needs to be passed through the liquid.

Rain water is usually contaminated with gases or chemicals which it absorbed as it fell. When rainwater reaches the ground it dissolves any soluble salts. Depending on which salts the water contains it may be classified as hard or soft.

Soft water

This is water which is free from dissolved calcium salts. Naturally occurring soft water is slightly acidic due to absorbed gases such as CO_2. Soft water tends to be more pleasant for washing in but has the major disadvantage of corroding pipework, lead pipes in particular.

Hard water

This is water which has fallen on, and filtered through chalk or limestone from which it dissolves small amounts of calcium and magnesium salts. The water may be either permanently or temporarily hard.

Permanent hardness   This is the result of water containing calcium or magnesium sulphates. Boiling has no effect on permanent hardness.

Temporary hardness   This is the result of the water containing calcium or magnesium hydrogen carbonates. The CO_2 dissolved in rainwater can attack limestone or chalk and convert the calcium carbonate and magnesium carbonate in the rock to soluble hydrogen carbonates. This temporary hardness can be removed by boiling the water; as a result CO_2 escapes into the air and calcium carbonate is precipitated as scale.

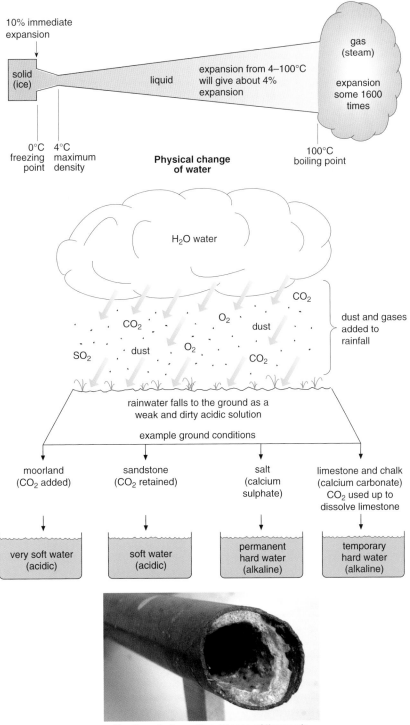

10% immediate expansion

solid (ice)

liquid

expansion from 4–100°C will give about 4% expansion

gas (steam)

expansion some 1600 times

0°C freezing point

4°C maximum density

100°C boiling point

Physical change of water

H_2O water

CO_2

O_2

dust

CO_2

SO_2

dust

O_2

CO_2

dust and gases added to rainfall

rainwater falls to the ground as a weak and dirty acidic solution

example ground conditions

moorland (CO_2 added)

sandstone (CO_2 retained)

salt (calcium sulphate)

limestone and chalk (calcium carbonate) CO_2 used up to dissolve limestone

very soft water (acidic)

soft water (acidic)

permanent hard water (alkaline)

temporary hard water (alkaline)

Section of pipe with evidence of limescale

Classification of water

Cold Water to the Consumer

Relevant British Standards
BS EN 806-2 and BS 6700

Throughout the United Kingdom, wholesome water (i.e. water fit for human consumption) is provided by the local water authorities to individual premises and various industries. When a supply of water is required, the water authority will supply water to a point just outside the property boundary line where, nowadays, a meter is usually installed to calculate the amount of water consumed. From this point the supply pipe is run into the premises, with precautions being taken to protect the pipe from movement, frost damage and corrosive soil.

Any pipe passing through or under a building must be ducted. This allows for its removal should the need arise. A consumer's stop tap is fitted where the supply pipe enters the building and should be fixed as low as possible with a drain-off cock immediately above it. (Note that older properties do not have meters.) The pipe is run from this point to feed the various systems of cold water supply. It is the responsibility of the installer to comply with the Water Supply Regulations 1999 when connecting to the supply main. These regulations have been designed in order to prevent wastage, contamination and erroneous measurement of water.

Wastage of water This could be the result of undue consumption, misuse or simply a faulty component, such as a leaking valve. To prevent this, water meters are being installed to register the amount of water used in serving the premises, for which the owner/occupier will eventually be invoiced. As a means of combating the problem of wastage, the installer must provide the dwelling with suitable overflow/warning pipes from cisterns to let the user know of a fault, and a means of isolation must be provided to allow its speedy shutdown. Allowance for thermal movement, frost protection, etc., must be made in order to prevent damage.

Water contamination The Water Supply Regulations include a list of five 'fluid categories' to identify the quality or condition of water, ranging from wholesome (fit for human consumption) to a level which presents a serious health risk. In very basic terms, the Regulations seek to ensure that once water has been drawn off for use, it should never be allowed to re-enter the water supply distribution network. There are several means by which water may be inadvertently contaminated. For example, it is possible that water may be sucked up from sanitary appliances (back-siphonage), such as baths and sinks, owing to peak flow demands creating a negative pressure on the supply pipe.

Contamination can also result from the use of unsuitable materials. Lead, for instance, is dissolved by water and as a result its use is prohibited nowadays. Even the solder used to join copper pipes, when used for hot or cold supplies, must be lead-free.

Erroneous measurement This refers to the discrepancy between the measurement or metering of water used and the quantity of water actually accounted and paid for.

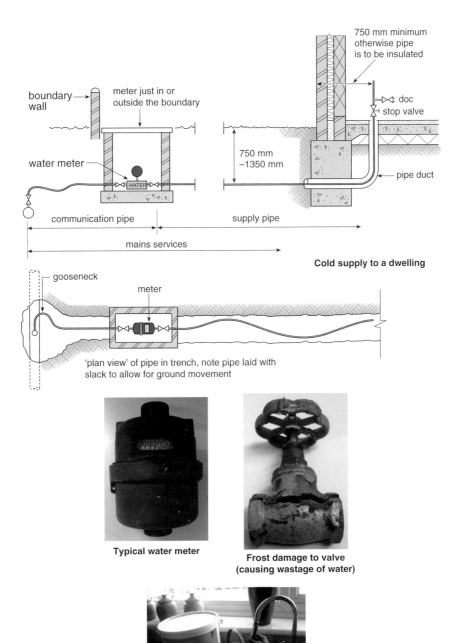

boundary wall

meter just in or outside the boundary

water meter

WATER

communication pipe

supply pipe

mains services

750 mm minimum otherwise pipe is to be insulated

doc stop valve

750 mm –1350 mm

pipe duct

Cold supply to a dwelling

gooseneck

meter

'plan view' of pipe in trench, note pipe laid with slack to allow for ground movement

Typical water meter

Frost damage to valve (causing wastage of water)

Hose connected to a tap without backflow protection (causing possible contamination)

Cold water to the consumer

Backflow Prevention

Backflow is a flow of water contrary to the intended direction of water flow. It can be caused due to **back pressure** or **back siphonage**.

■ *Back pressure* is the result of water pressure in the system being greater than that in the supply. Higher system pressures can be caused by the expansion of water in unvented dhw supplies, or in systems where a pump is used.

■ *Back siphonage* occurs as the result of negative pressures in the supply main, which may be caused by a major leak in the main or the fire services drawing off vast amounts of water.

Backflow prevention is achieved either by using a mechanical device or by a pipe arrangement which physically disconnects the supply from the system, maintaining an air gap. There are various backflow prevention devices, including single and double checkvalves and anti-vacuum valves. The Water Regulations list no less that 10 air gap configurations and 14 mechanical device combinations to combat the many scenarios where backflow could occur. The method selected would depend upon the severity of the risk; this is based upon the fluid risk category of the water that may be affected.

Fluid risk categories   Schedule 1 of the Water Supply Regulations identifies five fluid categories:

Category 1: No health risk. Wholesome drinking water
Category 2: Aesthetic quality is impaired due to, for example, being heated
Category 3: Slight health hazard due to contamination with substances of low toxicity
Category 4: Significant health hazard due to toxic substances, e.g. pesticides
Category 5: Serious health risk. Contains pathogenic organisms, e.g. from human waste.

Verifiable and non-verifiable checkvalves   Where a checkvalve incorporates a test point it is referred to as a verifiable check valve. A single check valve would be suitable where the fluid risk is only category 2. However for fluid risk 3, a double check valve would be required.

Reduced pressure zone valve (RPZ valve)   This is a comparatively new valve to the UK and is used to protect against fluid risk 4 applications, such as a fire sprinkler system filled with antifreeze in a commercial premises. It can only be installed by an accredited installer, approved by the water supplier. It should be tested for correct operation every year, with a test certificate issued. It should be noted that no drinking water should be drawn off downstream of the RPZ valve. It sometimes creates a drop in pressure and therefore is not suited to low pressure supplies.

Shown opposite are a few examples of backflow prevention methods, including air gaps and mechanical devices identifying the fluid risk they are suited to.

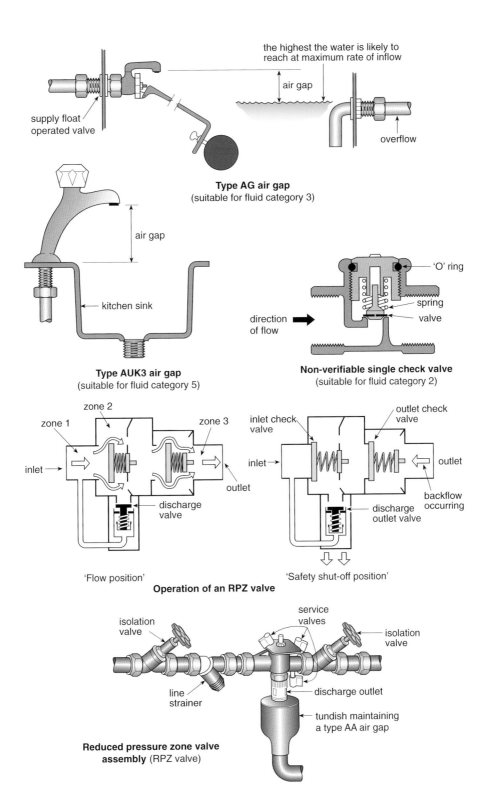

the highest the water is likely to
reach at maximum rate of inflow

air gap

supply float
operated valve

overflow

Type AG air gap
(suitable for fluid category 3)

air gap

kitchen sink

Type AUK3 air gap
(suitable for fluid category 5)

'O' ring

direction
of flow

spring

valve

Non-verifiable single check valve
(suitable for fluid category 2)

zone 2

zone 1

zone 3

inlet

outlet

discharge
valve

inlet check
valve

outlet check
valve

inlet

outlet

discharge
outlet valve

backflow
occurring

'Flow position'

'Safety shut-off position'

Operation of an RPZ valve

service
valves

isolation
valve

isolation
valve

line
strainer

discharge outlet

tundish maintaining
a type AA air gap

**Reduced pressure zone valve
assembly** (RPZ valve)

Assessing Water Efficiency in New Dwellings

Changes to part G of the Building Regulations, which came into force in 2010, identify that all new dwellings must have the water consumption of wholesome water limited so as not exceed 125 litres/person/day.

In order for the builder to obtain approval of their building plans they must demonstrate how this is to be achieved. This procedure is laid down in an industry guidance document entitled 'The Water Efficiency Calculator for New Dwellings', which is the government's calculation methodology for assessing water efficiency and is readily available through the internet.

In order to complete this calculation you need to know the specific flow rates and volumes of water used by the appliances within the new building. These are obtained from the manufacturer's product details and are simply entered onto the calculation table. If there are several different fitments and appliances used, then average flow rates need to be expressed. Savings from water recovery systems can be deducted from the calculated total to provide a total average volume of water consumption per person per day.

To find the average flow rate from the taps that is required to be entered onto the table, you select the largest of the following:

Average flow rate *or* Proportional flow rate

- The average flow rate is found by simply adding together the individual flow rates from all the individual taps and dividing by the number of taps.
- The proportional flow rate is found by taking the tap with the highest flow rate and multiplying this volume by 0.7

Example of Undertaking the Water Efficiency Calculation

The table opposite has been completed based upon the following:

Kitchen:

- 1 hot & 1 cold tap providing a combined average flow of 12 litres/min
- 1 washing machine using 6.3 litres/kg dry load
- 1 dishwasher using 1.6 litres/place setting

Bathroom and additional cloakroom:

- 3 hot & 3 cold taps providing a combined average flow of 8.4 litres/min
- 2 WCs flushing a combined average of 6 litres/min
- Bath volume 190 litres, with shower located above.

There is no water recovery system used within the dwelling.

Water Calculation for New Dwellings

Installation Type	Average Capacity or flow rate (l or l/min) 1.	Factor 2.	Fixed Use l/person/day 3.	l/person/day 1 × 2 + 3
Single WC (single flush)		4.42	0	
Single WC full flush -		1.46	0	
(dual flush) part flush -		2.96	0	
Several WCs (average)	6	4.42	0	26.52
Taps excluding the kitchen	8.4	1.58	1.58	14.85
Kitchen/Utility room taps	12	0.44	10.36	15.64
Bath (where shower present)	220	0.11	0	24.2
Shower (where bath present)		4.37	0	
Bath only		0.50	0	
Shower only		5.60	0	
Washing Machine	6.3	2.1	0	13.23
Dishwasher	1.6	3.6	0	5.76
Waste Disposal Unit		3.08	0	
Water Softener		1.00	0	
	Total flow calculated use from above =			100.2
	Multiply by a Normalisation factor of 0.91 =			91.182
	Plus (5.0 litres) for external use			5.0
	Total water consumption (litres/person/day)			96.182

This example clearly shows the water consumption falls within the maximum flow requirements; however, note that this is a small dwelling with no additional extras and the bath volume remains fairly small, with the shower present. Should a larger bath be required without the provision of a shower, or additional larger-volume sanitary fitments be sought, you might need to reconsider the design to include a shower, to bring down the installation-type maximum volume. Alternatively a system of water recovery would need to be considered. Where such a system is incorporated, such as grey water or rainwater recovery, then a further calculation would need to be completed to take this water contribution into consideration, for which further guidance from the industry guidance document should be sought.

Cold Water Systems

Relevant British Standards
BS EN 806-2 and BS 6700

Two distinct systems of cold water supply are in use – the direct and indirect systems – although modified systems are to be found, in which several appliances are on the mains supply and several fed from a cistern. It is essential that the plumber obtains advice and gives written notice of the design of a new system to the local water authority before commencing work. Failure to do this may mean a contravention of the Water Supply Regulations 1999. Whatever system is chosen, it must be designed to deliver cold water at the point of use at a temperature not exceeding 25°C.

Direct system

In this system, all the cold water in the house is fed 'directly' from the supply main. The water pressure is usually high at all outlets, so this system can have the disadvantage of being more prone to water hammer. In some areas of the UK the supply pressure is reduced at peak times. This can cause a negative pressure in the pipeline and loss of supply. Also, precautions need to be taken to prevent back-siphonage of foul water from appliances into the supply pipe.

The direct system is cheaper to install than the indirect, and does not require a roof space to accommodate the cistern. However, peak flow times must be considered and, above all, adequately sized pipes used to prevent a lack of suitable flow, should several appliances be used at once.

Indirect system

In this system, only one draw-off point (i.e. the kitchen sink) is fed from the mains supply pipe. All other outlets are supplied via a cold storage cistern, usually located in the roof space. Water pressure is usually much lower than with the direct system, but it will be maintained, even at peak times or during complete shutdown of the supply. Today in modern housing, with suitable precautions to storage cistern sizing and the prevention of water contamination, stored water is regarded as wholesome (fit for human consumption) so water from draw-off points other than the mains one is regarded as drinking water and, therefore, the same precautions must be maintained to prevent back-siphonage.

Cold supply to the domestic hot water (dhw) system

Unvented systems   These systems are fed directly from the mains supply pipe.

The biggest consideration is whether the supply main is large enough in diameter to provide a good flow rate should several appliances be operated at once. To prevent the hot water flowing back down the feed pipe, a check valve must be incorporated\break (see Unvented Domestic Hot Water Supply, page 108).

Vented systems   These require a supply via a cold feed cistern. The cold feed pipe is run separately from any cold distribution pipework to prevent hot water being drawn off when the cold supply is opened.

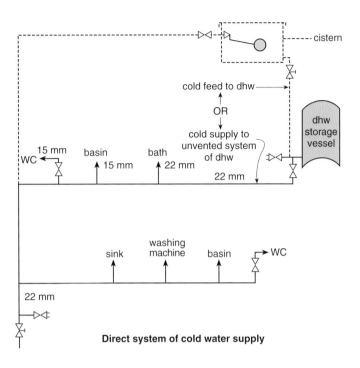

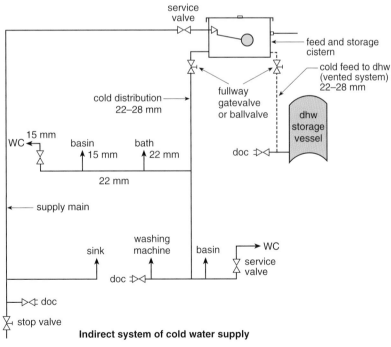

Cold water systems

Cold Water Storage

Relevant British Standards
BS EN 806-2 and BS 6700

When cold water needs to be stored to supply an indirect system of cold water, or to feed a system of dhw, it is held in a cistern which is usually located in the roof space.

The storage cistern should have a minimum capacity of 100 litres. If the cistern is also to act as a feed cistern for the hot water supply (being a combined storage and feed cistern), it should have a minimum capacity of 230 litres.

Cold distribution pipes from storage cisterns should be connected so that the lowest point of the water outlet is a minimum of 30 mm above the base of the cistern. This is to prevent sediment passing into the pipework. Connections of feed pipes to hot water apparatus from cisterns should be at least 25 mm above cold distribution pipes, if applicable. This should minimise the risk of scalding should the cistern run dry.

The float-operated valve (ballvalve) is fitted as high as possible and must comply with BS 1212, parts 2 or 3, thus maintaining an air gap and preventing back-siphonage. Overflow pipes should have a minimum internal diameter of 19 mm and in all cases be greater in size than the inlet pipe.

To prevent the ingress of insects, a tight-fitting lid must be provided, with a screened air inlet. Where a vent pipe passes through the lid, the pipe must be sleeved. Overflow warning pipes must also incorporate a filter or screen. Finally, the whole installation (cistern and pipes) must be insulated to prevent freezing.

Coupling of storage cisterns

In larger commercial properties it is desirable to have two or more cisterns coupled together instead of one large cistern. This is beneficial because one of them can be isolated and drained down, if required. Equally, on a smaller scale, lack of space in a house sometimes limits the size of a storage cistern. In such a case, two smaller cisterns can be joined together to give the required capacity. Different methods are used for the above examples. If the need arises to isolate one cistern, an isolating valve is fitted at each point in or out of the cistern, which can be shut off. For the purposes of a domestic house, it would be uncommon for one cistern to be isolated, so the mains supply is usually taken into one cistern and the cold distribution or feed pipe is taken out of the other. By designing it in this way the water in the second cistern would not become stagnant.

The washout pipe shown in the figure is only used on large cisterns (those holding over 2300 litres), for the purpose of draining down and cleaning out any sludge deposits, etc.

Cold supplies to be kept below 25°C

The Water Supply Regulations state that no cold water supplies should be warmed above 25°C. This requirement is to ensure water is not drawn off and wasted, but also reduces the growth of bacteria such as Legionella. Critical temperatures are between 25° and 50°C; therefore it may be necessary to insulate pipes to prevent them becoming too warm due to heat gain.

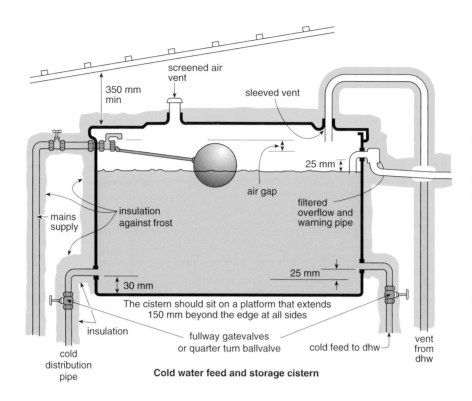

350 mm min

screened air vent

sleeved vent

25 mm

air gap

filtered overflow and warning pipe

insulation against frost

mains supply

30 mm

25 mm

The cistern should sit on a platform that extends 150 mm beyond the edge at all sides

insulation

fullway gatevalves or quarter turn ballvalve

cold feed to dhw

vent from dhw

cold distribution pipe

Cold water feed and storage cistern

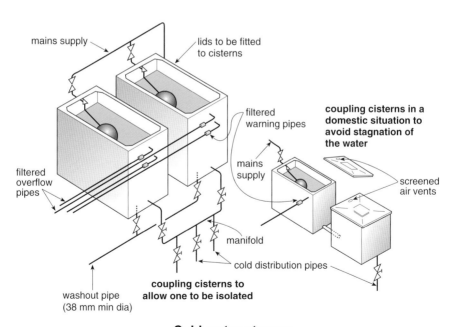

mains supply

lids to be fitted to cisterns

filtered warning pipes

coupling cisterns in a domestic situation to avoid stagnation of the water

filtered overflow pipes

mains supply

screened air vents

manifold

cold distribution pipes

washout pipe (38 mm min dia)

coupling cisterns to allow one to be isolated

Cold water storage

Water Treatment

Relevant British Standard
BS 7593

Before water is supplied to the consumer it is treated and purified by the water authority. When it arrives at the supply point, usually no more treatment is required. However, in hard water areas, where there are varying amounts of calcium salts in the water, it is sometimes desirable to treat the water to prevent excessive scale problems or provide a better liquid with which to wash (e.g. a laundry). The installation of a water softener or a water conditioner fulfils this purpose.

Water softeners

These soften hard water by passing it through a pressure vessel containing zeolites or a resin which absorbs the calcium in the water. After time, the zeolites become exhausted because they become saturated with calcium; they thus need to be regenerated with common salt. This is done automatically, by a system of backwashing which is timed to operate at around 3 AM via a timeclock or flow metering system, thus causing no inconvenience to the householder. Before installing the softener inlet connection, a branch pipe should be taken from the mains to provide a hard water drinking supply.

The installation of a water softener is quite straightforward, the connections being made in the way shown in the figure.

Water conditioners

Water conditioners do not soften water, they just stabilise the calcium salts which are held in suspension. There are two basic types of water conditioner: those that use chemicals and those that pass water through an electric or magnetic field. The calcium salts, if viewed through a microscope, appear star-shaped, and it is in this form that they can bind together.

Chemical water conditioners use polyphosphonate crystals. These dissolve in the cold water and bind to the star-shaped salts, making them circular. These polyphosphonate crystals are placed in the storage cistern or into specially designed containers fitted into the pipeline. Periodically the crystals must be replaced.

Electronic and magnetic water conditioners are devices fitted in the pipeline which pass a low current of a few milliamperes of electricity across the flow of water. This tends to alter the structure of the hard salts, making them more round or solid in shape and therefore they do not stick together but should pass through the system. Electronic and magnetic water conditioners should be installed as close as possible to the incoming main supply. Some types of electronic conditioners are plugged into the mains electric supply, whereas others rely on the current produced by electrolysis.

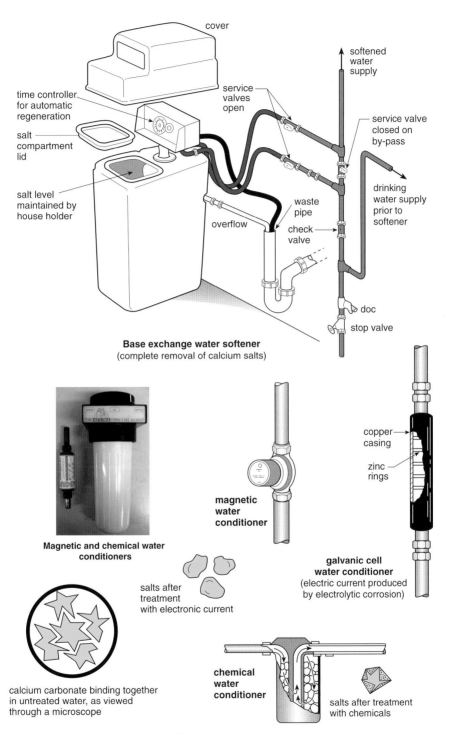

Base exchange water softener
(complete removal of calcium salts)

Magnetic and chemical water conditioners

magnetic water conditioner

galvanic cell water conditioner
(electric current produced by electrolytic corrosion)

copper casing

zinc rings

salts after treatment with electronic current

calcium carbonate binding together in untreated water, as viewed through a microscope

chemical water conditioner

salts after treatment with chemicals

Water treatment

Boosted Water Supplies

Relevant British Standards
BS EN 806-2 and BS 6700

There are two reasons why water may need to be boosted: (1) to give a better flow and pressure at the draw-off point in a domestic situation (see page 118, Connections to Hot and Cold Pipework); or (2) as a method of raising the water supply in high-rise buildings above the height that the mains will supply.

The pumps are usually fitted indirectly to the supply main. If fitted directly, a serious drop in the mains pressure may occur when the pumps are running. The indirect system consists of a suitably sized break cistern located at the inlet to the pumping set (see figure). Nowadays, 'packaged' pumping sets are installed consisting of dual pumps to overcome the problem of failure of (or the need to renew) one of the pumps. The second pump also assists at times of high demand on the system, cutting in as necessary. To prevent pump seizure and stagnation of water, the pumps should be designed to work alternately. Two types of system will be found: those using pressure-sensing devices and those using float switches.

Pressure-sensing devices   These include transducers or pressure switches which sense the drop in pressure in the pipeline. These come fitted to, and form part of, the packaged pumping set. To prevent the continuous cutting in and out of the pumps, a delayed action ballvalve or float switch is used in the high level cisterns. If draw-off points are required on the riser, a pressurised pneumatic storage vessel is sometimes incorporated to prevent the continual cutting in and out of the pumps; this consists of a vessel containing a rubber bag surrounded by a charge of air. When the pumps are running, water can enter and fill the bag, taking up volume and compressing the air. When the pumps are turned off, the compressed air forces the water back out into the pipeline, as and when required.

Float switches   These are devices which rise and fall with the water level. They are therefore located in cisterns or pipelines to sense a drop or lack of water within the system. If the high level cistern is of large capacity, it may be necessary to have a drinking water header to prevent stagnation, or a separate high level cistern for drinking water purposes.

To prevent the pumps running dry for any reason, a sensing device needs to be incorporated in the pipe feeding the pumps, e.g. an in-line sensor or a float switch fitted in the break cistern.

Water hammer can be a problem when fitting pumps to any system. Therefore a hydropneumatic accumulator should be incorporated if necessary. Packaged pumping sets incorporate these as standard. They are basically small pressurised pneumatic vessels, the air taking up the shock wave (for example see page 129).

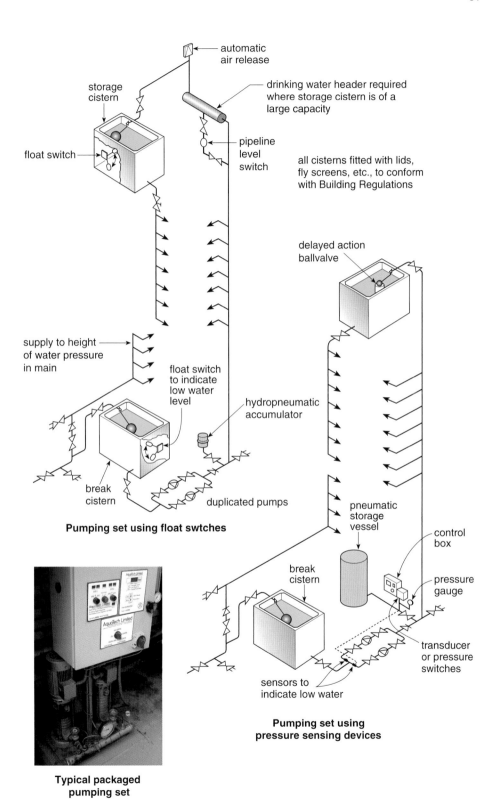

automatic air release

storage cistern

drinking water header required where storage cistern is of a large capacity

float switch

pipeline level switch

all cisterns fitted with lids, fly screens, etc., to conform with Building Regulations

delayed action ballvalve

supply to height of water pressure in main

float switch to indicate low water level

hydropneumatic accumulator

break cistern

duplicated pumps

Pumping set using float swtches

pneumatic storage vessel

control box

break cistern

pressure gauge

transducer or pressure switches

sensors to indicate low water

Pumping set using pressure sensing devices

Typical packaged pumping set

Boosted water supplies

2 Hot and Cold Water Supplies

Fire-fighting Systems

Relevant British Standards
BS 4422 and BS 5306

Sprinkler systems

On the outbreak of fire, a sprinkler system causes an automatic discharge of water to be sprayed, usually from sprinkler heads located near the ceiling. See the figure opposite for a typical arrangement for the pipe layout. See also page 100, Domestic sprinkler systems. In commercial premises, as distinguished below, there are two basic designs:

- The *wet-pipe system*, in which the sprinkler system is permanently charged with water.
- The *dry-pipe system*, in which the sprinkler system is charged with compressed air and is used in unheated buildings where the temperature may fall below 0°C. If they were to be charged with water these pipes would be liable to freeze.

The operating principle of a sprinkler head is that when the temperature around the head rises to a predetermined level, either a water filled glass bulb breaks, or a solder strut melts, allowing the valve to fall and open.

Hose reel systems

When hose reels are used, they should be sited in prominent positions adjacent to exits, so that the hose can be taken to within 6 metres of any fire. Two basic designs of hose reels are available: those which automatically turn on as the hose is reeled out, and those which need to be turned on at the wall. For the latter type, a notice must be provided near the reel indicating the need to turn on the supply. The hose reel must be adequate to supply a minimum of 0.4 litres per second.

The water supplies feeding sprinkler systems and hose reels need to be adequate (see particularly BS 5306); they are usually maintained via a system of boosted water supply.

Wet and dry risers

These systems are for the use of the fire brigade and consist of pipes, (100–150 mm in diameter) running up the building with one or two fire brigade hydrants on each floor. The purpose of this pipe is to save time running canvas hoses up the staircases should the building be on fire. The dry pipe system is used in buildings up to 60 metres in height (20 storeys) and is fitted with an inlet at ground level for the fire brigade to connect to the nearest hydrant. The wet riser is used in taller buildings because the mains pressure would be insufficient to rise to such great heights, and is charged with water under pressure by a booster pump capable of delivering 23 litres/s. The water pressure supplied to the hydrants should not exceed 690 kN/m^2, otherwise damage may occur to the hose. Therefore, a pressure relief valve is fitted at the hydrant; this opens if the pressure is too great, the discharging water returning to the break cistern.

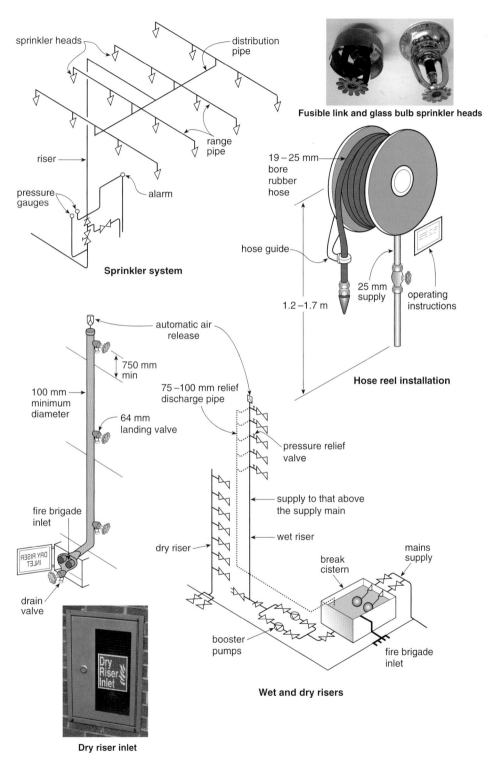

sprinkler heads

distribution pipe

range pipe

riser

pressure gauges

alarm

Sprinkler system

Fusible link and glass bulb sprinkler heads

19 – 25 mm bore rubber hose

hose guide

1.2 – 1.7 m

25 mm supply

operating instructions

Hose reel installation

automatic air release

750 mm min

100 mm minimum diameter

64 mm landing valve

fire brigade inlet

DRY RISER INLET

drain valve

75 – 100 mm relief discharge pipe

pressure relief valve

supply to that above the supply main

dry riser

wet riser

break cistern

mains supply

booster pumps

fire brigade inlet

Dry riser inlet

Wet and dry risers

Fire-fighting systems

Domestic Sprinkler Systems

Domestic sprinkler systems are relatively rare in the UK, but are becoming more and more sought after as a means of protection in the event of a fire. Used in conjunction with a smoke detector they are said to provide 98% protection. Should a fire break out in the home, the longer it burns undetected the more dangerous it becomes. Most people who discover a fire in the home do so too late and as a result many are trapped. The installation of a domestic sprinkler system should only be undertaken by an operative who has completed a recognised training course in design and installation. Water damage will clearly result from the activation of a sprinkler head, however it is approximately one hundred times less damage than could be expected from the fire services at a well-established fire. When considering the installation of a sprinkler system the local fire authority, water company and the fire insurer should be consulted.

Water supplies

One of the first things to consider is the water supply pressure and flow. It is vital that sufficient flow and pressure are available at the sprinkler head at all times, both uninterrupted and reliable. Failure to meet these basic requirements would mean the system is unsuitable. Good water flow is essential to meet the needs of cooling the combustible materials below their ignition temperature and a good pressure is needed to create an effective water spray for the sprinkler head. The water supply needs to maintain the manufacturer's required flow and pressure when discharging through two sprinkler heads at the same time. At least 60 L/min is required through any single sprinkler head, and 42 L/min through any two sprinklers operating at the same time. This would usually require the incoming supply pipe to be at least 25 mm nominal internal diameter. The water may be provided from a direct mains connection providing there is sufficient supply to provide both the domestic water and sprinkler system together plus an additional flow rate of 50 L/min. It is possible to have a system that incorporates a **demand valve** that closes off the domestic water supply should the sprinkler system be activated. Alternatively, a separate dedicated water supply would be needed for the sprinkler system. This could be via a second mains connection or by the use of a stored water supply. Where a stored supply is selected there must be sufficient storage to provide the minimum recommended flow and pressure by the sprinkler head manufacturer for a period of at least 10 minutes.

Sprinkler heads

These need to be of a design that is suitable for residential use. They need to be installed throughout all the habitable parts of the dwelling, each head covering not more than 12 m^2. The sizes available give a nominal diameter at the orifice outlet of between 10 and 20 mm. The temperature ratings for individual heads is given in the table opposite; the head selected should allow for 30°C above highest anticipated ambient temperature.

Sprinkler temperature ratings

Sprinkler type	Temperature range (°C)	Colour code
Fusible link	55–77	None
	80–107	White
Glass bulb	57	Orange
	68	Red
	79	Yellow
	93	Green

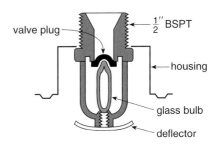

Typical glass bulb sprinkler head

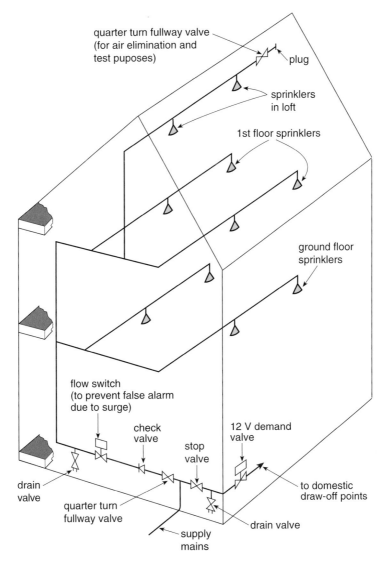

Domestic sprinkler system

Hot Water Systems (Design Considerations)

Relevant British Standards
BS EN 806-2 and BS 6700

When a supply of domestic hot water (dhw) is required, the designer has to consider many factors to ensure the most suitable system for the building in question.

The diagram below gives a brief guide to the system designs which are available. Generally, dhw systems can be divided into centralised and localised. The terms\ break *instantaneous* and *storage* simply indicate whether the water is heated only as required (instantly) or to a temperature indicated by a thermostat and held in a vessel until required (stored).

A **centralised system** is one in which the water is heated and possibly stored centrally within the building, supplying a system of pipework to the various draw-off points.

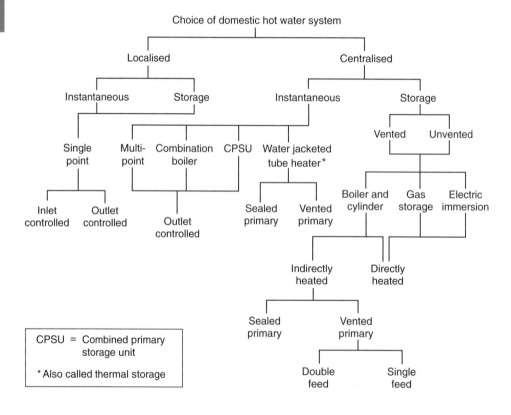

A **localised system** is one in which the water is heated locally to its needs, e.g. a single-point heater located above a sink. It may be chosen where a long distribution pipe would mean an unnecessarily long wait for hot water to be drawn off at the appliance.

In a centralised system, the water may be heated in the hot storage vessel itself, or it may be heated in a boiler or small gas circulator located in a more convenient position. Should this be the case, the water is fed to and from the boiler by what are known as primary flow and return pipes. The circulation of water can be achieved by *convection currents* being set up in the flow and return pipework, or by the use of a circulating pump. Water flows by convection currents as a result of expanding in volume when heated; modern systems utilising gas or oil boilers would require a pumped circulation system.

When water is heated it becomes less dense than cooler water. Because hot water rises, it is drawn off from the top of the storage vessel to supply the various draw-off points (taps). The cold feed is supplied low down in the vessel, thus preventing unnecessary cooling to the previously heated water. At the highest point in the system, a vent pipe is run up to terminate, with an open end just below the feed cistern lid. This pipe is to allow air to escape from the system upon initial filling and allows air in on draining down.

The vent pipe also acts as a fail-safe device should the cold feed become blocked, preventing the expanding water passing back up into the cistern. Should this occur, the water is forced over the vent and discharges into the cistern. The height to which a vent pipe is to rise above the water level in the cold cistern is found by allowing 40 mm for every 1 m head of water in the system, plus an additional 150 mm. For example, if the distance between the lowest point in the system and the water level in a cistern is 5 m, the vent pipe should be carried up above the water level by a minimum distance of: $5 \times 40 + 150 = 350$ mm.

The temperature at which the water is stored in the cylinder should not exceed 60°C. Failure to observe this limit may result in the user being scalded and scale build-up in hard water districts. Storage temperatures below 60°C are concerning as the Legionella bacteria survival time increases as the temperature drops below 60°C. At 60°C the bacteria is killed within a few minutes. Some means of controlling the temperature should therefore be provided. When a primary circuit is used with an independent boiler, the temperature is generally controlled by a cylinder thermostat positioned a third of the way up the base of the dhw cylinder operating a motorised valve; however, for older systems the water temperature may be controlled by a boiler thermostat or a thermostatic control valve located on the return pipe. These last two temperature control methods would no longer comply with the Building Regulations where an oil or gas boiler is installed.

The hot water supply to a bath needs to be regulated to ensure the temperature does not exceed 48°C; this is usually achieved by incorporating an in-line blending valve, mixing some cold with the hot water. Such a valve needs to be installed as close to the appliance as possible to prevent the colonisation of waterborne pathogens.

Direct Hot Water Supply (Centralised)

Relevant British Standards
BS EN 806-2, BS 5546 and BS 6700

Various fuels and systems of design can be used to heat the water in a centralised direct hot water system, including the following:

Electric water heating   A system which uses an immersion heater installed in the hot water storage vessel. This behaves in a similar way to the element in an electric kettle: when the desired temperature is achieved, sensed by a thermostat, the element is switched off. It is essential that the heater element extends to near the bottom of the storage vessel because it will not heat the water below it. The heater should be at least 50 mm from the base of the storage vessel to prevent convection currents disturbing any sediment. Sometimes two heater elements are used, one fitted at low level and one much nearer the top. The higher immersion heater is switched on if, for example, only enough hot water is required to fill a sink, whereas should enough hot water to fill a bath be required, the lower heater is switched on.

The electrical power supply to an immersion heater must come directly from the consumer unit (see page 280) to terminate close to the hot storage vessel with a double pole switch. It is from here that a 21 A heat-proof flex is run to the heater.

Gas storage heaters   Purpose-made vessels which have gas burners installed directly below the stored water. The system incorporates an open flue which must be discharged to the external environment, the flue passing up through the storage cylinder.

Boiler-cylinder system   A system in which a small boiler (e.g. a gas circulator) is located somewhere in the building and the hot water is conveyed via primary flow and return pipes. When using this system to heat the dhw, it is not possible to include radiators on the primary pipework, unless they are made of non-ferrous metals. This is because the water which passes to the boiler is being supplied by the feed cistern and is constantly passing through the pipework as it is drawn off at the taps, thus bringing in entrapped gases which will cause atmospheric corrosion to ferrous metals. Direct systems which use primary flow and return pipes should not be used in hard water areas because scale build-up in the pipework will block the flow.

Instantaneous systems   It is possible to heat the water directly by passing it through a heat exchanger. Several systems are available, e.g. the Multi-point, the water-jacketed tube heater, the combination boiler and the combined primary storage unit. The main disadvantage with instantaneous heaters is that only a limited number of draw-off points can be supplied at once, because of the restricted flow rate through the heat exchanger.

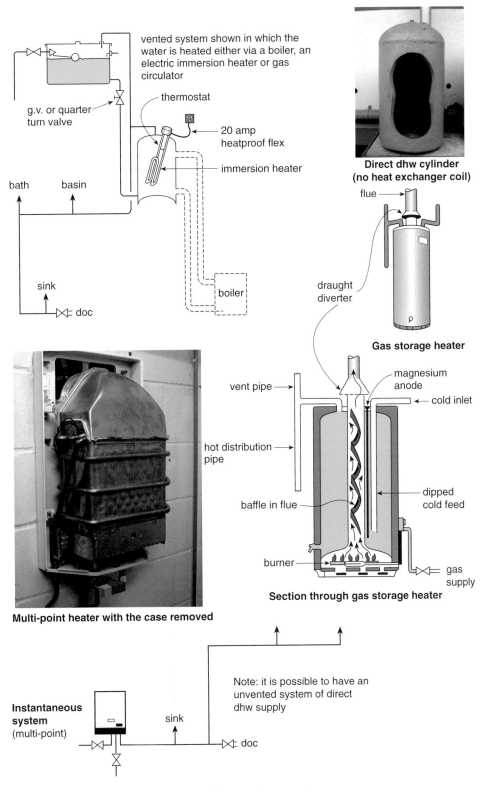

vented system shown in which the water is heated either via a boiler, an electric immersion heater or gas circulator

g.v. or quarter turn valve

thermostat

20 amp heatproof flex

immersion heater

bath

basin

sink

doc

Direct dhw cylinder (no heat exchanger coil)

flue

draught diverter

Gas storage heater

vent pipe

magnesium anode

cold inlet

hot distribution pipe

dipped cold feed

baffle in flue

burner

gas supply

Section through gas storage heater

Multi-point heater with the case removed

Instantaneous system (multi-point)

sink

doc

Note: it is possible to have an unvented system of direct dhw supply

Direct hot water supply

Indirect Hot Water Supply

Relevant British Standards
BS EN 806-2, BS 5546 and BS 6700

The indirect dhw system is probably the most common form of dhw and allows a boiler to be used for central heating purposes also. The storage vessel is the heart of these systems and consists of a special cylinder in which is fitted a heat exchanger. The heat exchanger allows water from the boiler circulating in the primary pipework to pass through, but not mix with, the water in the cylinder itself. Thus, in effect, it really consists of two systems which appear to join at the hot storage vessel.

Water is heated in the boiler and conveyed to the hot storage vessel via primary flow and return pipes by the use of a circulating pump; older systems utilised gravity circulation (convection currents). The water supplying this primary circuit can be taken from the f & e cistern, which is usually located in the roof space, or directly from the supply main in the case of a sealed system (see Part 3, Central Heating). The water, once in the primary pipework, is never changed except for maintenance purposes. Therefore, any calcium carbonate (limescale) will have been precipitated and any gases which came in with the fresh supply will have escaped from the water. Thus, the water in this state is somewhat neutralised; it will not cause excessive corrosion of steel radiators and is suitable for central heating purposes.

Because the primary water must not be changed, the water supply for domestic purposes needs to be taken from a separate supply, and, if using a feed cistern in the roof space, it must be separate from the f & e cistern to prevent mixing of the waters in each system.

As we have seen, most indirect systems are of the double feed design, using two cisterns in the roof space, and having a coil or annulus type heat exchanger fitted into the cylinder. But there is a second older type, known as a single feed indirect system (the design in the figure opposite being known as the Primatic). This system used a specially designed heat exchanger which allowed the primary circuit to fill up via a built-in feed pipe. In so doing it maintained an air break separating the primary and secondary waters. The Primatic system is no longer installed.

The figure shows how the expansion of the water in the primary circuit is taken up by moving the air in the top dome back through its cold feed pipe to the lower dome.

The primary circulation system must not be too large (having many radiators) because the excessive quantity of water that the system will contain would, when expanding, exceed that of the space available in the dome, thus forcing the air out. Also, a circulating pump must not be used on the primary circuit to the dhw cylinder for the same reason. Should the air be lost, the space would be filled instead with water and it would be converted, in effect, into a direct system of hot water supply; this would give rise to corrosion problems.

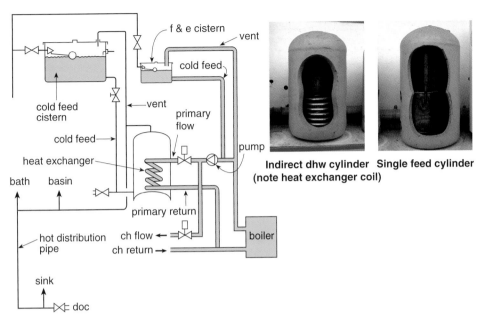

Indirect dhw cylinder **Single feed cylinder**
(note heat exchanger coil)

**Vented double feed indirect system
of hot water supply**

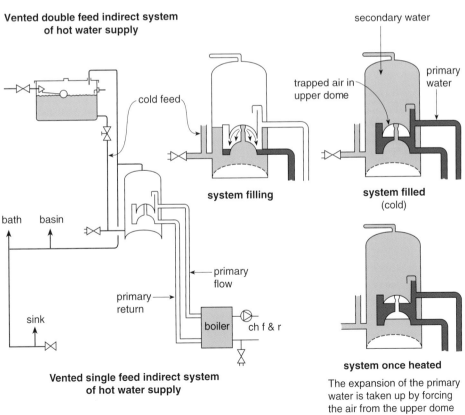

system filling

system filled
(cold)

**Vented single feed indirect system
of hot water supply**

system once heated

The expansion of the primary
water is taken up by forcing
the air from the upper dome
to the lower dome

Indirect hot water supply

Unvented Domestic Hot Water Supply

Relevant British Standards
BS EN 806-2 and BS 6700

When considering the installation of a system of unvented dhw (stored supply in excess of 15 litres), Part G of the Building Regulations should be observed. This identifies several requirements to be met by the local authority. Systems must be installed by an 'approved' installer who is registered with a recognised body and the system must be purchased as a *unit* or *package*. From April 2006, systems installed need to be self-certificated in order to meet the requirement of Part G.

A *unit* is a system in which all the component parts have been fitted by the\break manufacturer at the factory. A *package* is a system in which the temperature-activated controls are incorporated but all other components are fitted by the installer. In both cases this ensures that the safety devices, which are 'factory set', are installed with the system. The Regulations state that at no time must the water reach 100°C, which is ensured by the use of three safety devices: the thermostat (operating at 60°C); a high temperature thermal cut-out device (which locks out at 90°C); and a temperature relief valve (designed to open at 95°C).

To comply with water regulations, a check-valve must be fitted on the supply pipe to prevent a backflow of hot water down the supply main. Should the expansion vessel not function for any reason, the water, on expanding, will be forced out of the pressure relief valve, forming an additional safety device.

A pressure-reducing valve is fitted as a precaution to reduce excessive water pressures which may cause damage to the system. In order to ensure equal pressures at both hot and cold draw-off points, the cold supply pipe is sometimes branched off after this valve.

The advantages of these systems are the higher pressures obtainable at the draw-off points, the use of less pipework and the fact that less time is required for installation. The disadvantages are frequently overlooked. First, such a system can only be installed should the flow rate (volume of water) be sufficient to supply both hot and cold water at once, bearing in mind that several appliances could be running at the same time. Second, in hard water districts, the build up of scale around temperature and pressure relief valves could make those valves ineffective. Regular servicing of the system is therefore essential.

Some unvented dhw cylinders use a sealed expansion vessel to take up the expanding water. Others incorporate an inverted dome which traps a pocket of air, thus doing away with the need for a sealed expansion vessel. Unfortunately, high water turbulence within the cylinder sometimes causes the air trap to be lost, although one manufacturer has overcome this problem by using a floating baffle to cut down on turbulence. After a period of time the air may be absorbed into the water rendering the dome ineffective, resulting in the pressure relief valve opening. Consequently the cylinder will need draining down to recharge the dome with air.

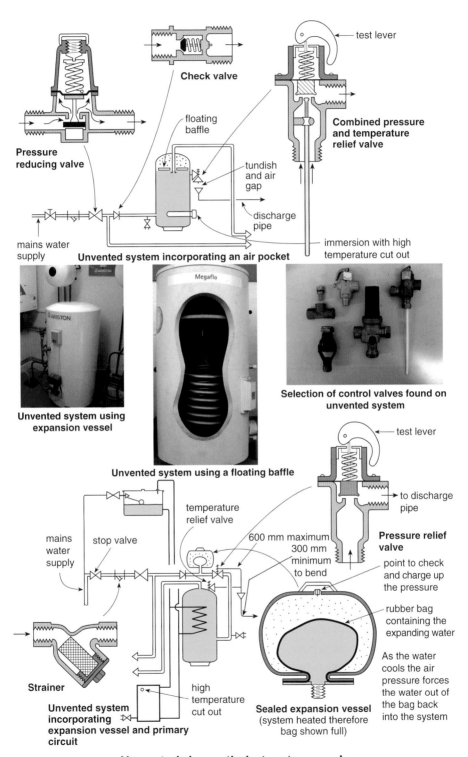

Check valve

test lever

Combined pressure and temperature relief valve

Pressure reducing valve

floating baffle

tundish and air gap

discharge pipe

mains water supply

Unvented system incorporating an air pocket

immersion with high temperature cut out

Unvented system using expansion vessel

Unvented system using a floating baffle

Selection of control valves found on unvented system

test lever

temperature relief valve

to discharge pipe

mains water supply

stop valve

600 mm maximum
300 mm minimum to bend

Pressure relief valve

point to check and charge up the pressure

rubber bag containing the expanding water

As the water cools the air pressure forces the water out of the bag back into the system

Strainer

high temperature cut out

Unvented system incorporating expansion vessel and primary circuit

Sealed expansion vessel
(system heated therefore bag shown full)

Unvented domestic hot water supply

Hot Distribution Pipework

Relevant British Standards
BS EN 806-2 and BS 6700

The hot water in the storage vessel needs to be preserved for as long as possible to save on fuel consumption. The cylinder and pipework should therefore be insulated. However, hot water can also be lost due to the circulatory flow of hot water, by convection currents, up the vertical vent pipe. To prevent this, the hot distribution should be run a minimum of 450 mm horizontally on leaving the top of the hot storage vessel.

When a tap is opened, before the hot water can discharge from the spout, the cold water in the pipe has to be drawn off and is invariably allowed to run to waste. This also causes a certain amount of inconvenience to the user. Where possible the system should be designed to provide a temperature of no less than 50°C within 30s, although this may not be achievable where instantaneous or combination boilers are used. The run of pipe to the appliance is referred to as a 'dead leg' and should, where possible, not exceed that indicated in the following table.

Maximum length of uninsulated hot distribution pipe

Internal bore of pipe (mm)	Length (m)
less than 10	20
11–19	12
20–25	8
Over 26	3

Where it is not possible to keep within these limits, the pipe should be thermally insulated, or some other method should be used to ensure that the hot water appears quickly at the tap. Either of two methods can be adopted to meet this requirement: a specially designed heat tracing tape can be used, which heats up as necessary, maintaining the water temperature; or a system of secondary circulation will need to be installed.

Secondary circulation An arrangement in which a pipe is run back to the dhw cylinder from the furthest point on the distribution, thus forming a circuit. Water can flow around this circuit usually by the use of a non-corrosive circulating pump, thus allowing hot water to be kept close to the draw-off points. The return pipe is connected within the top third of the cylinder to prevent the cooler water, lower down the cylinder, mixing with the hot water and reducing its temperature.

Hydraulic gradient When a tap is opened and water drawn from the system, the water level in any vertical pipe will drop. The amount by which it drops will depend on the size of the pipework and the flow rate being drawn off (see figure). In any cistern-fed system, be it hot or cold, if the pipework is less than perfect, when several appliances are running at once, one will be starved of water (usually the highest draw-off point). This is due to the above-mentioned drop in water level in the vertical pipe.

When installing a system of hot water supply, it is particularly important to consider this design concept, because any high connections in the vertical rise of the vent pipe will be starved of water unnecessarily.

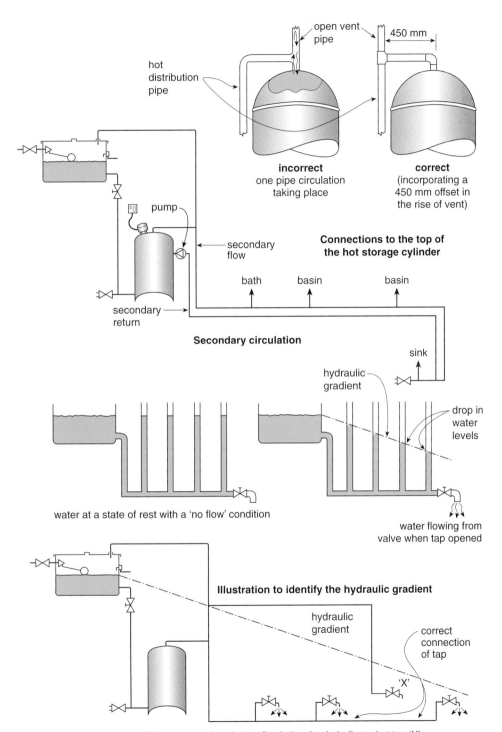

open vent pipe

450 mm

hot distribution pipe

incorrect
one pipe circulation taking place

correct
(incorporating a 450 mm offset in the rise of vent)

Connections to the top of the hot storage cylinder

pump

secondary flow

secondary return

bath   basin   basin

Secondary circulation

sink

hydraulic gradient

drop in water levels

water at a state of rest with a 'no flow' condition

water flowing from valve when tap opened

Illustration to identify the hydraulic gradient

hydraulic gradient

correct connection of tap

'X'

All taps opened and a 'no flow' situation is indicated at tap 'X'
(tap 'X' will only operate providing no other taps are open)

Hot distribution pipework

Heat Recovery Period

This is the amount of time required to heat up a quantity of water. The recovery time varies, depending upon the power rating of the heat source used, but basically the higher the power rating, the shorter the heating up period of the water.

Power of heat source and the heat recovery period are related by the following formula:

$$\text{Power} = \frac{\text{SHC} \times \text{kg} \times \text{temperature rise}}{\text{time in seconds}}$$

SHC = specific heat capacity of water (= 4.186 kJ/(kg k))

kg = the mass of water to be heated (mass of 1 litre = 1 kg)

Example: Find the power required to heat 100 litres of water in 2½ hours from 4°C to 60°C.

$$\text{Power} = \frac{\text{SHC} \times \text{kg} \times \text{temp rise}}{\text{seconds in } 2\frac{1}{2}\text{h}} = \frac{4.186 \times 100 \times 56}{9000} = 2.6 \text{ kW}$$

Note that the above example does not allow for heat loss and for the heating up of the hot storage vessel itself. It would be advisable to add, say, 10% for this; thus\ break 2.6 + 10% = 2.86 kW, and as a result a 3 kW heater would be chosen.

It will be seen that as the incoming temperature of the water has a bearing upon the recovery time, one can only estimate its value. Often the householder will require some guidance concerning the time it will take to heat the water using an existing immersion heater. This is also found using the above formula, although it will need transposing to suit this new situation, to give:

$$\text{Time (seconds)} = \frac{\text{SHC} \times \text{kg} \times \text{temp rise}}{\text{power (kW)}}$$

Example: Find out how long it will take to increase the temperature of 136 litres of water from 10°C to 60°C, using a 3 kW immersion heater. Ignore heat losses, etc.

$$\text{Time} = \frac{4.186 \times 136 \times 50}{3} = 9488 \text{ s or } 2.6 \text{ h} (2 \text{ h } 36 \text{ min})$$

Where the water is heated in a boiler, etc., away from the hot storage vessel, it will need to be conveyed via primary flow and return pipework. Circulating this water to and from the storage vessel can be achieved by gravity circulation or speeded up by the use of a circulating pump (see page 148, Fully Pumped System). The greater the circulation pressure from the heat source to the storage vessel, the shorter will be the heat recovery period. Where gravity circulation is chosen to convey the hot water, it should be understood that the greater the circulating head, the greater will be the circulating pressure. Where only a poor- or low-circulating head can be achieved, larger pipe sizes or the inclusion of a circulating pump will be required. It must be noted that for gas or oil installations gravity circulation is no longer permitted in the design of the system.

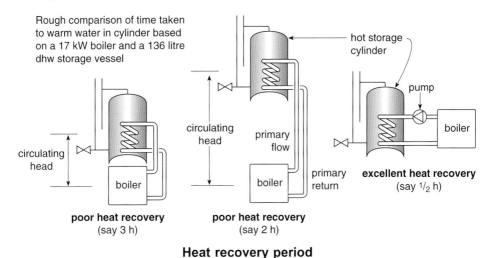

Rough comparison of time taken to warm water in cylinder based on a 17 kW boiler and a 136 litre dhw storage vessel

circulating head

boiler

poor heat recovery
(say 3 h)

circulating head

primary flow

boiler

primary return

poor heat recovery
(say 2 h)

hot storage cylinder

pump

boiler

excellent heat recovery
(say $1/2$ h)

Heat recovery period

The heat recovery period is also influenced by the design and dhw cylinder used. All modern systems use BS 1566 cylinders. These have no less than 5–6 turns within the heat exchanger coil and the cylinder includes factory-fitted insulation foam. High performance cylinders can also be purchased with an even greater number of coils. The following chart shows the number of boiler cycles and approximate time required to raise a typical 136 litre cylinder up to the required temperature.

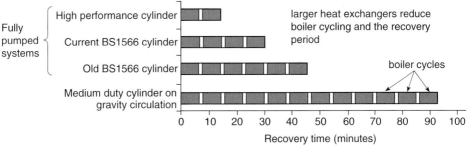

Fully pumped systems

High performance cylinder

Current BS1566 cylinder

Old BS1566 cylinder

Medium duty cylinder on gravity circulation

larger heat exchangers reduce boiler cycling and the recovery period

boiler cycles

Recovery time (minutes)

High performance cylinder
(note the many coils to provide rapid heat recovery)

Instantaneous Domestic Hot Water Supplies (Centralised)

Relevant British Standards
BS EN 806-2, BS 5546 and BS 6700

With instantaneous dhw systems, the principle is to pass the cold water through a heat exchanger, such as a coil of pipework passing through a heat source, which heats the water by the time it comes out the other end. There is a limit to the speed at which the water can be heated; therefore, the flow rate (volume) of water needs to be minimised; failure to minimise the flow rate will result in an insufficient heat-up. Because of this reduced flow rate of water passing through the heat exchanger it is not possible to supply several outlets at once; as a result these systems are unsuitable where there is to be high demand for hot water. With these systems you only heat the water as and when it is required; therefore, a saving can be made in fuel consumption. Instantaneous heaters include multi-points, water-jacketed tube heaters, CPSUs and combination boilers. (See page 154 for combination boilers.)

The multi-point This consists of a gas burner located beneath a heat exchanger. When hot water is required, the water, in passing through the heater, causes the gas valve to open, and this is ignited by a pilot flame. The gas valve opens as the result of a reduction in the water pressure on one side of a diaphragm in the pressure differential valve caused by water passing through a Venturi. Attached to the diaphragm is the push rod which opens the gas line. On shutting the water supply, the pressure in the differential valve equalises and the diaphragm is sprung shut, closing off the gas supply.

The water-jacketed tube heater Sometimes referred to as a thermal storage system, this is a system in which the water to be heated is passed through a stored supply of central heating water. It is like an indirect system of dhw in reverse; the domestic water flows through the heat exchanger and not the primary water. One end is connected to the cold supply main, the other directly to the taps. When the water is heated, a small amount of expansion occurs and is taken up in a small expansion vessel located in the cold supply line fitted to the unit or within the unit itself via an expansion chamber. Because the water in the heat exchanger can potentially become as hot as the water in the primary storage cylinder, it must be noted that water, upon leaving the unit, passes through a thermostatic mixing valve in which the water is cooled from the cold supply to give a temperature no hotter than 60°C.

The combined primary storage unit (CPSU) This system of instantaneous dhw is the same as a water-jacketed tube heater, the only difference being that everything is contained within the boiler and no separate hot water storage vessel is required.

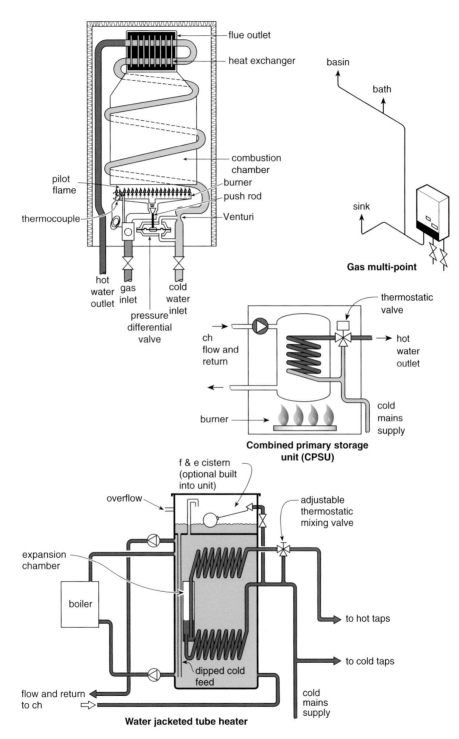

flue outlet

heat exchanger

basin

bath

combustion chamber

pilot flame

burner

push rod

thermocouple

Venturi

sink

Gas multi-point

hot water outlet

gas inlet

cold water inlet

pressure differential valve

thermostatic valve

ch flow and return

hot water outlet

burner

cold mains supply

Combined primary storage unit (CPSU)

f & e cistern (optional built into unit)

overflow

adjustable thermostatic mixing valve

expansion chamber

boiler

to hot taps

to cold taps

dipped cold feed

flow and return to ch

cold mains supply

Water jacketed tube heater

Instantaneous domestic hot water supplies

Localised Hot Water Heaters

Relevant British Standards
BS EN 806-2, BS 5546 and BS 6700

Two distinct types of localised dhw heaters will be found: the instantaneous and storage types. In each case the heater will serve only one sink, or two if fitted in close proximity to the heater.

Instantaneous single points

These heaters are fuelled either by gas or electricity and heat the water only when required. They are usually fitted with a swivel spout and located directly above the sanitary appliance, the water flow usually being inlet controlled. The gas heater works on the same principle as the multi-point (see page 114).

With electric instantaneous heaters, the water is allowed to flow into the heater, where it is surrounded by an electric heating element. Because of the small volume of water surrounding the element, the water quickly heats up as it is drawn through the heater. The temperature of the water will be directly related to the power rating of the appliance and the water flow rate. The water flowing through the heater is sensed by the pressure or flow switch located on the inlet supply, which in turn makes the electrical contacts to the immersion heater element.

Storage type single points

These heaters are located either above or below a sink or similar appliance and have a capacity not exceeding 15 litres. The stored water is heated by an electric element to, for example, 60°C, and on expanding the water is allowed to push up and discharge from the discharge spout. It is important to make the client aware of this dripping spout. When cold water enters the base of the unit it forces the hot stored water out. Obviously the discharge of hot water is limited and will soon start to cool, but it will be sufficient for small quantities of draw-off. When installing these heaters below an appliance, a special design of terminal fitting (tap) needs to be used – one which allows water to flow through the heater but at the same time allows the water to expand (see figure).

Some designs of single points incorporate a small expansion vessel which enables an outlet control valve to be used, eliminating dripping outlet spouts.

**Instantaneous gas
water heater**
outlet controlled

Storage type electric water heater
inlet controlled

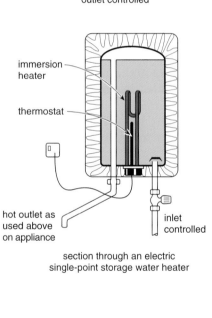

immersion
heater

thermostat

hot outlet as
used above
on appliance

inlet
controlled

section through an electric
single-point storage water heater

electrical
element

electrical
supply

outlet
controlled

inlet

section through an electric
instantaneous water heater

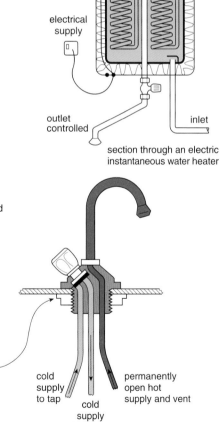

Specialised design of tap to allow for
the expansion of water from the
storage heater when installed
below the appliance

cold
supply
to tap

cold
supply
to heater

permanently
open hot
supply and vent

Localised hot water heaters

Connections to Hot and Cold Pipework

Relevant British Standards
BS EN 806-2 and BS 6700

Connections to showers

The hot and cold supplies to a shower will need to be of equal pressure. Where the supply is directly from the service main, provision must be made to ensure no back-flow occurs. This is usually achieved by ensuring that the shower head cannot discharge below the overspill level of the appliance or by incorporating a double check-valve assembly into the pipeline.

Connections via a storage cistern will need to be such that an adequate pressure is achieved; in general, a minimum distance of 1 metre from the underside of the cistern to the shower head should be maintained. The pressure can be increased by the use of a booster pump fitted into the pipeline in which a small self-contained unit, designed to give a greater head of water, is used. The only proviso is that at least 150 mm of initial head is available to allow the flow switch to operate and start the pump when the supply is opened.

It is possible to use a flow-activating button where no head at all is available for some designs of pump. Booster pumps may be fitted before or after the shower-mixing valve, although in general they should be installed in such a location as to ensure that it is constantly flooded with water. Where a shower booster which draws more than 12 litres/min is considered, the water authority may need to be advised regarding its use.

To ensure that the shower is never starved of water, the cold supply to the shower mixing valve should be independent of other draw-off points and its connection to the storage cistern should be below that of the cold feed to the dhw cylinder.

The temperature of the water to the shower is regulated by means of manual or thermostatic control. With the manually controlled valve, a dial on the control head is turned to open or close either the hot or cold porthole size, thus restricting the flow. Thermostatic mixing valves are fitted with a temperature-sensing device which is designed to expand due to heat and should maintain a constant outlet temperature, opening or closing the portholes automatically as required.

Connections to bidets

There are two types of bidet: those with pillar taps to give an over-the-rim type discharge, thus maintaining an air gap; and those with a submerged nozzle which discharges a spray of water upwards from the base of the appliance. Those with an ascending spray are not permitted to be connected directly to the supply main and must have their hot and cold supplies run via separate distribution pipes, independent of other draw-offs. This can be achieved as shown in the figure. Note that a check valve and an additional separate vent pipe from the hot distribution pipe to the appliance are required.

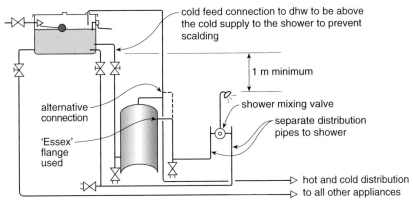

cold feed connection to dhw to be above the cold supply to the shower to prevent scalding

1 m minimum

shower mixing valve

separate distribution pipes to shower

alternative connection

'Essex' flange used

hot and cold distribution to all other appliances

Storage fed shower

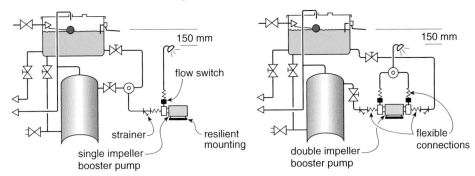

150 mm

flow switch

strainer

resilient mounting

single impeller booster pump

150 mm

flexible connections

double impeller booster pump

Installation of shower booster pumps

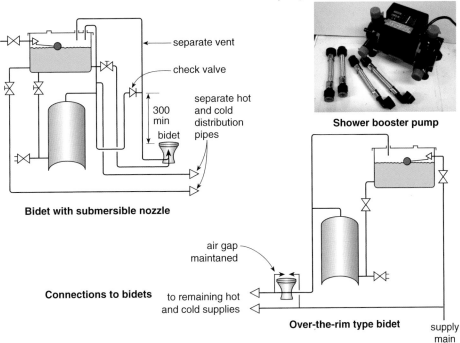

separate vent

check valve

300 min

bidet

separate hot and cold distribution pipes

Shower booster pump

Bidet with submersible nozzle

air gap maintaned

Connections to bidets

to remaining hot and cold supplies

Over-the-rim type bidet

supply main

Connections to hot and cold pipework

Installation of Pipework 1

Relevant British Standards
BS EN 806-2 and BS 6700

Pipe supports

There are many designs of pipe support bracket and the one chosen will depend upon the material nature of the pipe, the cost allowed for the job, and upon circumstances; for example, it would be pointless to use plastic pipe clips in schools or hospitals, etc., where they could very easily be damaged. Whatever pipe support is chosen, the fixing must be secure to prevent damage and the possible development of air locks. As a guide, the general recommended pipe support spacings are given in the following table, but one must remember that one clip too many is better than one clip too few and, in many cases, plumbers have to use their own judgement.

Maximum spacings for internal pipework (m)

Pipe size		Copper pipe		Steel pipe		Plastic pipe	
(mm)	(in)	horizontal	vertical	horizontal	vertical	horizontal	vertical
15	$\frac{1}{2}$	1.2	1.8	1.8	2.4	0.6	1.2
22	$\frac{3}{4}$	1.8	2.4	2.4	3.0	0.7	1.4
28	1	1.8	2.4	2.4	3.0	0.8	1.5
35	$1\frac{1}{4}$	2.4	3.0	2.7	3.0	0.8	1.7
42	$1\frac{1}{2}$	2.4	3.0	3.0	3.6	0.9	1.8
54	2	2.7	3.0	3.0	3.6	1.0	2.1

Design considerations

If the pipe is to run through structural timbers such as floor joists, it is essential that the structural members are not weakened. Notches and holes should be as small as practicable, but should also allow for pipe expansion and contraction. The size and position of a notch or hole need to be considered and should not exceed the dimensions in the figure.

Example: For a joist 200 mm deep and 3 m long (measured from centre line of the bearing) any notch must have a maximum depth of H ÷ 8

Therefore notch depth $200 \div 8 = 25$ mm

and be at least 0.07 L from its bearing; therefore $0.07 \times 3000 = 210$ mm and no greater than L ÷ 4 from its bearing; therefore $3000 \div 4 = 750$ mm.

Note: Joists ≤100 mm must not be cut or drilled and the maximum depth of joist to be considered is 250 mm, therefore a 300 mm joist is only regarded as being 250 mm.

In vented systems the pressure is usually quite poor in comparison with mains supply pipework. Therefore, it is essential to run the pipework with no dips or high spots, which may allow a trap of air to form, causing a blockage (air lock). To this end, pipes should be run horizontal, or to an appropriate fall, allowing the air to escape from the system.

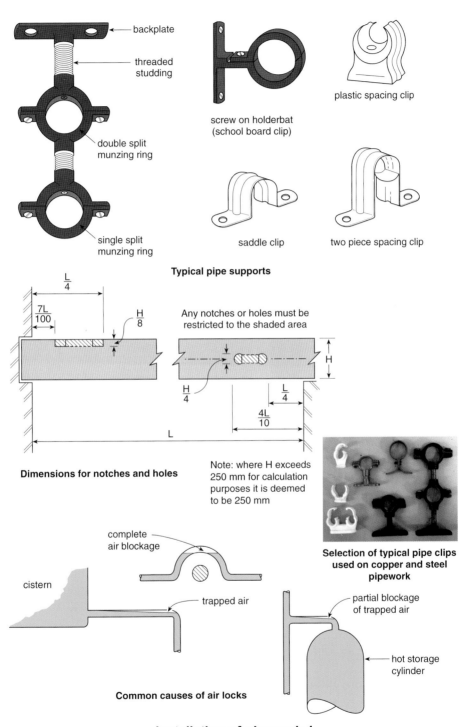

backplate

threaded studding

double split munzing ring

single split munzing ring

screw on holderbat (school board clip)

plastic spacing clip

saddle clip

two piece spacing clip

Typical pipe supports

Dimensions for notches and holes

Any notches or holes must be restricted to the shaded area

Note: where H exceeds 250 mm for calculation purposes it is deemed to be 250 mm

Selection of typical pipe clips used on copper and steel pipework

complete air blockage

cistern

trapped air

partial blockage of trapped air

hot storage cylinder

Common causes of air locks

Installation of pipework 1

Installation of Pipework 2

Relevant British Standards
BS EN 806-2 BS 5422 and BS 6700

Thermal insulation

When installing any pipework, steps need to be taken to prevent the pipe contents from *freezing* in the case of cold water supplies and *losing heat* in hot water systems.

Therefore, some form of insulation may be required. Insulation materials entrap small air pockets which are a poor conductor of heat. Should a pipe freeze, damage can occur to the pipework; therefore, insulation is imperative in exposed locations. Roof spaces and garages must be regarded as exposed. The insulation requirements depend on the thermal conductivity of the insulation material. The following table gives a guide to the minimum insulation for frost protection in housing.

Minimum thickness of insulation material

Outside pipe diameter (mm)	Indoor installations (e.g. roof space)		Outdoor installations (including below ground)	
	Flexible foam and expanded plastic (mm)	Loose-fill (mm)	Flexible foam and expanded plastic (mm)	Loose-fill (mm)
0–15	22	89	27	100
16–22	22	75	27	100
23–42	22	75	27	89
43–54	16	63	19	75
Flat surfaces	13	38	16	50

It should be noted that smaller pipes require a greater thickness of material because they are more prone to freezing up.

Accessibility of pipework

It is a water regulation that all water pipes and fittings be readily accessible for inspection and repair. Often the designer/installer has no wish to see exposed pipework; therefore, a duct or chase is made in the wall, etc., and enclosed by a cover to allow movement and ease of removal. This cover may be covered at the choice of the client with plaster or a tile finish (see figure for examples). Note that thermal insulation material may also be required in exposed or unheated dwellings, although not shown in the examples. It is only possible to encase the pipe in the floor screed where it forms part of a heating element of a closed circuit of under-floor radiant heating.

When passing pipes through floors or walls, the pipe should be sleeved to allow movement, and the space between the pipe and sleeve 'fire stopped' to prevent the passage of smoke and flame.

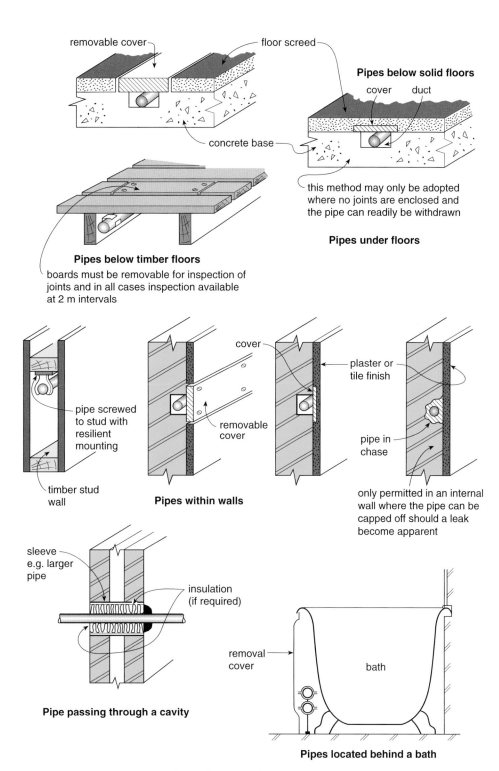

Pipes below solid floors

this method may only be adopted where no joints are enclosed and the pipe can readily be withdrawn

Pipes under floors

Pipes below timber floors

boards must be removable for inspection of joints and in all cases inspection available at 2 m intervals

pipe screwed to stud with resilient mounting

timber stud wall

Pipes within walls

plaster or tile finish

pipe in chase

only permitted in an internal wall where the pipe can be capped off should a leak become apparent

sleeve e.g. larger pipe

insulation (if required)

removal cover

bath

Pipe passing through a cavity

Pipes located behind a bath

Installation of pipework 2

Sizing of Hot and Cold Pipework

Relevant British Standards
BS EN 806-2 and BS 6700

For everyday plumbing in a domestic dwelling, the installer uses a rule of thumb (tried and tested) method which consists of 15–22 mm pipe on the supply main and a 22–28 mm pipe for the hot and cold distribution pipework. In each case the larger size is chosen if there are many draw-off points, the pipe slowly reducing in size as necessary to each appliance. For bigger systems requiring many outlets over a large area or several floors, the main distribution pipe run will need to be sized correctly to ensure sufficient pressure and flow at the draw-off points, without excessive noise problems.

Shown opposite is a completed table, which acts as a key to the figure below it. An explanation of the stages carried out to choose the pipe sizes is as follows.

Column 1 This is the pipework which is being sized; note that the system is broken into various sections.

Column 2 The flow required is found by making an assessment of the probable maximum demand of water, in litres per second, at any given time, because it is very unlikely that all the sanitary appliances will be used at once. To perform this assessment, a method has been devised based upon the theory of probability, in which a loading unit rating is given to each type of sanitary appliance.

Sanitary appliance	Loading unit
WC cistern	2
Bath	10
Wash basin	$1\frac{1}{2}$
Sink and washing machine	3
Shower	3
Bidet	1

By multiplying the number of each type of appliance by its loading unit and adding together the results, the total loading units for the system will be found.

To convert loading units to flow rate (litres per second), the conversion table opposite is used.

Thus, in our example, where each floor has one bath, two basins and two WC cisterns, the number of loading units is:

$$\text{Bath:} \qquad 1 \times 10 = 10$$

$$\text{Wash basin:} \quad 2 \times 1\frac{1}{2} = 3$$

$$\text{WC cistern:} \quad 2 \times 2 = \underline{4}$$

$$\text{Total:} \qquad \qquad 17 \text{ loading units}$$

From the conversion table opposite, 17 loading units gives a flow rate of 0.4 litres/s. Note that for pipe section A all five floors are being served; therefore the total loading units for this section will be: $17 \times 5 = 85$, which converts to a flow rate of 1.1 litres/s.

Cold distribution pipe serving five flats

1	2	3	4	5	6	7	8	9	
Section	Flow rate	Suggested pipe size	Velocity	Loss of head	Effective pipe length	Frictional head	Progressive head	Actual head	Notes
	(l/s)	(mm)	(m/s)	(m)	(m)	(m)	(m)	(m)	
A	1.1	28	1.8	0.2	×16.5	=3.3	3.3	3	Undersized
A	1.1	35	1.25	0.07	×18.2	=1.27	1.27	3	✓
B	0.92	28	1.5	0.15	×4.5	=0.68	1.95	6	✓
C	0.8	22	2.2	0.43	×4.0	=1.72	3.67	9	Possible noise
C	0.8	28	1.3	0.12	×4.5	=0.54	2.49	9	✓
D	0.6	22	1.6	0.27	×4.0	=1.08	3.57	12	✓

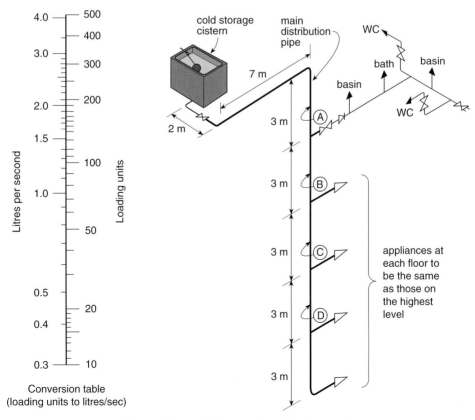

Pipe sizing of hot and cold pipework

Column 3 An assumption is made at this stage that a particular pipe size is correct, and the table will confirm (or reject) its possible use.

Columns 4 and 5 The velocity and loss of head can simply be read from the graph opposite. To use, take a horizontal line from the flow in litres per second to intersect the pipe diameter; from this point the readings can be taken.

Column 6 The effective length of pipe run is found by adding the actual net length of pipe to the length of pipe due to frictional loss.

Equivalent pipe lengths, due to frictional loss through copper fittings

Pipe o.d (mm)	Elbow (m)	Tee (m)	Stopcock (m)	Check-valve (m)
15	0.5	0.6	4.0	2.5
22	0.8	1.0	7.0	4.3
28	1.0	1.5	10.0	5.6
35	1.4	2.0	13.0	6.0
42	1.7	2.5	16.0	7.9
54	2.3	3.5	22.0	11.5

In the example given, the effective length for section A, with an assumed pipe diameter of 28 mm, is:

$$\text{Actual length} = 12.0 \text{ m}$$

$$\text{Three elbows} = 3.0 \text{ m}$$

$$\text{One tee} = \underline{1.5 \text{ m}}$$

$$\text{Total} = 16.5 \text{ m}$$

Column 7 The frictional head is found by multiplying the loss of head by the effective pipe length (i.e. column 5 × Column 6).

Column 8 The progressive head is the sum total of the frictional heads for each section above the section in question.

Example: The progressive head for section C will be sections A + B + C

$$\text{Therefore } 1.27 + 0.68 + 0.54 = 2.49 \text{ m}$$

Column 9 The actual head is the total head available; it is standard practice to measure this vertically from the underside of the storage cistern to the end of the section of pipe in question.

In conclusion, one estimates a suggested pipe diameter and completes the table for the section to prove its suitability for use. The pipe size is correct provided that the progressive head does not exceed the actual head and the velocity does not exceed 2 m/s in cold water pipework and 1.5 m/s in dhw systems, thus limiting noise transmissions.

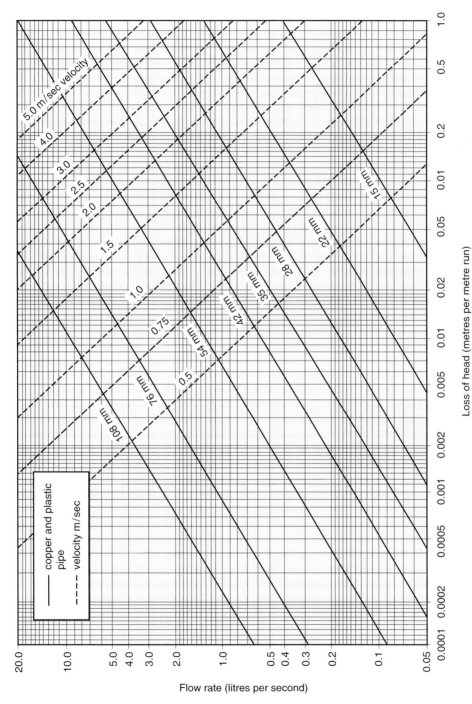

Pipe sizing of hot and cold pipework

(Reproduced with permission of The Institute of Plumbing and Heating Engineering)

Noise Transmission in Pipework

Relevant British Standards
BS EN 806-2 and BS 6700

Impulsive noise (water hammer)

A hammering noise which occurs in high pressure water pipes; the noise is caused by surges of pressure. There are two basic noise types:

(1) A noise which consists of a sudden loud bang and is often caused by a loose stopcock jumper or washer which quickly flips shut onto the seating. It may, on the other hand, be caused by pipes which have not been fixed correctly and flap about; in which case the noise is caused by sudden back surges of pressure, perhaps created by the rapid closing of a tap.

(2) Oscillation or ballvalve murmur which consists of a series of bangs or rumbles generated in the pipeline. The noise is created by a float-operated valve quickly opening or closing, this being caused by ripples or waves which form on the surface of the water in a storage cistern. To overcome this problem, a larger ballfloat is often used or a damping plate fitted to the float or ballvalve lever arm; alternatively, baffle vanes are fitted in the cistern to prevent waves forming.

One method often employed to cure water hammer is to shut down slightly the incoming stop-valve. This does not reduce the pressure but it reduces the flow rate of the water.

Flow noise   A general noise caused by the water flowing through the pipework. If the water velocity is kept below 3 m/s in high-pressure pipework and below 2 m/s in\break low-pressure pipework, this flow noise will not be significant. Sudden changes of direction and minimal downstream pressures can cause cavitation, which is turbulence, causing air bubbles to form and collapse. Flow noise can result unnecessarily when one uses roller pipe cutters and fails to remove the internal burr.

Pump noise   When a booster pump has been provided, noises should not be generated by water pressure and flows, providing the pump is correctly sized. However, the noise of the motor running may give concern, in which case isolation of the pump from the building is the only answer. Noise transmission from pumps can be reduced by using rubber-type connections to the pipework and installing the pump on a resilient mounting.

In large buildings, excessive water noise problems may be overcome by the installation of a hydropneumatic accumulator. This consists of a rubber bag into which the water can enter; the air surrounding the bag is charged to just below the system working pressure. Should a shock wave occur in the pipework, the pressure surge is taken up by the cushion of air.

Splashing noise   When water drops into cisterns the falling water can be somewhat noisy. To prevent this, a collapsible silencer tube can be used, consisting of a polythene bag. It is sometimes possible to discharge the water onto an inclined plate, which breaks the waterfall and inhibits the resulting splashing sound.

Movement noise   When pipes, especially hot pipes, expand and contract they need to move and as a result must not be restricted (see pages 120–123 which identify good pipework installation practices).

Entrapped air bubbles and boiler noise This is the result of poorly installed pipework allowing air to be entrapped in dhw systems. Sometimes, owing to scale build-up in the flow from a boiler, a boiling noise (kettling) is generated, which is caused by steam forming and condensing. A similar noise is also caused where a flame impinges onto the heat exchanger, causing local hot spots.

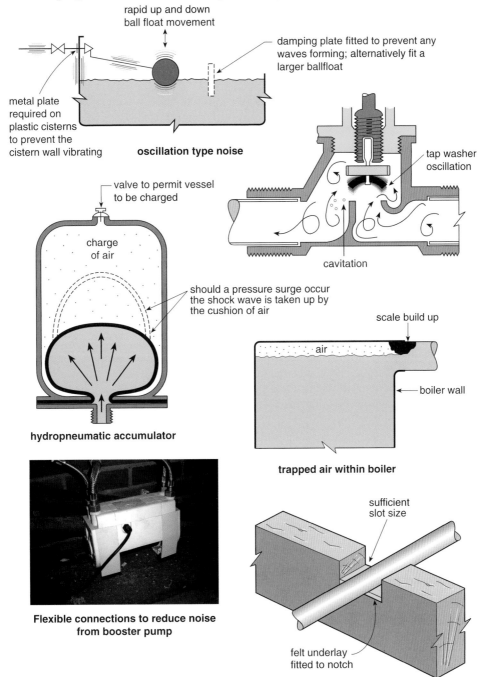

rapid up and down
ball float movement

damping plate fitted to prevent any
waves forming; alternatively fit a
larger ballfloat

metal plate
required on
plastic cisterns
to prevent the
cistern wall vibrating

oscillation type noise

tap washer
oscillation

valve to permit vessel
to be charged

charge
of air

cavitation

should a pressure surge occur
the shock wave is taken up by
the cushion of air

scale build up

air

boiler wall

hydropneumatic accumulator

trapped air within boiler

sufficient
slot size

**Flexible connections to reduce noise
from booster pump**

felt underlay
fitted to notch

Noise transmission in pipework

Commissioning of Hot and Cold Supplies

Relevant British Standards
BS EN 806-2 and BS 6700

When a system of hot or cold water supply has been installed, it should be inspected and tested appropriately. The following checklist should be followed:

(1) **Visual inspection**   All pipework should be inspected to ensure it is fully supported and free from jointing compounds, flux, etc. The feed or storage cistern should be cleaned and made free of swarf. All valves should be closed to allow filling up in stages.

(2) **Testing for leaks**   The system should be slowly filled in stages. Air should be expelled by opening the highest draw-off point on the section being tested. The~testing is carried out in stages so that any leaks can be identified easily. Before opening the isolation valve to a cistern, the float-operated valve seating should be temporarily removed to allow any grit, etc., in the pipe to flow into the cistern, thus preventing the blockage of the small hole in the seating itself. When the valve is reassembled, the water level should be adjusted as necessary.

Sometimes it is desirable to test the installation using a hydraulic test pump, giving a test pressure of one and a half to two times the system working pressure. This is achieved as shown in the figure, the test pressure being maintained for at least 1 hour. Pipework which is to be encased must be tested in this fashion prior to the connection of the water supply to prevent the need for costly removal of the encasement cover.

(3) **System flushing and disinfection**   All systems, large or small, will require the flushing through of the pipeline to remove flux residuals, wire wool, etc., from inside the pipe – usually achieved by opening the tap or draining off for a period of time. Domestic hot water systems should be flushed both cold and hot; this will give a better removal of the deposits from within the system. The BS requires that all systems other than those of private domestic dwellings be disinfected before being put into use. This is carried out after any initial flushing and is achieved by dosing the system with a measured quantity of sodium hypochlorite solution. The procedure outlined over the page should be observed.

(4) **Performance tests**   Every terminal fitting, e.g. draw-off point, float-operated valve, should be checked for suitable flow rate (volume of water) and pressure, and that it is operating to give suitable performance. For example, it would be pointless having an appliance with 3 bar pressure if the water only passed through a pinhole; it would take for ever to fill.

The performance test must be carried out under probable flow demands, i.e. several appliances should be opened at once. Generally the pressure and flow performance tests are carried out by visual inspection, but it is possible to use a pressure gauge and flow measuring device to ensure that the performance meets the required specification. A check should also be made at this stage for noise transmissions in the pipeline caused by rapid closure of valves.

All dhw thermostats will need to be inspected for correct operation and adjusted to maintain the required water temperature, not exceeding 60°C.

(5) **Final system checks** After any problems have been resolved, cistern lids should be secured, insulation material applied, and labels fixed to valves for identification purposes, as necessary. The job should be left clean and tidy.

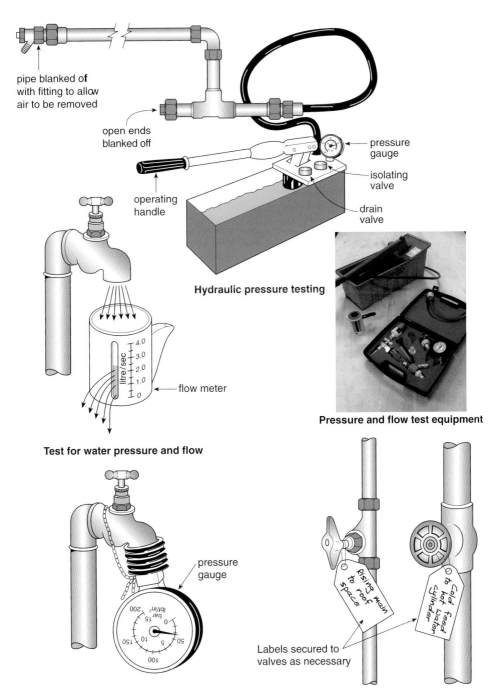

pipe blanked off with fitting to allow air to be removed

open ends blanked off

pressure gauge

isolating valve

operating handle

drain valve

Hydraulic pressure testing

litre/sec

4.0
3.0
2.0
1.0
0

flow meter

Pressure and flow test equipment

Test for water pressure and flow

pressure gauge

lbf/in²
bar

Labels secured to valves as necessary

Rising main to roof space

to Cold feed hot water cylinder

Commissioning of hot and cold supplies

Disinfection of Hot and Cold Water Systems

Relevant British Standards
BS EN 806-2 and BS6700

In relation to industrial and commercial hot and cold water systems used for potable water supplies it is a requirement that the system is maintained in a condition fit for its purpose. As a result these water systems need to be disinfected in the following circumstances:

- Where it is a new installation or has had a major extension
- Where it is suspected that contamination may have occurred
- Where a system has not been in regular use
- Where water is stored in a way that could create the risk of legionnaire's disease.

Disinfection is carried out by thoroughly flushing and filling the system with a\break measured quantity of chlorinated water at 50 ppm. Chlorination is achieved using a sodium hypochlorite solution (similar to household bleach). The solution may be of various strengths, usually between 5 and 10%, therefore the quantity added to the system needs to be carefully considered. Where unknown it is best to assume it to be 10%. Thus if we wish to create a concentration of 50 ppm we would require $50 \div 1\,000\,000 \times 100 \div 10 = 0.0005$ parts of sodium hypochlorite to one part water, which equates to **1 litre to every 2 000 litres** of water in the system.

Assessment of system volume   This can be calculated by using the formulae identified on page 34. To assist, the following table identifying water volumes in copper tubes may be of use.

Water volume (litres per meter run) for copper tube to BS EN 1057

Tube size	Litres/m	Tube size	Litres/m	Tube size	Litres/m
15 mm	0.145	35 mm	0.835	66.7 mm	3.247
22 mm	0.320	42 mm	1.232	76 mm	4.197
28 mm	0.539	54 mm	2.091	108 mm	8.659

Safety   Prior to undertaking any process of disinfection you must give notice to all parties, including the water authority who may have concern for the water discharge. Locate and mark all outlets with an appropriate warning notice advising of the operation taking place. In order to avoid toxic fumes being generated no other chemicals, such as toilet cleaner, should be added to the water discharge until the process has been completed. COSHH data sheets should be reviewed and appropriate personal protective equipment used (i.e. goggles and gloves) when mixing the solution. Where necessary it is possible to remove the chlorine from the water by mixing with a neutraliser solution, such as sodium thiosulphate, added at a rate of:

System volume $(m^3) \times ppm(mg/L)$ chlorine $\times 2 =$ No. of grams required

Disinfection procedure for gravity fed systems

1. Undertake the safety procedures described and position warning notices
2. Thoroughly flush new systems to remove flux residuals and swarf
3. Calculate the capacity of the cistern and add to this water 1 litre of sodium hypochlorite for every 2000 litres (to check dose is correct see below).
4. Working from the cistern, open each draw-off point until the disinfection solution is detected, usually by smell. Then close this fitting and move to the next outlet, slowly progressing around the system, drawing the solution to all points. As the chlorinated water is drawn off from the cistern it will be necessary to maintain the 50 ppm concentration.
5. Once the entire system is full of chlorinated water at the correct dose the system is left for 1 hour.
6. After the hour has elapsed the chlorine level should be checked in the cistern and at selected outlets. Where this level is less than 30 ppm you need to re-dose and leave for a further hour.
7. Upon successful completion the chlorine can be neutralised before draining.
8. Finally the system is thoroughly flushed with clean water until the residual chlorine level at the outlets is no higher than that present in the supply mains (0.2 ppm).

Disinfection procedure for mains fed systems

Where no storage cistern is available to introduce the sodium hypochlorite solution it will be necessary to introduce a temporary branch connection. This allows for a pump, checkvalve and cistern, as shown below, to be introduced, thereby allowing the solution to be added.

During dosing the supply would need to be isolated. It may be possible to add the solution to the highest point in the system, eliminating the need for a pump.

Checking the chlorine concentration

This is achieved by using a simple colourimetric test. The procedure involves filling a small clear plastic tube with a sample of water to be checked. To this is added a special tablet. The tube is shaken to dissolve the tablet and the colour of the water is inspected, which, when compared with the following table, gives an indication of the chlorine level.

Water colour	Chlorine level (ppm)
Clear	0
Light pink	0.2–1
Dark pink	1–5
Red	5–10
Purple	10–20
Blue	20–30
Grey/green	30–50
Yellow/brown	Over 50
Colour develops then clears	Excessive

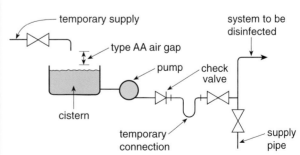

Temporary connection for dosing mains supplies

Maintenance and Servicing Schedule

Relevant British Standards
BS EN 806-2 and BS 6700

No system can be guaranteed for ever, but its life expectancy can be greatly improved by identifying faults before they have a chance to cause inconvenience. Planned preventive maintenance, regularly carried out, will not only help to ensure that the system performs as it was intended, but may also prevent costly damage to equipment and buildings. A maintenance schedule is generally drawn up, and should be observed, giving guidance on what to look for when fulfilling the terms of a servicing contract.

Shown is a typical schedule as used when inspecting a system of hot and cold water supplies.

Maintenance schedule

Date of Inspection: _____ Inspected by: _____ Remarks: _____ _____ _____	Inspection carried out at:		
Component	**Remarks**	**Inspected**	**Notes**
Meters	* Read meter and check water consumption for early signs of wastage * Confirm in correct working order		
Meter and stop valve \quad chamber	* Ensure ease of opening to access covers * Clean out as necessary		
Earth bonding	* Check for alterations and suitable earth bonding maintained		
Water analysis	* '6 month' chemical and bacteriological analysis of drinking water systems where bulk storage exceeds 1000 litres		
Inspection covers and ducting	* Ensure ease of opening to covers and clean out service ducts as necessary * Check for signs of leakage from pipework and surrounding ground or surface water * Check for the accumulation of gas		
In-line control valves	* Operate and confirm easy and effective operation * Labels clearly identify their purpose and are securely attached * Emergency valve keys readily available		
Terminal valves	* Check for suitable operation and effective closing * Remove scale build-up and clean sprayheads of shower mixers, etc. * Check timing delay of self-closing taps * Adjust water levels to float operated valves * Check for suitable pressure and flow at outlets		

Component	Remarks	Inspected	Notes
Pipework	* Check supports and inspect for loose fittings * Check provision is maintained for expansion and contraction * Check for soundness of pipework * Inspect for signs of corrosion * Inspect insulation material for soundness * Inspect fire stopping to ensure that it is maintained		
Storage cisterns	* Confirm the cleanliness of vessels * Look for signs of leakage * Check for stagnant water (e.g. dust on surface of water) * Check condition of cistern supports * Confirm operation of overflow * Ensure lid and insulation are sound		
Pumps	* Check operation of any pump(s) fitted and ensure noise levels are mimimal		
Pressure and temperature relief valves	* Open test lever to confirm valve not stuck down * Check discharge pipe not blocked		
Pressure-reducing valves	* Check pressures downstream of valve		
Pressure vessels	* Inspect for corrosion and leakage * Drain vessels of water and measure gas pressure; adjust if necessary		
Filters	* Remove gauze/mesh trap and clean out		
Electrical components	* Check operation of all controls to include thermostatic devices * Check suitability of wiring to IEE standards		

Any system that has not been fully maintained may fail to meet the requirements of an insurance policy should a system failure result in damage to the property.

Operative taking water flow, pressure and temperature measurements

Part 3
Central Heating

Plumbing, 4th Edition. R. D. Treloar.
© 2012 Blackwell Publishing Ltd. Published 2012 by Blackwell Publishing Ltd.

Domestic Central Heating

Relevant British Standards
BS EN 12828 and BS 5449

The term *central heating* refers to a method of heating a building from a central heat source; it is this which distinguishes it from various forms of localised heating such as gas/coal fires or electric storage heating located within a room. The plumber is only concerned with wet systems of central heating; however, the warm air system is very often connected to primary flow and return pipework to a hot storage vessel. A design of central heating can also be found which uses such materials as electric cable embedded in the walls, designed to warm the structure; the concept is the same as that of radiant systems heating (see page 176).

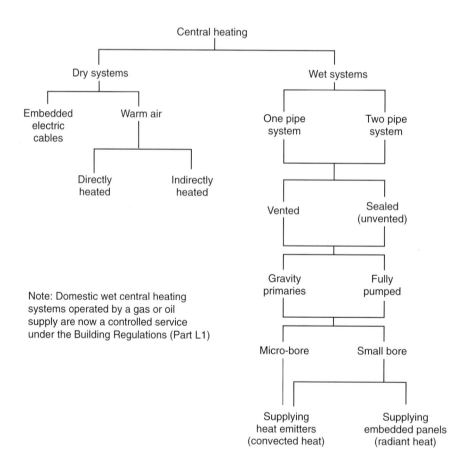

Note: Domestic wet central heating systems operated by a gas or oil supply are now a controlled service under the Building Regulations (Part L1)

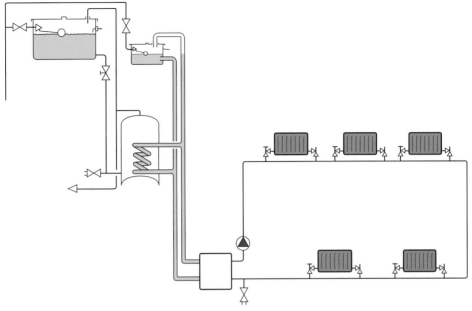

**Small bore 'one pipe' vented heating system
with gravity primaries to the dhw**

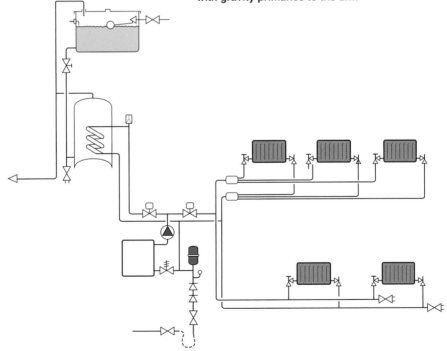

**Fully pumped 'two pipe' sealed heating system
with micro-bore to first floor radiators**

Domestic central heating

Wet Central Heating Systems

Relevant British Standards
BS EN 12828 and BS 5449

Originally, systems were designed to circulate hot water by convection currents; because these systems had a very low circulating pressure, however, larger pipe sizes were required. More modern systems use a pump to speed up this circulation. Because of this increased flow rate, it is possible to use smaller pipes which contain less water and thus heat up more quickly. Installing a pump also enables freedom in system design; for example, the water can be pumped to circulate below the level of the boiler as it does not rely upon convection currents. The water which is heated up in the boiler is often also used to circulate, either by pump or by convection currents, to an indirect hot water storage vessel, giving a domestic hot water supply.

One pipe system   This system reduces installation costs, less pipework being required. It does, however, have some disadvantages when compared to the more expensive two pipe systems. First, the first heat emitter in the system passes its cooler water back into the main flow pipe; this results in the heat emitters at the end of the heating circuit being cooler than those at the beginning; therefore, careful balancing of the system is essential. Second, because of its design, the pump only forces water around the main flow pipe and not through the individual radiators, these being heated only by convection currents. For this reason, the heat emitter used must offer only a very little resistance to the natural upward flow of hot water.

Two pipe system   With this system, the water is not only pumped around the circuit but also through the radiators, giving them a much shorter heating up period. Balancing out the heat to each radiator proves to be reasonably simple. It is not uncommon for the first radiator in the system to have its lock-shield valve just fractionally opened when balancing; this is due to the minimal frictional resistance to the flow through this heat emitter.

Two pipe reversed return system (three pipe system)   This is a special design of the two pipe system in which the length of each heating circuit to each heat emitter in the system is about the same. When the cooler water leaves the first heat emitter in the system, it does not simply join the return pipe and travel back to the boiler, as in the two pipe system; instead, it travels to the furthest point in the system and upon receiving the return water from the last heat emitter, runs back to the boiler return connection. This ensures that frictional resistance to the water flow is the same to each radiator. Therefore, although the three pipe system can prove a little more complicated and expensive to install, balancing of the system proves to be a simple task.

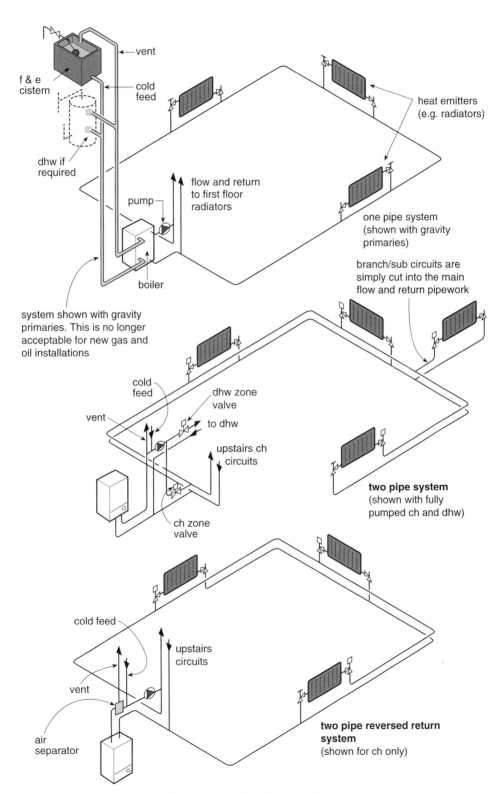

vent

f & e
cistern

cold
feed

heat emitters
(e.g. radiators)

dhw if
required

flow and return
to first floor
radiators

pump

one pipe system
(shown with gravity
primaries)

branch/sub circuits are
simply cut into the main
flow and return pipework

boiler

system shown with gravity
primaries. This is no longer
acceptable for new gas and
oil installations

cold
feed

dhw zone
valve

to dhw

vent

upstairs ch
circuits

two pipe system
(shown with fully
pumped ch and dhw)

ch zone
valve

cold feed

upstairs
circuits

vent

air
separator

two pipe reversed return
system
(shown for ch only)

Wet central heating systems

Central Heating Components 1

Relevant British Standards
BS EN 442, BS EN 215 and BS 2767

Heat emitters

There are two main types of heat emitter: the radiator and the convector. Radiators, despite their name, give off very little radiant heat; they simply expose a large hot surface to the room, convection currents are set up, and the room is heated.

Radiators will be found in several designs (see figure). One of these, the convector radiator, has fins or plates welded to its back side which improve its ability to warm the air, exposing a greater surface to the air flow. The plates are heated by conduction from the hot radiator surface. **Convectors** have a relatively small finned heating surface through which hot water passes. A series of closely attached fins warm up by conduction, through which the air passes and is warmed. Two basic designs of convector will be found: those relying on the setting up of natural convection currents and those which use a fan to assist air circulation.

Radiator valves

These are valves fitted to the heat emitters: generally one valve is fitted to each end. The valves are identical except that one is fitted with a lockshield head to prevent unauthorised people from tampering with the regulated flow of water. The lockshield valve is used only when balancing out the heating system, ensuring an equal distribution of hot water; if for any reason a radiator has to be removed from the wall, the valve can be shut down. The other radiator valve has a wheel head which can be used to turn the heater on and off.

Modern systems use thermostatic radiator valves (TRVs) to open and close the hot supply automatically when the room requires heat. Either the valve is fitted with a built-in heat sensor, or the sensor can be fitted in a better position away from the valve. Remote sensors prove useful if, for example, the radiator valve is often covered with a curtain. As the sensor heats up, a volatile liquid expands and is forced into the bellows chamber; this causes it to expand and exert a pressure on a pin, closing the valve.

When installing thermostatic radiator valves in a system, it is essential that not all the heat emitters are fitted with a means of thermostatic control as rooms fitted with a room thermostat, designed to provide boiler interlock (see page 158), must be fitted with a manually operated valve. One manufacturer utilises a wireless TRV valve in conjunction with a control box, thus providing boiler interlock without the need for a separate room thermostat. When all the TRVs are closed the boiler will be shut down. Most modern TRVs are bi-flow, meaning that they can be fitted on either the flow or return pipe. However, a check may need to be made as the valve may be inoperable or subject to undue noise. Should a radiator need to be removed at any time, it is generally necessary to remove the temperature sensing head and secure the pressure pin down with a special manual locking nut; otherwise, should the temperature drop in the room, the valve may open, resulting in the discharge of water onto the floor.

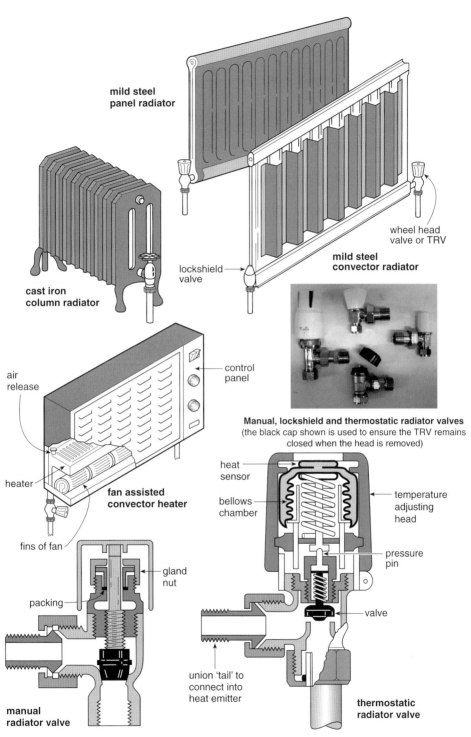

mild steel panel radiator

cast iron column radiator

lockshield valve

wheel head valve or TRV

mild steel convector radiator

air release

control panel

heater

fan assisted convector heater

fins of fan

Manual, lockshield and thermostatic radiator valves
(the black cap shown is used to ensure the TRV remains closed when the head is removed)

heat sensor

bellows chamber

temperature adjusting head

pressure pin

gland nut

packing

valve

union 'tail' to connect into heat emitter

manual radiator valve

thermostatic radiator valve

Central heating components 1

Central Heating Components 2

Feed and expansion (f & e) cistern A small cistern located at the highest point of a vented system of central heating. The cistern is designed to act initially as the fill-up point for the system, but its prime function is to allow the water in the system to expand. For this reason, on initial fill-up, the water level must be adjusted to just above the cold feed outlet into the system. The cistern must be placed in such a location that it will not be affected by the position and head of the circulating pump. A minimum dimension of the maximum head developed by the pump, divided by three, needs to be maintained between the water level and the pump to prevent undue water movement in the f & e cistern (see figure). When a c.h. system is connected to an indirect dhw storage cylinder, the f & e cistern should ideally be located just below the water level of the dhw storage cistern; thus, if a split in the heat exchanger occurs and the c.h. and dhw systems mix, the fault will be identified by water discharging from the f & e cistern overflow.

Circulating pumps These devices, sometimes called accelerators, are fitted to the pipework to assist water circulation. Basically, pumps utilise a circular veined wheel which draws water in through its centre and throws it out by centrifugal force. The water velocity should not be too fast as noise will be generated. Generally, the velocity should not exceed 1 m/s for small bore systems and 1.4 m/s for micro-bore systems. The duty load of a pump should overcome the resistance of the index circuit (the circuit offering the greatest frictional resistance to flow). The location of a pump should, if possible, be such that it gives a positive pressure within the circuit, thus ensuring no air is drawn into the system via micro-leaks (e.g. air being sucked in through radiator gland nuts).

Automatic air release valve A specially designed valve which enables air to escape from the system by allowing a float to rise and fall with the water level in the system. If water is present, the float will be held up, forcing a washer against the outlet seating. When these valves are fitted to heating circuits, they must only be installed in 'positive' flow pipework; otherwise, if installed on the negative (low pressure) side of a pump, air will be drawn into the system when the pump is running.

Anti-gravity valve A valve fitted vertically in the pipeline, designed to overcome unwanted gravitational circulation in central heating pipes. During the summer months when the boiler is used to heat the dhw, the radiators will sometimes get hot due to gravitational circulation. The valve will only open when pressure is created by a pump. The pressure exerted by convection currents is insufficient to cause the valve to lift. Modern systems employing fully pumped circuits incorporating motorised valves do not require these valves to be fitted.

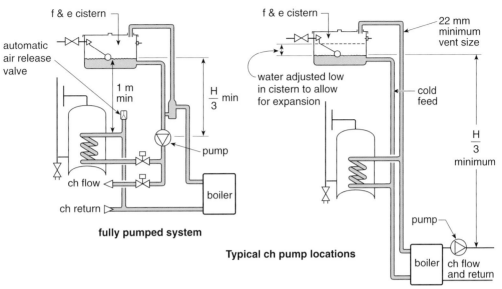

automatic air release valve

f & e cistern

1 m min

$\frac{H}{3}$ min

pump

ch flow

ch return

boiler

fully pumped system

f & e cistern

water adjusted low in cistern to allow for expansion

22 mm minimum vent size

cold feed

$\frac{H}{3}$ minimum

pump

boiler | ch flow and return

Typical ch pump locations

pumped ch, gravity primaries
(found on older systems or systems utilising solid fuel)

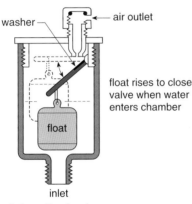

washer

air outlet

float rises to close valve when water enters chamber

float

inlet

Automatic air release

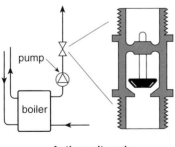

pump

boiler

Anti-gravity valve
valve opens only when pump is running

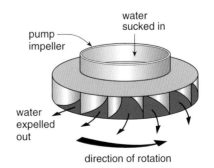

Domestic circulating pump

water sucked in

pump impeller

water expelled out

direction of rotation

Central heating components 2

Heating Controls

Motorised valve   A valve with an electrically operated motor fitted on top which opens or closes the pipeline automatically. The power supply operating the valve is regulated by either a thermostat or a time clock. There are two basic types of motorised valve, these being a *zone valve* and a *diverter valve*. The zone valve simply opens and closes the waterway and is fitted in a straight run of pipe, whereas a diverter valve is fitted at a 'tee' connection and sends the flow of water either one way or the other; this valve can be wired to give priority to either the domestic hot water or the heating circuit. Many diverter valves are designed to have a midway position, allowing water to flow in both directions at the same time. Systems using one zone valve are sometimes referred to as 'C' plan systems; where two zone valves are used, the system is called the 'S' plan, and where a mid-position three port valve is incorporated it is called the 'Y' plan. See page 288, Central Heating Wiring Systems.

Programmer   This is a device consisting of a time clock which automatically switches the boiler and pump on and off and other controls which enable the user to override the time clock settings. The programmer also allows users to control the system so that it only heats the water they require; for example, so that it only heats such domestic hot water as may be required during the summer months. Programmers can be of two designs: (1) the basic design brings on both the heating and dhw at the same time; (2) the full programmer has independent time control for all circuits, i.e. both heating and hot water. The basic programmer fails to comply with the Building Regulations.

Room thermostat   This is a thermostat designed specifically to control the temperature within a building. When the desired room temperature is achieved, the thermostat breaks a switch contact turning off a pump, boiler or some other control and thus preventing the flow of heat to the room. A room thermostat should not be located where it would be affected by extreme temperatures; for example, in cold draughts or near any heat source. A good location would be in a living room or lounge at a height of about 1.5 m providing no additional heat source is used, e.g. fire.

Cylinder thermostat   This thermostat is designed to control the temperature of the dhw. It is clamped to the outside wall of a hot storage cylinder a third of the way up from the base, and makes or breaks the electrical circuit, usually to a motorised valve, allowing or preventing the circulatory flow of primary water from the boiler. The thermostat should be set so that it provides 60°C water at the top of the cylinder.

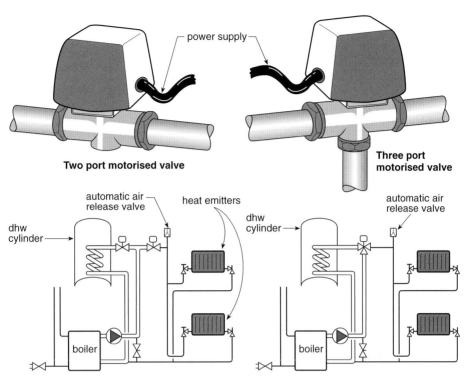

power supply

Two port motorised valve

Three port motorised valve

dhw cylinder

automatic air release valve

heat emitters

dhw cylinder

automatic air release valve

boiler

boiler

Fully pumped systems incorporating motorised valves

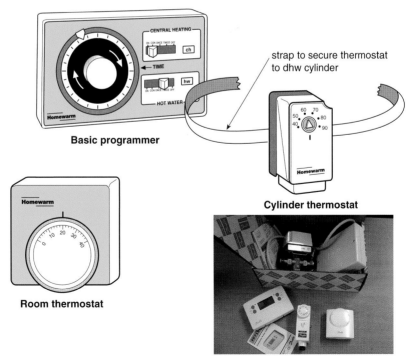

Basic programmer

strap to secure thermostat to dhw cylinder

Cylinder thermostat

Room thermostat

**Heating controls package
(bought as a complete set)**

Heating controls

Fully Pumped System

Relevant British Standards
BS EN 12828 and BS 5449

This is a system which operates fully under the influence of a pump. It does not rely upon convection currents to circulate hot water to the hot storage vessel and can therefore have its boiler above or below the hot storage vessel. The motorised valves may be fitted either on the flow or return pipework and are generally wired up to the cylinder and room thermostats and made to close automatically when the desired temperature within the cylinder or room has been reached.

When designing a vented fully pumped system supplied via an f & e cistern, care must be taken to locate the circulating pump in such a position as to ensure no positive or negative pressure at the vent pipe which could lead to pumping water over or sucking air in from this open-ended pipe. The position at which the cold feed enters the system is regarded as the neutral point and from this point to the pump it will be under a sucking, or negative, influence. From the pump back to the cold feed inlet, a pushing or positive force will occur. The ideally designed system should have a positive flow, pressurising the system. If the vent pipe connection is within 150 mm of the cold feed connection, the vent will also be located at the neutral point and as a result no problems should occur regarding pumping water over or sucking in air.

An **air separator** is sometimes used. This allows cold feed and vent connections to be closely grouped; it also causes a turbulent of water flow in the pipe run which allows the formation of air bubbles which can simply rise up and out of the system, eliminating unnecessary corrosion problems.

All domestic c.h. and dhw systems heated by oil or gas should use a fully pumped design in order to comply with the Building Regulations (Part L1). Systems heated by solid fuel, however, may use gravity dhw in order to overcome the problem of removing the excessive heat from the boiler. It should be noted that a shorter heat recovery time can be achieved with a fully pumped system. This can be shortened further by taking the primary flow into the lowest connection of the dhw cylinder heat exchanger coil. This allows for a greater heat transference by conduction and convection due to the greater heat difference between the primary and secondary waters. The greater the heat difference, the greater the heat transference. This concept is shown in the bottom illustrations opposite.

Reversed circulation

This sometimes occurs in fully pumped systems in which some radiators get hot when not required (see figure). This is the result of a pressure difference between the two tee fittings at A and B when valve 'X' is closed. A water flow is set up, by-passing the closed motorised valve. This situation is simply avoided if the flow to the radiator circuits is split after the flow to the dhw cylinder and rejoins before the dhw return connection.

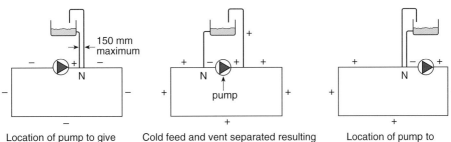

Location of pump to give negative pressure (system works but is undesirable)

Cold feed and vent separated resulting in water pumping over the vent. Face the circulator (pump) the other direction and air would be sucked in through the vent

Location of pump to give positive pressure

Illustration showing principle of correct pump location

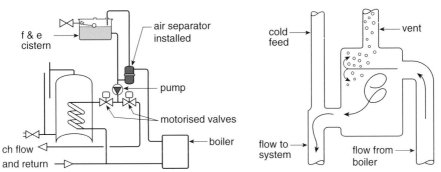

Fully pumped system using an air separator

Three tapping air separator
(the cold feed is introduced as shown)

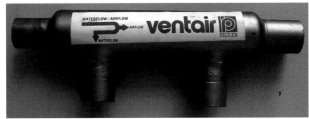

Four tapping air separator

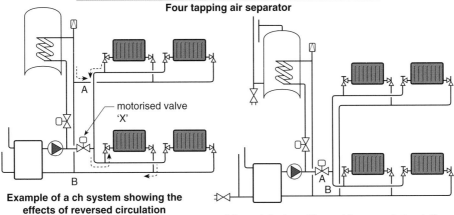

Example of a ch system showing the effects of reversed circulation

A layout designed to avoid reversed circulation

Fully pumped system

Sealed (Closed) Systems

Relevant British Standards
BS EN 12828, BS 4814 and BS 5449

These are hot water heating systems in which the water supplied to the system is fed from the supply main via a temporary supply pipe; connected to the circuit is a double check-valve assembly or some other means of preventing back-siphonage. Once the system is filled with water, the inlet supply is shut off; thus all the water is entrapped in the circulating pipework. Note that the temporary fill connection should be removed after filling. When the boiler is fired, the water heats up and expands and because this water (which is expanding) cannot be taken up in a feed and expansion cistern, it generates a pressure on the internal pipework. Eventually this pressure acts upon a diaphragm and compresses the air and nitrogen gas located in a sealed expansion vessel, thus taking up the expansion of the water.

To allow for water replacement, which may be necessary because of leakage or venting, etc., either the temporary hose has to be replaced or the system is charged up to a pressure slightly in excess of the expansion vessel pressure. In the latter case the make-up water will be taken up in the vessel itself, but care must be taken to ensure sufficient capacity in the expansion vessel to allow for the expanding water. Alternatively, a make-up cistern can be located above the highest point of the system which in turn can be filled either manually or connected to the supply main. The expansion vessel should be connected to the system on the inlet side of any pump, thus preventing the exertion of positive pressures on the diaphragm; also, the vessel should be located on the cooler return pipe to give a longer life.

A pressure gauge is installed to indicate the fill and system pressure. This should ideally be located close to the expansion vessel and fill connection. At no time must the temperature of these systems be allowed to exceed 100°C: a high temperature cut-out device must therefore be installed in addition to the normal boiler thermostat. To prevent excessive pressures building up within the system, a pressure relief valve is fitted to open at 2–3 bar pressure. Any discharging water should be safely conveyed to a suitable drain point, via a tundish (a funnel-shaped pipe which must maintain an effective air gap).

It must be noted that because pressure is created in this type of system, higher water temperatures can be reached because water boils at a higher temperature under pressure. Therefore, convector heaters or heating panels are often chosen in preference to panel radiators to prevent anyone from being scalded. The advantages of sealed systems over more conventional systems include:

- Less pipework is necessary on installation
- The pumping of water over vent pipes or drawing air into the system as in fully pumped systems is eliminated
- Higher water temperatures can be achieved
- The boiler can be positioned anywhere, even in the roof space, as no header cistern is required.

Sealed system controls are commonly found incorporated within a combination boiler.

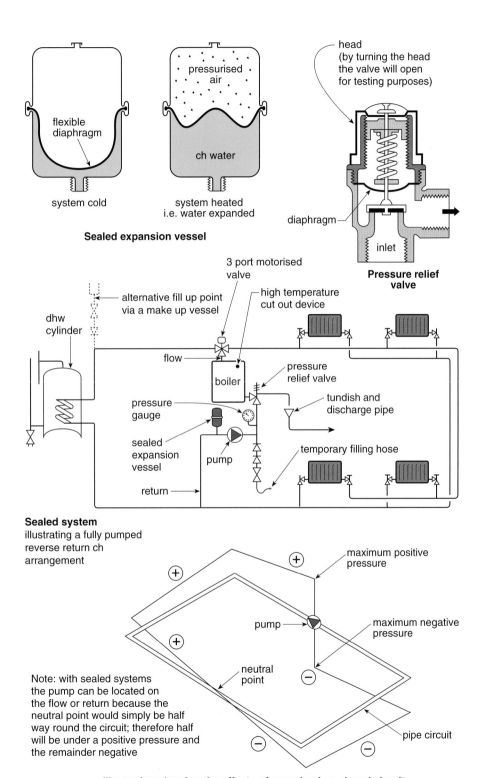

Sealed expansion vessel

flexible diaphragm

system cold

pressurised air

ch water

system heated i.e. water expanded

head (by turning the head the valve will open for testing purposes)

diaphragm

inlet

Pressure relief valve

3 port motorised valve

alternative fill up point via a make up vessel

high temperature cut out device

dhw cylinder

flow

boiler

pressure relief valve

tundish and discharge pipe

pressure gauge

sealed expansion vessel

pump

temporary filling hose

return

Sealed system
illustrating a fully pumped reverse return ch arrangement

maximum positive pressure

pump

maximum negative pressure

neutral point

pipe circuit

Note: with sealed systems the pump can be located on the flow or return because the neutral point would simply be half way round the circuit; therefore half will be under a positive pressure and the remainder negative

Illustration showing the effects of pumping in a closed circuit

Sealed system

Boilers

Relevant British Standards
BS EN 12828, BS 5449 and BS 6798

Solid fuel boilers   These include those which burn such materials as wood, straw and coal. The design can vary tremendously from the 'pot' burner, in which the heat exchanger (area containing the water) surrounds the combustion chamber, to a design which burns smokeless fuels such as anthracite pellets (hard coal) fed automatically into the burner via a hopper, and giving off heat as it rises through a series of waterways.

Gas boilers   The heat exchanger of a gas boiler is a close network of waterways through which the hot gases pass. The sooting-up of this heat exchanger is not a concern because of the cleanliness of the fuel and its combustion process.

Oil boilers   These can be found in several designs although only those using a pressure jet burner are produced today. Oil burners can produce excessive carbon deposits (soot), therefore the heat exchanger needs to be designed so that it can easily be cleaned out. It consists of a chamber surrounded by the waterway, the heat being directed onto the walls by a series of baffles. See also p. 250.

Electric boilers   Two types of electric heater are used: (1) A storage heater uses cheap rate night-time electricity to heat an element which warms up a series of refractory blocks. During the day when the heat is required, a fan blows air around a closed circuit which in turn warms and blows onto a water-filled heat exchanger. (2) The second type of electric heater is of a fairly new design and consists of a copper tube surrounded by a series of electric heating elements. It is installed singularly or in banks of two to three heaters and is capable of supplying a wet central heating system of considerable size.

Boiler noises and design considerations   Noises created in the boiler are often due to the formation of scale, especially in the flow pipe; air is entrapped by the scale and a kind of boiling noise (often called kettling) ensues, caused by steam forming and condensing. The scale can generally be removed by treating the system with a descaling solution. With solid fuel boilers and low-water-content boilers, it may be necessary to have a heat leak from the boiler to allow the dissipation of residual heat. A heat leak may be a boiler bypass for solid fuel or a heat emitter; sometimes the dhw cylinder can be used, providing that the circuit cannot be closed, allowing the heat to circulate and thus escape from the boiler.

In the specification of any boiler, one needs to consider the heat *input* and heat *output*. The heat input is the result of the fuel being consumed, whereas the heat output is the energy produced and available for use. For example, a 75% efficient gas boiler which has a 17 kW input will only provide 12.75 kW output to the system. With the exception of condensing boilers, when commissioning the system, after the initial warm up, the water should never be allowed to flow back to the boiler below 55°C, as condensation will form in the heat exchanger, resulting from the combustion of the fuel, and will cause corrosion problems, shortening the life of the boiler appreciably.

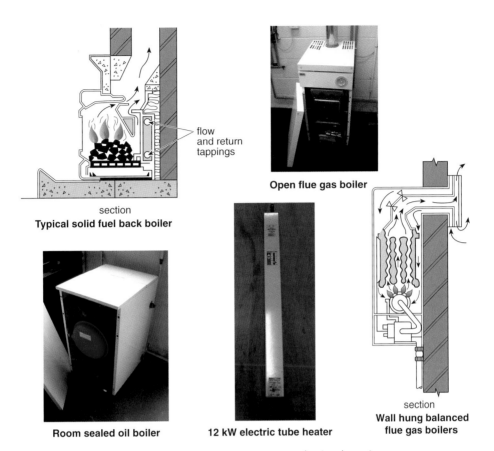

flow
and return
tappings

section
Typical solid fuel back boiler

Open flue gas boiler

Room sealed oil boiler

12 kW electric tube heater

section
**Wall hung balanced
flue gas boilers**

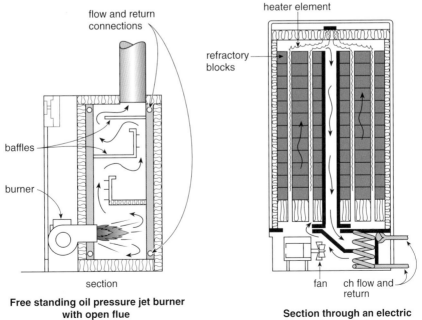

flow and return
connections

heater element

refractory
blocks

baffles

burner

section
**Free standing oil pressure jet burner
with open flue**

fan   ch flow and
return

**Section through an electric
storage boiler**

Boilers

Combination Boiler (Combi)

This is a specially designed boiler which is used to heat up the domestic hot water instantly, as and when required, and also to serve a system of hot water central heating. The combination boiler reduces installation costs because no feed and storage vessels are required for the supply of water; also, by omitting the storage of domestic hot water, this boiler saves money which might have been spent heating the water unnecessarily.

Combination boilers are only suitable for energy-efficient homes in which the demand for hot water is limited. There is a limit both to the volume of water and the speed with which it can be heated up, and also, during time of dhw demand, there is no c.h. When considering whether or not to install a combination boiler, one needs to ensure that the flow rate of water is sufficient to supply the volume of water needed and possibly, in the case of direct mains supply systems, to allow for all hot and cold draw-off points. Therefore, if several taps are opened at once it may lead to some appliances being starved of water.

There are many variations of combi, all working on different design concepts. One such system operates as follows:

(1) Should the central heating system call for heat, the pump is energised. This starts the water flowing. As the water passes through a Venturi, a pressure differential occurs in the deficiency valve causing the gas valve to open.
(2) Gas flows through the main burner and is ignited.
(3) The water is rapidly heated in the low-water-content heat exchanger and can only circulate around the boiler through the dhw heat exchanger. The expansion of the water is taken up in the sealed expansion vessel.
(4) As and when the temperature of the water reaches 55–60°C, the thermostatic element expands, causing the hot water system valve to close and the heating system valve to open; this allows water to flow around the heating circuit.
(5) The closing of the hot water system valve also causes a rod to rise and activate a micro-switch. This notifies the boiler control box that higher temperatures can be achieved which are manually determined by the setting of the flow temperature selector, on the control panel, and range from 60 to 90°C.

Should domestic water be required the following operation takes place:

(1) When a hot draw-off point is opened, water flows through the differential pressure valve; this causes the diaphragm to lift and activate a micro-switch, energising the pump.
(2) The main burner ignites and the boiler functions, as in (1), (2) and (3) above.
(3) As the cold water passes over the thermostatic element, it keeps the element cool, ensuring that it does not expand, causing the heated water to flow out to the heating circuit. Therefore the central heating hot water only circulates through the dhw heat exchanger and around the boiler.
(4) As the cold water passes over the dhw heat exchanger, it is rapidly heated before being discharged through the hot tap.

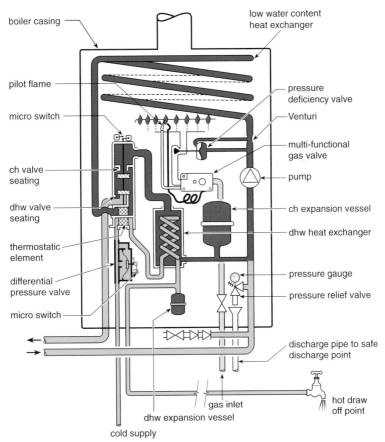

boiler casing

low water content
heat exchanger

pilot flame

pressure
deficiency valve

micro switch

Venturi

multi-functional
gas valve

pump

ch valve
seating

ch expansion vessel

dhw valve
seating

dhw heat exchanger

thermostatic
element

pressure gauge

differential
pressure valve

pressure relief valve

micro switch

discharge pipe to safe
discharge point

gas inlet

hot draw
off point

dhw expansion vessel

cold supply

**Inside view of a condensing
combination boiler**
Note the plate heat exchanger to the bottom
left and sealed expansion vessel above it.
The condensing boiler principle is explained
on the next page.

Combination boiler

High Efficiency or Condensing Boiler

This is a design of boiler which can be more efficient than the more traditional boiler. The efficiency of a typical non-condensing boiler is around 75%, whereas with condensing boilers it can be over 90%. This increased efficiency is due to the extraction of heat from the otherwise wasted flue gases. Most boilers have a single combustion chamber enclosed by the waterways of the heat exchanger through which the hot gases can pass. These gases are eventually expelled through the flue, located at the top of the boiler, at a temperature of around 180°C.

Condensing boilers, on the other hand, are designed first to allow the heat to pass through the primary heat exchanger; then through a secondary heat exchanger. Some designs simply use larger or a more effective heat exchanger. These can reduce the flue gas temperature to about 55°C. This reduction of temperature causes the water vapour (formed during the combustion process) to condense and, as the droplets of water form, fall by gravity to collect at the base of the flue manifold. The remaining gases are expelled to the outside environment through a fan-assisted balanced flue. The condensation produced within the appliance should be drained as necessary into the waste discharge pipework or externally into a purpose-made soakaway. The condensate produced is slightly acidic (about the strength of a fizzy drink) and therefore will corrode metal pipework. For this reason, plastic materials need to be used for the condensate drain.

It is only possible for a condensing boiler to work to these very high efficiencies if the flow and return pipework is also kept below 55°C. These low f & r temperatures need to be maintained for the heat transference to occur from the flue to the water (i.e. heat is transferred from hotter to cooler materials).

For a c.h. system to work with radiators and dhw primary circuits, flow temperatures need to be 75–80°C, therefore the boiler, although highly efficient, is not in condensing mode. The appliance only works in its condensing mode, during initial heat-up. To achieve a system which will function in its condensing mode, the installer needs to consider a suitable system of radiant heating. This will be identified and discussed later in the book.

Since April 2005 all domestic gas and oil boilers used in habitable dwellings need to be of the high efficient/condensing type. Some exceptions do exist, however (see page 160).

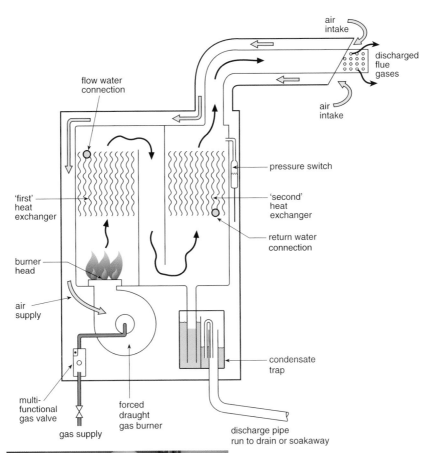

air
intake

discharged
flue
gases

air
intake

flow water
connection

pressure switch

'first'
heat
exchanger

'second'
heat
exchanger

return water
connection

burner
head

air
supply

condensate
trap

multi-
functional
gas valve

forced
draught
gas burner

gas supply

discharge pipe
run to drain or soakaway

Inside view of a condensing boiler
Note the large heat exchanger

Condensing boiler

Domestic Heating and the Building Regulations

Wet central heating systems and domestic hot water supplies now fall within the requirements of the Building Regulations (Part L1), which means that what you put into a dwelling must be approved by the Local Authority Building Control Officer. Appropriately qualified and approved plumbing operatives can self-certificate their own installation through their regulatory body, which alleviates the need for costly building approval for the customer; however, the system installed must meet the criteria laid down. It is essential to use the best possible design in order to conserve the consumption of fuel and prevent any waste of energy. The minimum central heating system specification to achieve is as follows:

- A boiler with an efficiency of greater than 90%.
- A dhw cylinder, if applicable, meeting the requirements of maintaining a good recovery time with factory fitted thermal insulation.
- The following minimum set of heating and hot water controls:
  - *Independent time control for c.h. and dhw* is achieved by the use of a 'full' programmer, or separate time control for both dhw and c.h. Also separate c.h. zones are required for systems over $150\,\text{m}^2$.
  - *Boiler interlock* must be provided.
  - *TRVs* on all radiators, except in rooms with a room thermostat.
- Suitable commissioning and handing over in accordance with the manufacturer's instructions.

Boiler choice is covered in greater depth on the following page. Good cylinder design will be achieved if the requirements of BS 1566 are met for indirect cylinders, or BS 7206 for unvented hot water storage units. In relation to thermal storage units, the performance needs to meet the specification of the Waterheater Manufacturer's Association.

Boiler interlock   Boiler interlock is a condition where the boiler will not fire unless a thermostat, connected to the system remotely from the boiler, is calling for heat. The room thermostat and cylinder thermostat provide the best examples of boiler interlock. However, another example would be a pipe thermostat. For example, where a frost thermostat has been included in the system to prevent freezing of the pipework; the frost thermostat overrides the other controls and brings on the boiler as the outside temperature falls to a temperature of around 0°C. This would allow the boiler to fire and it would continue to operate until it shut down as the result of the cylinder or room thermostat, depending upon how the system was wired up. This would result in the system becoming unnecessarily hot. The pipe thermostat however would switch off the boiler, or even prevent it from firing, when the pipe it is fitted to is around 5–8°C.

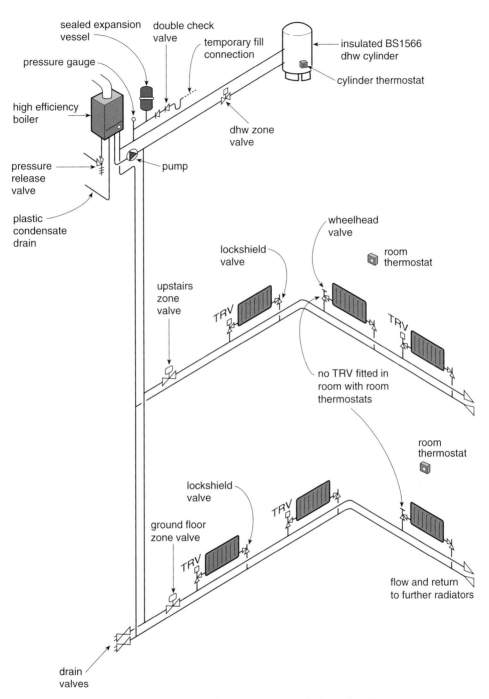

**Fully pumped sealed heating system with two heating zones
as required for dwellings over 150 m²**

Domestic Boilers Requirements

Where a solid fuel boiler is installed it simply needs to meet the efficiency requirements of the Heating and Testing Approval Scheme (HETAS).

However, for a domestic gas- or oil-fuelled boiler the Building Regulations have identified that the type of boiler installed as a replacement, or for a new system, needs to be of the high efficiency/condensing type. This kind of boiler will, as a result, produce condensation and water vapour in the form of a plume gas or cloud, seen discharging from the terminal, which could result in unacceptable installation problems. Therefore, there is a procedure that can be followed to determine if an exception is justified and a boiler with lower efficiency can be installed, which will not give rise to the same problems.

Exceptions to the rule for non-condensing boilers

Using the form opposite, obtainable from Building Control, the task is to find the *lowest cost option* for installing a high efficiency boiler. You must assume the building is empty and disregard any furniture/fittings or customer preferences. Where no feasible installation position can be found an explanation is simply required, completed by the installer, in box Y of the form. However, this decision must not be based on customer preference and you must have good justification for not complying with the law, such as a block of flats where the plume gas would be a nuisance. Should a problem exist, the data is gathered from the following procedure and all the points are added together.

1. In box A enter the points from the following table:

Points for property and fuel type

Building design	Natural gas	LPG	Oil
Flat	710	660	830
Mid terraced	640	580	790
All other types	590	520	760

2. If the boiler is a replacement, does it need to be fitted in a different room? If so insert points indicated in box B.
3. Is an extended flue (more than 2 m) needed? If so insert points in box C.
4. Is a condensate pump or soakaway needed? If so insert points in box D.
5. If the total exceeds 1000 then an exception is justified.

The form must be signed by a competent person, and must be kept with the completion certificate as it may be needed when the house is sold.

L1 ASSESSING WHERE NON-CONDENSING BOILERS COMPLY

CALCULATION AND DECLARATION FORM

This form may be used to show that a non-condensing boiler is reasonable provision for the purposes of complying with Part L of the Building Regulations.

1   Full address of property assessed: _____

      _____

      _____

      Postcode: _____

2   Dwelling type (tick one only)   Flat ☐ Mid-terraced ☐ End-terraced ☐ Semi-detached ☐ Detached ☐

3   Existing boiler fuel (tick one only)   Natural gas ☐ LPG ☐ Oil ☐ Solid fuel ☐ None ☐

4   New boiler fuel (tick one only)   Natural gas ☐ LPG ☐ Oil ☐

5   Existing boiler type (tick one only)   Wall mounted ☐ Back boiler ☐ Floor standing ☐ None ☐

6   Existing boiler position (tick one only)   Kitchen ☐ Utility room ☐ Garage ☐ Living room ☐ Bedroom ☐ Other ☐ None ☐

7   In the lowest cost option is the new boiler positioned in a different room from the existing boiler position?   Yes ☐ No ☐ Inapplicable (no existing boiler) ☐

8   If YES to section 7, state new boiler position   Kitchen ☐ Utility room ☐ Garage ☐ Living room ☐ Bedroom ☐
         Other:

9   Determine points for property type and new boiler fuel from the Table on the reverse of this form and insert in box A   **Box A**

10 New boiler position in a different room from the existing boiler? (see 7) If YES insert 350 in box B   **Box B**

11 Extended flue (longer than 2m) necessary? If YES insert 200 for gas boilers, or 350 for oil boilers, in box C   **Box C**

12 Condensate pump or soakaway necessary? If YES insert 100 in box D   **Box D**

13 ASSESSMENT SCORE   **TOTAL of points in boxes A + B + C + D**   **Box T**

14 Declaration (tick one box only)

Box W   I declare that the boiler to be installed is oil fired and will be installed before 1st April 2007, OR

Box X   I declare that the boiler is being replaced under the original manufacturer's or installer's guarantee, within 3 years of the original installation date, OR

Box Y   I declare that there are no feasible condensing boiler installation options (as defined by the assessment procedure) because:

_____

_____

_____

Box Z   I declare that I have considered all feasible boiler installation options in the property above, and the option defined in boxes A to D produces the lowest total T.

Signed _____   Date _____

Name (in capitals) _____   Status (agent or installer) _____

Competent person scheme _____   Competent person registration number _____

Notice to householder:

1   Where Box W has been ticked, a non-condensing oil boiler may be installed before 01 April 2007.

2   Where box X is ticked, a like-for-like replacement boiler is reasonable.

3   Where Box Y has been ticked or box Z has been ticked and the assessment score in section 13 exceeds 1000, this document may be used as evidence that installation of a condensing boiler has been assessed as impractical or uneconomic. **Nevertheless you may choose to exceed the Building regulations requirement** if a suitable installation option can be found. Condensing boilers are more efficient and therefore save on fuel costs and cause less harm to the environment. You may be eligible for a grant that defrays some of the additional costs – contact your local energy efficiency advice centre, or the energy efficiency helpline of your gas or electricity supplier (phone number on back of bill).

4   You should retain this form. It may be required when you sell your home.

Points for property type and fuel

Building type	Natural gas	LPG	Oil
Flat	710	660	830
Mid-terrace	640	580	790
Others (end-terrace semi-detached, or detached)	590	520	760

Central Heating System Protection

Pump overrun

In order to prevent a boiler overheating, particularly with low water content boilers, it may be necessary to allow the pump to continue running for a short period after the boiler has shut down. This removes the residual heat in the boiler, which would otherwise boil and possibly cause an ejection of water from an open vent. It may also cause damage to the boiler and will certainly cause unnecessary boiler noises. The boiler manufacturer will specify if boiler overrun is required. They may also include a pump and bypass for this purpose.

Central heating bypass

Several heating system applications may require the inclusion of a bypass; for example to provide a route of water flow where a pump overrun, described above, is required. Other reasons include:

- To assist in meeting the 10°C design temperature drop across the system when balancing (see page 182) thus allowing the water to flow directly back into the boiler.
- To allow water flow where thermostatic radiator valves are installed and have closed down. Pressure would otherwise build up in the system.

The size of the bypass should be 15 mm in diameter for boilers up to 18 kW, and 22 mm for boilers exceeding 18 kW. Prior to current Building Regulations the type of bypass valve fitted in a domestic dwelling was usually of the manually adjusted, lock-shield headed type. Today, automatic bypass valves should be installed; these automatically open as the pressure builds up in the system.

Frost protection

The minimum requirement for any form of frost protection will be the application of thermal insulation material both to the pipes and f & e cistern, where applicable. It is possible to add an antifreeze solution, where exposed pipework or controls may not be fully protected; alternatively the boiler can be made to fire up, thereby heating the water. It is essential where this method is employed that the boiler is fully electrically interlocked with some form of thermostatic control to prevent unnecessary wastage of fuel. Two general methods are used: set back control or the use of a frost thermostat.

Set-back control This is a design where the heating system is left on, under 24-hour control, via the use of a room thermostat. During the daylight hours the thermostat is set to provide the desired temperature for the environment. At night, or when the building is out of use, the temperature is allowed to fall no lower than a set point, typically 8°C. Where the building is this cold inside it is likely to be much colder outside, so the boiler brings on the heating to maintain the temperature.

Thermostatic control   This is a system where a thermostat is located at the coldest point of the system, usually close to the boiler where it has been installed in an outbuilding. The thermostat is wired up to override all other electrical controls and as a result will bring on the boiler and pump to warm the heating system. Because the system may be designed with zone valves that may be in the closed position, it is essential that the wiring considers opening either the heating or dhw circuit if necessary to enable the water to flow. When a frost thermostat electrical contact is made, due to a low outside temperature it will remain closed until such time as the environment warms up; where the frost thermostat is outside this may not be for some time. So boiler interlock is eventually made when the room or cylinder thermostat is satisfied. This situation would have heated the pipework unnecessarily high. To prevent this, a 'pipe thermostat' set to break the electrical circuit at around 2–3°C should be incorporated in the circuit. Once the pipe thermostat is satisfied, the boiler will shut down. The pipe thermostat must be located on the pipe in the most exposed location, usually on the return to the boiler.

Frost thermostat
Located near the coldest
part of the system

Pipe thermostat
Located typically on the return
pipe to the boiler

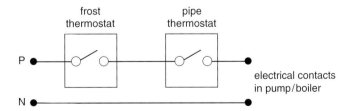

The frost thermostat will make contact in cold weather conditions. Once the pipe thermostat is satisfied the pump/boiler will not run, therefore 'boiler interlock' is provided.

Wiring diagram

Advanced Central Heating Control

Weather compensation

This is a system that varies the boiler flow temperatures dependent upon the inside load requirements and the outside weather condition. When a system is designed (see page 166) it is sized upon an outside temperature of −1°C. For much of the time that the central heating is on, this outside temperature is not experienced. For example, the outside temperature may be 10°C but the heating system will still be on, assuming the time control and thermostat are calling for heat. In this case, the system is oversized by 11°C and the system will not work to its maximum loading. To save on fuel consumption, an electric control unit makes an adjustment and sends the water around the system at a cooler temperature. This is often achieved by the use of a three port modulating valve that allows water to bypass the boiler itself if necessary. The flow temperature is continually monitored as the outside temperature rises and falls. It is essential when installing this design of system to make the customer aware of the rising and falling temperatures that will be experienced by the heating system, as felt by touching the various radiators, etc., otherwise you would soon be getting complaints from the end user stating that there is a problem with the system, when in fact it is doing just what you desire, thereby saving on fuel. The capital installation cost for this control system is high, therefore is rarely used in the domestic situation.

Optimum start

This is a system that varies the heating start-up time each day in order to reach the set room temperature by the desired time, as set by the occupant of the building. Basically the building owner sets the programmer to have the room up to a desired temperature by a set time. The optimiser then brings on the boiler automatically some time before this.

By referral to the graph opposite it can be seen that the end user requires the heating to be at 21°C by 7.30 am. The following morning the control unit decided to automatically bring on the heating at 5.30 am, allowing 2 hours to raise the temperature of the dwelling. If by 7 am the room temperature is achieved the control unit discovers that it has made an error and has brought on the heating system too early. Therefore, on the next occasion it brings on the heating system ½ hour later. Thus it adjusts automatically the time at which the system needs to come on. The outside and inside temperatures will also have an influence as to the time it will take for the desired room temperature to be achieved; therefore, as with the weather compensating system above variables will exist. For the first few weeks in the life of a new system designed in this way the computer is gathering data and continually alters the heating start up times until it gets it right. After a while the system knows that when a specific temperature of 'x'°C is required at a specific time in the morning it needs to allow 'y' amount of time to bring the system up to temperature.

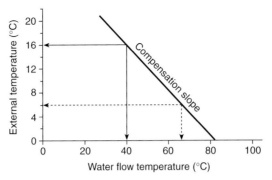

Note: as the external temperature increases, cooler water is made to flow around the ch system

Weather compensation

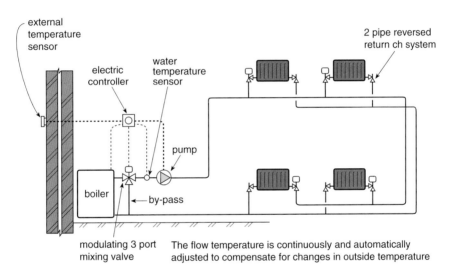

The flow temperature is continuously and automatically adjusted to compensate for changes in outside temperature

External temperature control

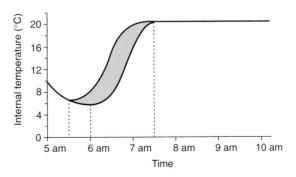

Optimum start

Radiator and Boiler Sizing

Relevant British Standards
BS EN 12828 and BS 5449

Correct heat emitter sizing can be achieved using a c.h. calculator such as the *Mears*, a computer program or a mathematical calculation, as shown here. First one must find the rate of *heat loss from the room*, which occurs in both of the following ways:

1. *Heat loss due to air change and natural ventilation.* This is found using the formula:

$$\begin{array}{ccccccccc} \text{Volume of} & & \text{Air change} & & \text{Temperature} & & \text{Ventilation} & & \text{Rate of} \\ \text{room} & \times & \text{rate} & \times & \text{difference} & \times & \text{factor} & = & \text{heat loss} \\ \text{(m}^3\text{)} & & \text{(per hour)} & & \text{(°C)} & & \text{(0.33 W/m}^3\text{ °C)} & & \text{(W)} \end{array}$$

   The following *air change rates* should be allowed for: two per hour in kitchens, bathrooms and dining areas, and one and a half per hour for other rooms.

2. *Heat loss through the building fabric.* This is found using the formula:

$$\begin{array}{ccccccc} \text{Surface area} & & \text{Temperature difference} & & \text{U value} & & \text{Rate of heat} \\ \text{(m}^2\text{)} & \times & \text{(°C)} & \times & \text{(W/m}^2\text{°C)} & = & \text{loss (W)} \end{array}$$

The temperature difference This is the difference between the internal and external environment. The external temperature is usually taken to be -1°C, although a colder outside temperature may be allowed for in exposed locations. The internal temperature is to the client's needs, usually based on the following: 21°C for living, dining, bath and bed sitting rooms and 16–18°C for kitchen, hall, WC and bedrooms. Note that where the dwelling adjoins another (e.g. semi-detached) one assumes a temperature difference of 6°C.

The U value

This is found from tables, such as those listed in the Chartered Institution of Building Services Engineers, or Institute of Plumbing and Heating Engineering Design Guides. It is possible to use the following approximate values without knowledge of the correct U value.

Appro+ximate U value through building fabric

Construction	W/m² °C	Construction	W/m² °C
External solid wall	2.0	Ground floor, solid	0.45
External cavity wall	1.0	Ground floor, wood	0.62
External cavity wall (filled)	0.5	Intermediate floor, heat flow up	1.7
External timber wall	0.6	Intermediate floor, heat flow down	1.5
Internal wall	2.2	Flat roof	1.5
Single glazing	5.7	Pitched roof (100 mm insulation)	0.34
Double glazing	3.0	Pitched roof (no insulation)	2.2

Heat emitter sizing If one finds the total rate of heat loss (in watts) from a room and installs a heater giving the same heat output, the temperature will be maintained. To allow a cold room to warm up requires the heater to be increased in size by a small percentage (usually 15%) although this is not applicable if the heating is on for 24 hours a day. For the bungalow in the figure, find the heat emitter requirements for the lounge and the bedroom. The answer is on page 168.

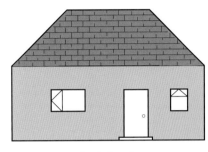

front elevation

Notes:
All dimemensions in metres
Solid external brick wall
Solid floor
Single glazed windows
Double glazed doors
100 mm insulation in roof space
3 kW to be allowed for dhw
Room heights 2.4 m

3 Central Heating

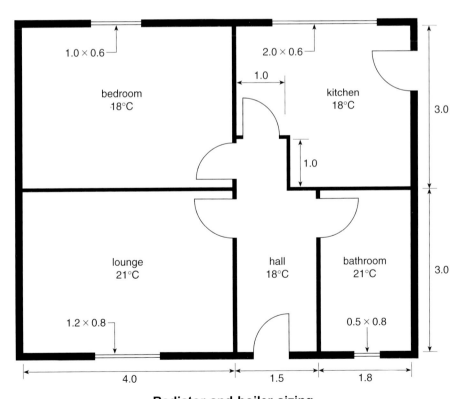

Radiator and boiler sizing

Heat emitter sizing (cont'd)

Having found the heat input required in watts, one simply refers to a manufacturer's radiator cataloge, as in the following example, to find the size of heat emitter required. From the schedule, the lounge will require a 1600 mm long × 590 mm high single convector, or a 960 mm × 590 mm double convector may be chosen. The bedroom will reqire a 1280 mm long × 590 mm single convector or a 800 mm × 590 mm double convector radiator.

Heat requirements. Location: lounge

Fabric loss element	Area L × B = (m²)		Temperature difference (°C)		U value (W/m² °C)	Heat loss rate (W)
Window	1.2 × 0.8 = 0.96	×	22	×	5.7	120.38
External walls	7 × 2.4 = 16.8 − 0.96 = 15.84	×	22	×	2.0	696.96
Internal walls	7.0 × 2.4 = 16.8	×	3	×	2.2	110.88
Floor	4.0 × 3.0 = 12.0	×	22	×	0.45	118.80
Roof	4.0 × 3.0 = 12.0	×	22	×	0.34	89.76
					Sum of fabric loss =	1136.78
Ventilation loss						
	volume × air change	× temperature difference × factor				
	3 × 4 × 2.4 × 1.5	× 22		× 0.33	=	313.63
		Sum fabric loss + ventilation loss			=	1450.41
		Plus 15% for intermittent heating			=	217.56
		Total rate of heat loss			=	1667.97

Heat requirements. Location: bedroom

Fabric loss element	Area L × B = (m²)		Temperature difference (°C)		U value (W/m² °C)	Heat loss rate (W)
Window	1.0 × 0.6 = 0.6	×	19	×	5.7	64.98
External walls	7.0 × 2.4 = 16.8 − 0.6 = 16.2	×	19	×	2.0	615.6
Internal walls	no heat losses	×	n/a	×	n/a	0.0
Floor	4.0 × 3.0 = 12.0	×	19	×	0.45	102.6
Roof	4.0 × 3.0 = 12.0	×	19	×	0.34	77.52
						860.7
Fabric gain[a]						
Internal walls	4.0 × 2.4 = 9.6	×	3	×	2.2	−63.36
					Sum of fabric loss =	797.34
Ventilation loss						
	Volume × air change	× temperature difference × factor				
	3 × 4 × 2.4 × 1.5	× 19		× 0.33	=	270.86
		Sum of fabric loss + ventilation loss =				1068.20
		Plus 15% for intermittent heating =				160.23
		Total rate of heat loss =				1228.43

[a]There is a heat gain to the bedroom from the lounge.

Sample section from a radiator schedule

Convector radiators				Height 23 in (590 mm)			Tappings $4 \times \frac{1}{2}$ in		
	Heat emission			Length			Heat emission		
Order code	Btu/h	W	Price £	in	mm	Order code	Btu/h	W	Price £
23 SC 12	1662	487	57.99	18.9	480	23 DC 12	2969	870	78.45
23 SC 16	2235	655	67.67	25.2	640	23 DC 16	4009	1175	98.07
23 SC 20	2805	822	78.94	31.5	800	23 DC 20	5047	1479	117.68
23 SC 24	3378	990	90.22	37.8	960	23 DC 24	6087	1484	137.29
23 SC 28	3951	1158	101.50	44.1	1120	23 DC 28	7128	2089	156.91
23 SC 32	4521	1325	112.78	50.4	1280	23 DC 32	8169	2394	176.52
23 SC 36	5094	1493	124.05	56.7	1440	23 DC 36	9209	2699	196.13
23 SC 40	5668	1661	135.33	63.0	1600	23 DC 40	10250	3004	215.75
23 SC 44	6237	1828	157.89	69.3	1760	23 DC 44	11291	3309	235.36
23 SC 48	6811	1996	168.64	75.6	1920	23 DC 48	12328	3613	274.59
23 SC 52	7384	2164	198.38	81.9	2080	23 DC 52	13369	3918	315.31
23 SC 56	7954	2331	220.42	88.2	2240	23 DC 56	14409	4223	329.89
23 SC 60	8527	2499	280.90	94.5	2400	23 DC 60	15450	4528	366.54

Boiler sizing

The required boiler output is determined as follows:

(1) Total up the fabric requirements for all rooms
(2) Add half the sum total of ventilation heat losses. The total ventilation heat loss is not included because some air change will be the result of warm air passing from one room to another; thus it is, in effect, a heat gain.
(3) Add the domestic hot water heat requirements (e.g. 1 kW for every 50 litres)
(4) Add a 20% margin for heat loss from pipes and initial warm up.

Given the fabric and ventilation heat losses from the bungalow previously identified, and allowing for a 150 litre dhw cylinder, the boiler output would need to be 8½ kW, as the calculation here shows.

Boiler size for bungalow

Room	Vent loss (W)	Fabric loss (W)
Lounge	313.63	1136.78
Bedroom	270.86	797.34
Kitchen	294.94	794.40
Hall	124.15	130.33
Bathroom	188.18	709.32
Total	1191.76	3568.17 +
Half vent loss = 1191.76 ÷ 2 =		595.88
		4164.05
dhw requirement →		3000.00
		7164.05
20% margin →		1432.81
		8596.86 W ($8\frac{1}{2}$ kW)

Whole House Boiler Sizing Method

When replacing an existing boiler, one of several practices could be adopted, including:

- Like for like replacement, without confirming correct size (*bad practice*)
- Use a calculator or computer programme (*generally oversizes boiler*)
- Calculate the dwelling's heat load (*accurate but time-consuming*)
- Whole house boiler sizing method (*simple, quick and reasonably accurate*).

The whole house boiler sizing method is a simple and quick method that can be employed to calculate the size of a replacement domestic oil or gas regular boiler (not a combination boiler) up to 25 kW. It can only be used for buildings of simple shape, or where the building can be sub-divided easily.

Completion of the worksheet on the following page is fairly self explanatory, however the following should act as a guide:

Step 1   Take the internal measurements (in metres) to provide the overall length and width (it is possible to take the external measurement and simply deduct the width of the external walls). Also measure the room heights and list the number of floors. Enter this data into the shaded boxes.

Step 2   Enter the data from above into the shaded boxes, including the number of external walls. *Note: a mid-terrace house has no external walls along two sides of the measured length and an end-terraced or semi-detached house has only one external wall along this length. Also note the whole wall is regarded as external where it is attached to an unheated garage.* With the data entered, calculate the total external wall area.

Step 3   Using the values from Tables 1, 2 and 3 (below), calculate the window heat losses. Note: where very large windows are encountered use the actual window area.

Step 4   Calculate the roof and floor areas, entering the data from Table 4. Where the dwelling is a top floor flat the floor area is zero; if the dwelling is a bottom floor flat the roof area is zero.

Step 5   Add boxes A, B, C and D together and multiply by the factor from Table 5.

Step 6   Taking the floor area, room height and number of floors data from above calculate the volume and multiply by the factor from Table 5.

Step 7   Finally, add together boxes E and F and add a further 2000 W for the dhw to give the final boiler output size. At this stage add on any additions, or simply add radiator outputs for extensions to the basic box shape dwelling.

Table 1 Window factors	
Detached	0.17
Semi-detached	0.2
Mid-terrace	0.25
Flat	0.25

Table 2 Window U-values	
Double glazed wood/plastic	3.0
Double glazed metal frames	4.2
Single glazed wood/plastic	4.7
Single glazed metal frames	5.8

Table 3 Wall U-values	
Filled cavity wall	0.45
Unfilled cavity wall	1.6
Solid wall 220 mm	2.1
Solid brick 343 mm	1.68

Table 4 Roof U-values	
Pitched <50 mm insulation	2.6
Pitched 50–75 mm insulation	0.99
Pitched >75 mm insulation	0.44
Flat uninsulated	2.0
Flat 50 mm insulation	0.54

Table 5 Location factors	
North and Midlands	29
Northern Ireland	26.5
Scotland	28.5
South East and Wales	27
South West	25

Assess the dwelling shape

A. Simple rectangular dwelling
Use worksheet alone.

B. Extension and loft conversion
Use worksheet and add on
radiators sizes in section 7.

C. Non-rectangular dwelling
Divide into sections and
repeat calculations.

3 Central Heating

1. Take three measurements (in meters)

Length ▭   Room height ▭

Width ▭   Number of floors ▭

2. Calculate TOTAL external wall areas

| | No. of ext. walls | | | | | Room height | No. of floors | Total ext. wall area m2 |

Width ▭ X ▭ = ▭
+ = ▭ X ▭ X ▭ = ▭
Length ▭ X ▭ = ▭

Number of ext. walls

3. Calculate wall and window heat losses

Total ext. wall area ▭ **Table 1** X ▭ = Window area ▭ **Table 2** X ▭ = ▭ **A Window heat loss**

▭ − ▭ = Wall area ▭ **Table 3** X ▭ = ▭ **B Wall heat loss**

Total ext. wall area   Window area   Wall area

4. Calculate floor and roof heat losses

Length ▭ X Width ▭ = Roof area ▭ **Table 4** X ▭ = ▭ **C Roof heat loss**

▭ − ▭ = Floor area ▭ X 0.7 = ▭ **D Floor heat loss**

Length   Width

5. Add up fabric heat losses

A + B + C + D = ▭ **X** **Table 5** ▭ **=** ▭ **E Total fabric heat loss (W)**

6. Calculate ventilation heat loss

Floor area ▭ X Room height ▭ X No. of floors ▭ = Volume ▭ X 0.25 X **Table 5** ▭ ▭ **F**

Ventilation heat loss (W)

7. Calculate boiler output (in kW)

E + F = ▭ + Water heating (W) **2000** = ▭ + ▭ = ▭ **BOILER OUTPUT**

Add in any extension
From separate worksheet or radiators sizes

Divide by 1000 to get kW

Pipe and Pump Sizing

Relevant British Standards
BS EN 12828 and BS 5449

The size of the pipe and pump required to serve the heat emitters in domestic situations is generally based on rule of thumb and general experience, which in most cases works sufficiently well. However, for the larger job or where efficiency is paramount, the size of the pipe and pump may be calculated.

Shown opposite is a completed example of a small heating system and the calculation table. Note that the pipe size is indicated in the calculation table; the pump size is deduced from its results (see page 175). The stages and interpretation of the results may be explained as follows:

Column 1   This is the section of pipework which is being sized; note that the system is broken down into various sections.

Column 2   The total required heat emitter values are inserted here, taking account of all the heat losses and intermittent heating (see pages 166–169 on radiator sizing).

Column 3   An allowance of 10–25% of the heat emitter size is given to allow for heat loss from the pipes due to standards of insulation, pipe runs, etc. I have allowed 20%. Example in section A, 20% of 11.1 = (20 ÷ 100 × 11.1) = 2.22.

Column 4   This is the sum total of columns 2 and 3; thus in section A 11.1 + 2.22 = 13.32, this being the total heat requirement for the section.

Column 5   A value determined by multiplying the specific heat capacity of water (4.186 kJ/kg °C) by the system design temperature drop. I have assumed 75°C flow and 65°C return; thus 75−65 = 10°C. Therefore, by calculation, 4.186 × 10 = 41.86.

Column 6   Flowrate is found by dividing column 4 by column 5; thus in section A 13.32 ÷ 41.86 = 0.32, this being the flow, in litres per second, required to maintain the required heat emissions.

Column 7   The actual pipe size required is found by referral to the table on page 174. The flow rate is given within the table for various diameters of pipe. When using the table, ideally a pipe size should be chosen which gives a pressure loss within the range 200–400 pascals/metre (Pa/m) for economic reasons. (Where the pressure loss is much less than 200 Pa/m the pipe is likely to be oversized, which increases the installation costs. Conversely, where the pressure loss is much greater than 400 Pa/m the pipe would be undersized, a larger pump would be required and running costs would increase.)

Example: In section A where a flow rate of 0.032 L/s is required, a pipe size of 22 mm or 28 mm may have been chosen. Neither actually falls within the 200–400 Pa/m pressure loss band and both are at equal distances from the desired pressure loss range. I have selected the 22 mm pipe size for economic reasons because it is cheaper to purchase and easier to install.

Pipe sizing to c.h. system and indication of system working pressure

1	2		3	4		5	6	7	8		9	10
Section	Total heat emitter value (W)		20% addition for heat loss from pipes (W)	Total heat requirements (W)		Specific heat capacity × temp. drop	Flow rate (l/s)	Pipe size (mm)	Pressure loss (Pa/m)		Total effective length (m)	Pressure required (Pa)
A	11.1	+	2.22	13.32	÷	41.86	0.32	22	520	×	30	15600
B	5.5	+	1.1	6.6	÷	41.86	0.16	22	160	×	27	4320
C	2.0	+	0.4	2.4	÷	41.86	0.06	15	180	×	10	1800
D	3.5	+	0.7	4.2	÷	41.86	0.1	15	460	×	20	9200
E	1.8	+	0.36	2.16	÷	41.86	0.05	15	140	×	15	2100
F	5.6	+	1.12	6.72	÷	41.86	0.16	22	160	×	38	6080
G	2.6	+	0.52	3.12	÷	41.86	0.07	15	240	×	19	4560

3 Central Heating

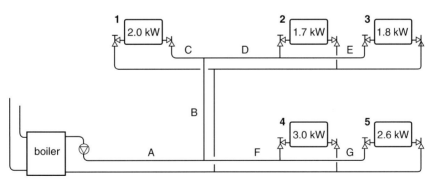

Schematic illustration of a two pipe c.h. system

Effective pipe lengths : section A – 30 m
(including flow section B – 27 m
and return) section C – 10 m
section D – 20 m
section E – 15 m
section F – 38 m
section G – 19 m

Pipe and pump sizing

Flow of water at 75°C in copper pipes

Pressure loss Pa/m	Flow rate in litres/s					
	Pipe size					
	12 mm	15 mm	22 mm	28 mm	35 mm	42 mm
90.00	0.021	0.040	0.118	0.239	0.430	0.725
92.50	0.022	0.041	0.120	0.242	0.437	0.735
95.00	0.022	0.042	0.122	0.246	0.444	0.748
97.50	0.022	0.042	0.124	0.250	0.450	0.759
100.00	0.023	0.043	0.125	0.253	0.457	0.769
120.00	0.025	0.047	0.139	0.281	0.506	0.852
140.00	0.028	0.052	0.152	0.306	0.551	0.928
160.00	0.030	0.056	0.164	0.330	0.594	1.00
180.00	0.032	0.060	0.175	0.352	0.635	1.07
200.00	0.034	0.064	0.186	0.374	0.673	1.13
220.00	0.036	0.067	0.196	0.394	0.710	1.19
240.00	0.038	0.071	0.206	0.414	0.745	1.25
260.00	0.039	0.074	0.215	0.433	0.779	1.31
280.00	0.041	0.077	0.224	0.451	0.812	1.37
300.00	0.043	0.080	0.233	0.469	0.844	1.42
320.00	0.044	0.083	0.242	0.486	0.874	1.47
340.00	0.046	0.086	0.250	0.503	0.904	1.52
360.00	0.048	0.089	0.258	0.519	0.913	1.57
380.00	0.049	0.092	0.266	0.535	0.962	1.62
400.00	0.050	0.094	0.274	0.551	0.990	1.66
420.00	0.052	0.097	0.282	0.566	1.02	1.71
440.00	0.053	0.099	0.289	0.581	1.04	1.75
460.00	0.055	0.102	0.297	0.595	1.07	1.80
480.00	0.056	0.104	0.304	0.610	1.10	1.84
500.00	0.057	0.107	0.311	0.624	1.12	1.88
520.00	0.059	0.109	0.318	0.637	1.15	1.92
540.00	0.060	0.112	0.324	0.651	1.17	1.96
560.00	0.061	0.114	0.331	0.664	1.19	2.00
580.00	0.062	0.116	0.338	0.677	1.22	2.04
600.00	0.064	0.119	0.344	0.690	1.24	2.08

(From CIBSE Guide Section C4, reproduced with permission of the Chartered Institution of Building Services Engineers.)

Column 8   Once the pipe size has been chosen, the pressure loss is recorded (see the table above). Thus in section A where the flow rate is 0.318 and a 22 mm pipe is selected, the pressure loss is 520 Pa/m.

Column 9   The total effective length is the run of flow and return pipework, including the actual pipe length and any additional length due to fittings. See page 126, Pipe Sizing of Hot and Cold Pipework, for a worked example. In this example I have given the effective pipe runs and they are indicated on the figure.

Column 10   This is the actual pressure required for each section and is found by multiplying column 8 by column 9; thus in section A, $520 \times 30 = 15600$ Pa.

Pipe sizing

In specifying the pump performance, both the maximum flow rate (litres/s) and the maximum pressure (Pa) need to be stated.

- The maximum flow rate can clearly be seen in the completed example against the largest pipe size, to which 0.32 L/s is indicated.
- The maximum pressure is found by adding together the pressure from each section of pipework in a circuit. In our example there are three possible circuits to consider.

The circuit with the greatest pressure drop is known as the *index circuit* and in theory if you can circulate around this pipeline you can circulate around any part of the system. Calculation to find the index circuit in the system on p. 173:
1st circuit, to radiator no. 1 (sections A, B and C)
2nd circuit, to radiator no. 3 (sections A, B, D and E)
3rd circuit, to radiator no. 5 (sections A, F and G)

1st circuit		*2nd circuit*		*3rd circuit*	
section A:	15600	section A:	15600	section A:	15600
section B:	4320	section B:	4320	section F:	6080
section C:	1800	section D:	9200	section G:	4560
Total	21720 Pa	section E:	2100	Total	26240 Pa
		Total	31220 Pa		

Thus we can conclude by identifying the index circuit as that from the boiler to radiator no. 3, requiring a pressure of 31220 Pa (or 31.2 kPa). Note that the index circuit is not always the circuit with the greatest actual pipe length; this is dependent upon the amount of fittings used.

The final pump selection can now be made. Our example requires a pump/circulator with a maximum pressure of 31.2 kPa and a maximum flow of 0.32 L/s. By referral to the performance curve of a pump (given in the manufacturer's data), we can select a pump which meets our requirements, the pressure and flow falling within the curve itself. (Note: $1 \text{kPa} = 1 \text{kN/m}^2$.)

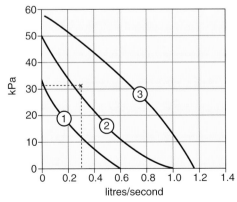

By referring to the performance curve the pump chosen will need to be selected on the number three setting

Radiant Heating

This is a system of heating designed to raise the temperature of a room space by the emission of infrared energy, which is basically thermal radiation. Radiant heat passes directly through air and will only heat the more solid surfaces upon which it falls. Radiant heat panels are usually mounted in the floor, walls or ceiling of a room. The panels are heated electrically or by circulating hot water or hot air through them. Unlike other forms of central heating, the effectiveness of radiant heating does not rely on the efficient circulation of air due to direct contact with the heat source.

For an equal state of comfort in a room, systems relying mainly on convection currents (such as a radiator type system) must provide a higher air temperature within the room because the cold surfaces of walls and windows, etc., remove heat from the human body that can only be replaced by the surrounding/ambient air. Radiant heat on the other hand warms up the floors, walls, windows, etc., and thus reduces the heat lost from the human body. Therefore, lower air temperatures within the room are maintained; this also provides a greater feeling of freshness, and with the reduction of convection currents in the room, cold draughts and dust problems are reduced to a minimum.

Radiant heating can save fuel, unless the heating of the room is intermittent, as in, for example, a building which lowers its air temperature at night and requires it to be rapidly raised in the morning. The design of radiant heating systems is quite straight-forward, requiring only a coil of embedded pipe running through the surface of a floor, wall or ceiling; the only important requirement is that the surfaces of the radiant heaters should not be metallic. A continuous pipe, usually plastic with no joints, is placed into a plastic floor panel positioned over the whole floor surface. This provides the correct spacing, usually at 100–200 mm intervals, as required. See the following page for further design details. Fitted behind the heater should be thermal insulation to give heat emission only into the room. With this system, one major disadvantage is the danger of a pipe leakage which can prove difficult to find and expensive to repair.

In the following table is given the recommended surface temperatures of walls, etc., fitted with embedded pipe panels. Notice how panels located in the ceiling can be used to give off a higher radiant heat emission, and therefore give a quicker room heating period. The water temperature flowing through the pipes should be between 40 and 55°C. To achieve the higher surface temperatures used in the ceiling, the heating pipes are placed closer together.

Panel location	Water temperature	Surface temperature (°C)	Heat given off (%)
Ceiling	55	40	65
Floors/walls	40	24	50

For a radiant system to operate efficiently and effectively it needs to be on constantly, thus preventing the building fabric from cooling. Because the temperature of water flowing through the pipework is greatly reduced, compared with the more traditional system using radiators with flow temperatures around 75–82°C, these systems can be designed with a condensing boiler, which will increase the efficiency of the system even more.

3 Central Heating

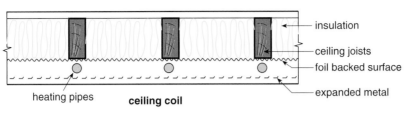

ceiling coil

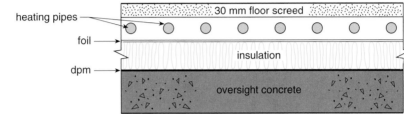

solid floor heating coil

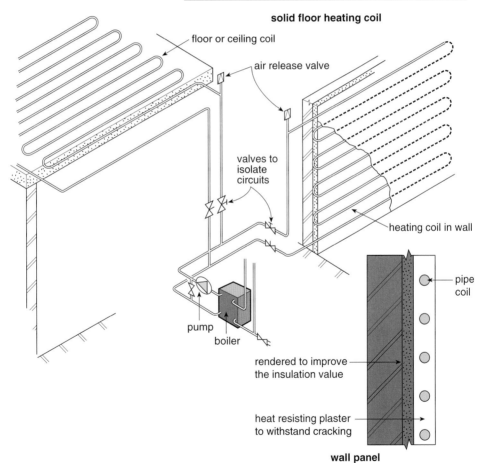

Radiant heating

Underfloor Central Heating Sizing

Underfloor central heating is a form of radiant heating, as described on the previous page. In order to correctly estimate the piping requirements for a specific installation, the manufacturer specifications should ideally be sought, in which they would state the maximum output performance of their system based upon the temperature of the water flowing through the pipework. This would usually be no higher than $100\,W/m^2$ for solid floors, and $70\,W/m^2$ for timber floors. The choice of floor covering would also have a bearing upon the final heat output; a floor covering with a greater thermal resistance will reduce the output from the underfloor heating. The following table shows typical thermal resistances from the four basic floor finishes.

Floor finish	Typical thermal resistance
Screed or ceramic tile	0.02 W/m^2
Vinyl	0.075 W/m^2
Standard carpet	0.1 W/m^2
Thick carpet or timber	0.15 W/m^2

Heat loss requirements

A heat loss calculation for the room is made first. The method used to make this calculation has previously been described on page 166. When completing this calculation the heat loss through the floor can be ignored, as the floor temperature is warmer than the room temperature. However, in reality a degree of heat loss does occur and this needs to be considered when making the final boiler output selection and 10% should be added at this stage.

With the room heat loss calculated, the heat requirement needs to be found by dividing the heat loss by the total floor area available. Note: in most cases the available floor area is the total floor area, however pipework would not be laid beneath kitchen units, for example, therefore the available area might be less.

With the heat requirement found, one simply refers to the table opposite, looking specifically at the room temperature required. This will provide an indication as to the pipe spacing needed and water flow temperatures required in order to give the desired output. It also gives the approximate floor temperature that will be obtained. Note: the table shown allows for standard carpet covering.

Example: A living room at 21°C, with heat loss from the room of $950\,W$ and the room size at $12\,m^2$. The total heat requirement would be $950 \div 12 = 79.17\,W/m^2$. Therefore referral to the table opposite suggests pipes spaced at $100\,mm$ apart with a water flow through them at 50°C or pipes spaced at $200\,mm$ apart with a water flow through them at 55°C.

Note: underfloor heating is not suitable for buildings with intermittent heating, high heat losses or where large amounts of floor area have fitments fixed to the floor, e.g. cupboards or shelving.

Heat output

Heat output for solid floors (with standard carpet floor covering)				
Room temperature	Output (W/m²)	Pipe spacing (mm)	Water flow temperature (°C)	Floor temperature (°C)
18°C	59	100	40	24
	75	100	45	25
	92	100	50	26
	109	100	55	28
	48	200	40	23
	62	200	45	24
	75	200	50	25
	89	200	55	26
21°C	48	100	40	26
	65	100	45	27
	82	100	50	29
	99	100	55	30
	40	200	40	25
	54	200	45	26
	67	200	50	27
	81	200	55	28
22°C	45	100	40	26
	62	100	45	28
	79	100	50	29
	95	100	55	31
	37	200	40	26
	51	200	45	27
	65	200	50	28
	78	200	55	29
Heat output for suspended or floating floors (with standard carpet floor covering)				
18°C	34	300	40	21
	44	300	45	22
	53	300	50	23
	63	300	55	24
	72	300	60	25
21°C	28	300	40	24
	38	300	45	25
	47	300	50	26
	57	300	55	26
	67	300	60	27
22°C	26	300	40	25
	36	300	45	26
	45	300	50	26
	55	300	55	27
	65	300	60	28

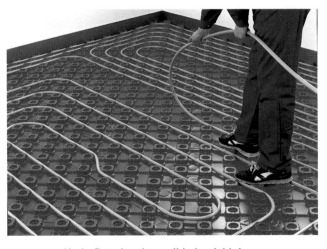

Underfloor heating coil being laid down

Warm Air Heating

Relevant British Standard
BS 5864

This is a central heating design in which warm air is blown through a system of ductwork to the rooms to be heated, assisted by a fan. The air in this system is either heated directly in a special boiler (in which the air circulates around the combustion chamber), or indirectly, in which case the air passes over a water-filled heat exchanger.

The direct system tends to have a shorter heating-up period, owing to the fact that the air is heated directly and no heat will have been lost from the flow and return pipe to the warm air heat exchanger. With this system, because a flue is required, it is not always possible to site the boiler in a desired central position; with the indirect system, however, the boiler can be sited away from the warm air heat exchanger (see figure).

The indirect system is generally more expensive to install but has the advantage of also heating water for domestic purposes, the boiler used being the same as that in a water-filled central heating system. Special boilers can, however, be purchased for the direct system which will allow water to circulate through the boiler for dhw purposes.

Once the air has been heated it is passed through a system of ductwork. The delivery air temperature at the room register/diffuser should not exceed 60°C.

Air from the heated rooms should be returned to the heater for reheating. If possible, the return air grill should be positioned opposite the warm air inlet diffuser; for example, if the warm air inlet is on one side of the room the return air outlet should be located on the opposite wall. No return air grills are positioned in bathrooms, kitchens or WC compartments because of the large amounts of condensation and possible odours which can be drawn into the system. The return air is either passed back to the heater by a system of ductwork, or it is drawn back to the hall and eventually to the heater through a duct.

Should the method be chosen which allows the return air to flow to the hall, the air is simply allowed to flow through grills in the internal walls, but note that some thought must be given to their siting in bedrooms, etc., for privacy. To ensure good comfortable room conditions, up to 25% of fresh air is often mixed with the return air: the heater air inlet manifold is fitted with a damper to regulate the proportions of fresh and return air. In large industrial buildings, often only fresh air is used from outside the building and no return air used at all.

Note that to reduce the spread of smoke in the event of fire, no grille should be positioned more than 450 mm above floor level.

Building Regulations compliant

It must be understood that a warm air unit is not a boiler and therefore falls outside the tight constraints for domestic gas boilers and central heating systems, as required under part L of the Building Regulations. This applies to heaters with or without an integral circulator.

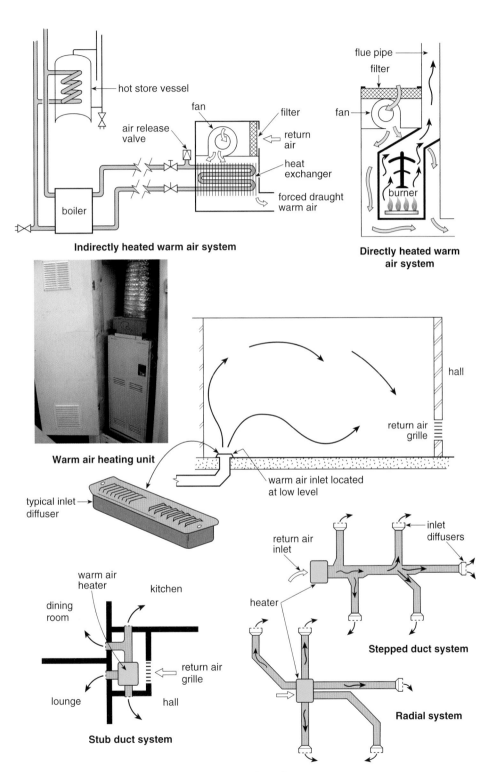

Indirectly heated warm air system

Directly heated warm air system

Warm air heating unit

Stub duct system

Stepped duct system

Radial system

Warm air heating

Commissioning of Wet Central Heating Systems

Relevant British Standards
BS 5449 and BS 7593
BS EN 12828 and PAS 33

On completion of a c.h. installation the system should be commissioned as follows:

(1) The pump should be removed and replaced with a suitable piece of pipe to bridge the gap. The system is now filled with water, any air being vented out as necessary from all high points and the system checked for any leaks.

(2) The system should be drained out, and should receive a flush through to remove any wire wool, etc. The pump is now replaced and the system refilled, as in (1) above. The boiler is now made to ignite and the system brought into operation by turning up any thermostats, etc. A check should be made of the boiler thermostat to confirm that it is working.

(3) At this stage the boiler can be commissioned for correct operation. See pages 232 and 264: commissioning of gas and oil appliances.

(4) Close all lockshield valves and go round the system balancing the heat emitters to each room. This involves slowly reopening the lockshield valve to give a mean water temperature to each heater. Ensure that the design temperature drop across the system is maintained at the boiler; this should be carried out using clamp-on thermostats at the flow and return pipes to achieve approximately 75°C flow and 65°C return temperatures, although this may vary, dependent on the system type/design. This will ensure maximum efficiency and give a longer life to the boiler.

(5) Check the operation of the programmer, room and cylinder thermostats, etc., and any motorised valves fitted to the system, and ensure that the pump switches off as required.

(6) Recheck for leaks. Turn off the boiler and drain the system while it is still hot. This assists the removal of flux residuals.

(7) Refill and vent the system, adding an inhibitor if applicable.

Having cured any problems, and secured cistern lids and insulation material, the installer should attach to the boiler a card identifying the date of installation and the name of the installer. Labels should also be fixed to any valves, etc., for identification purposes. The job should now be tidied up in readiness for handover to the owner.

(8) Handing over: The working of the system should be demonstrated to the user and the best methods of economic and efficient usage explained. All documentation supplied should be left with the owner/occupier who should be made aware of the need for a regular service contract to ensure that the equipment is maintained in an efficient and safe operating condition.

(9) Finally, all completion certificates should be issued and the approval body notified of the installation.

Power flushing

When a new boiler is connected to an existing central heating system it is often a requirement of the boiler manufacturer that the existing system is flushed using a power flushing unit. These are highly effective units that, when completed correctly, in accordance with the supplier details, rid the system of scale, black sludge and corrosion deposits. The system should be flushed with the existing boiler still in position prior to the installation of the replacement appliance.

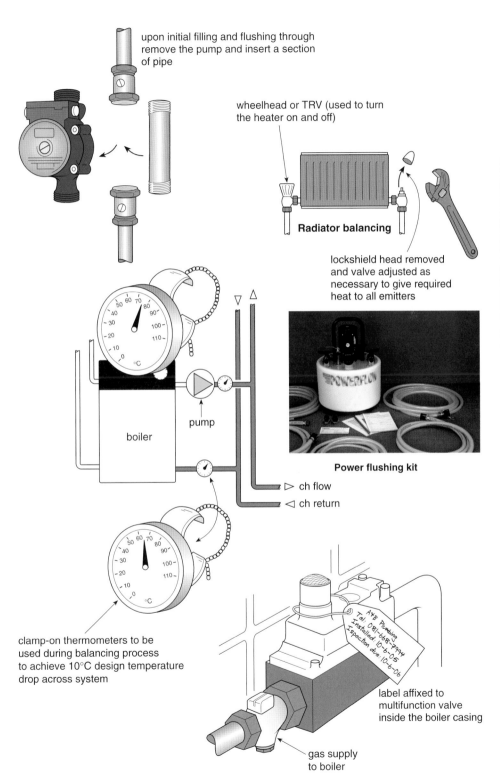

upon initial filling and flushing through remove the pump and insert a section of pipe

wheelhead or TRV (used to turn the heater on and off)

Radiator balancing

lockshield head removed and valve adjusted as necessary to give required heat to all emitters

Power flushing kit

pump

boiler

▷ ch flow

◁ ch return

clamp-on thermometers to be used during balancing process to achieve 10°C design temperature drop across system

A&B Plumbing
Tel. 081-668-7994
Installed 10-6-05
Inspection due 10-6-06

label affixed to multifunction valve inside the boiler casing

gas supply to boiler

Commissioning of wet central heating systems

Part 4
Gas Supplies

Plumbing, 4th Edition. R. D. Treloar.
© 2012 Blackwell Publishing Ltd. Published 2012 by Blackwell Publishing Ltd.

Properties and Combustion of Natural Gas

Properties of natural gas

Constituent	Chemical formulae	Approximate % by volume
Methane	CH_4	90.0
Ethane	C_2H_6	5.3
Propane	C_3H_8	1.0
Butane	C_4H_{10}	0.4
Carbon dioxide	CO_2	0.6
Nitrogen	N_2	2.7

Natural gas occurs below ground and is a by-product of the breaking down of decaying vegetation, etc., which is subsequently trapped beneath impervious land strata. The gas is non-toxic and odourless; the smell is added to the gas by the supplier to assist the identification of a leak. The relative density of natural gas is approximately 0.58, so it is lighter than air and will rise as a result. Natural gas is primarily made up of hydrocarbons which are a combination of hydrogen and carbon atoms. Both these atoms can be burnt in the presence of oxygen (O_2) and when burning, their chemical composition will change. For example, if we looked at methane burning in a sufficient quantity of oxygen to give complete combustion, the following reactions would occur:

Methane (1 volume) CH_4	+	Sufficient oxygen (2 volumes) $2 O_2$	gives off	Carbon dioxide (1 volume) CO_2	and	Water vapour (2 volumes) $2 H_2O$

Note: None of the atoms are ignored. Nothing has actually been consumed; its chemical state has simply been changed. The two gases given off are completely harmless and safe to breathe in, being present in the surrounding air. If we looked at the same volume of gas with an insufficient quantity of oxygen, incomplete combustion would result, as shown:

Methane (1 volume) CH_4	+	Insufficient oxygen (1 volume) O_2	gives off	Water vapour (1 volume) H_2O	and	Unburnt fuel $CO + H_2$

Apart from being inefficient, this situation is also highly dangerous, as the gas carbon monoxide (CO) is highly toxic; in fact, 0.5% volume in the air can prove fatal within a few minutes.

There are two fundamental requirements for the ignition of gas: a supply of oxygen and a gas temperature of approximately 704°C. The gas must be mixed with the oxygen to give a ratio of gas in the air of between 5% and 15%. Upon ignition the flame will burn through the gas/air mixture at a speed of 0.36 m/s. The manufacturer designs the burner to mix the gas with the air before leaving the burner head, and combine further in the reaction zone, allowing complete mixture and warming of the fuel.

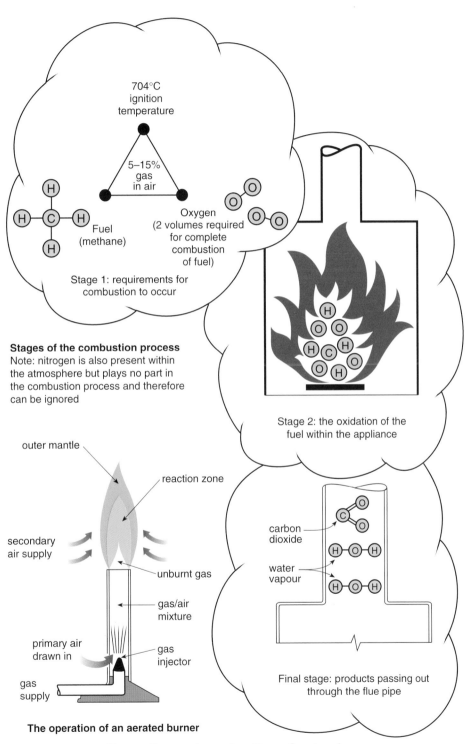

704°C ignition temperature

5–15% gas in air

Fuel (methane)

Oxygen (2 volumes required for complete combustion of fuel)

Stage 1: requirements for combustion to occur

Stages of the combustion process
Note: nitrogen is also present within the atmosphere but plays no part in the combustion process and therefore can be ignored

Stage 2: the oxidation of the fuel within the appliance

outer mantle

reaction zone

secondary air supply

unburnt gas

gas/air mixture

primary air drawn in

gas injector

gas supply

carbon dioxide

water vapour

Final stage: products passing out through the flue pipe

The operation of an aerated burner

Properties and combustion of natural gas

The Law Relating to Gas Installation Work

Any person carrying out any work in relation to gas fitting must be competent to do so. The Gas Safety (Installation and Use) Regulations 1998 clearly define what can and cannot be done.

You may do the work yourself, in your home, providing you comply with the Regulations, but if you wish to do the work for gain (i.e. be employed by someone) you must be a member of a class of persons approved by the Health and Safety Executive (Regulation 3.3). This in effect means you must be a registered operative as identified by the Gas Safe Register. If you are an employee, your employer must be registered and you must be a named operative with the company. If you work for yourself, then you yourself must register.

All gas fitting operatives are required by law to undertake assessments in the kinds of gas work that they are involved in. These assessments are conducted by certified bodies who comply with the rules of the Nationally Accredited Certification Scheme for Individual Gas Fitting Operatives (ACS). Gas fitting also involves other regulations, therefore other documents such as British Standards, Building Regulations, and manufacturers' information must be complied with. It would not be possible for a book such as this to identify all the Regulations which need to be observed, but the following lists briefly a few of those from the Gas Safety (Installation and Use) Regulations 1998.

Regulation No. 8   No person shall alter the building in any way so as to affect the safe operation of the gas fitting. Builders and double glazing people, etc., may inevitably break this law, but are still liable to prosecution.

Regulation No. 10   Where any pipework is to be removed, a suitable cross bonding wire must be incorporated to prevent the production of a spark or electric shock due to a fault in the electrical supply (see Part 6 Electrical Work, page 276).

Regulation No. 19   No pipes should be run in a cavity wall, except when passing from one side to the other. Where this is the case, a sleeve should be incorporated and sealed to prevent the passage of gas. No pipe must be installed in an unventilated shaft, duct or void.

Regulation No. 29   The installer must leave with the owner/occupier all instructions provided by the manufacturer.

Regulation No. 30   Open flued appliances are not to be installed in bath/shower rooms. In addition, bedrooms should also be avoided to satisfy the law and where used will require additional safety measures.

Regulation No. 33   When an appliance has been connected to the gas supply it must be tested in accordance with the manufacturer's instructions (i.e. commissioned).

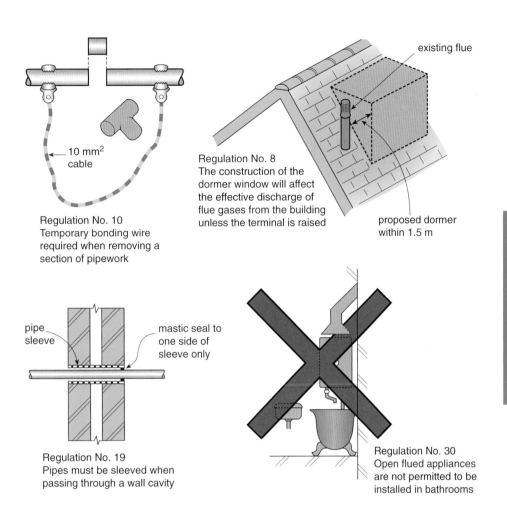

Regulation No. 10
Temporary bonding wire
required when removing a
section of pipework

10 mm² cable

Regulation No. 8
The construction of the
dormer window will affect
the effective discharge of
flue gases from the building
unless the terminal is raised

existing flue

proposed dormer
within 1.5 m

pipe sleeve

mastic seal to
one side of
sleeve only

Regulation No. 19
Pipes must be sleeved when
passing through a wall cavity

Regulation No. 30
Open flued appliances
are not permitted to be
installed in bathrooms

4 Gas Supplies

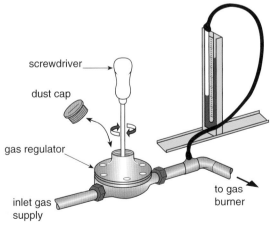

screwdriver

dust cap

gas regulator

inlet gas
supply

to gas
burner

Regulation No. 33
Burner pressure to be
adjusted in accordance
with manufacturer's
instructions

Gas safety regulations

Gas Supply to the Consumer

Relevant British Standards
BS 6400 and BS 6891

When a supply of natural gas is required, provided it is available in the local district, a service pipe is run to the building (see figure). Whenever possible, a special meter box should be installed. Three designs will be found: the semi-concealed, the surface-mounted or the sunken box. The meter may be located inside a garage or the dwelling itself, although these locations are not recommended because the supplier will have difficulty in obtaining meter readings.

The service pipe should not be routed below the foundations or through unventilated voids, and where the service pipe has to pass through a wall, the pipe must be sleeved to allow movement. The sleeve should be made good with mortar in the wall, and the gap between the pipe and sleeve sealed with a mastic sealant to prevent the passage of gas.

The supply gas pressure may be far in excess of that required by the consumer and is therefore reduced to a supply pressure of 20 mbar, and in no cases exceeds 30 mbar. This is achieved by the installation of a meter governor, which adjusts automatically to provide an adequate supply of gas at a constant pressure of 21 mbar ± 2 mbar. The meter governor is set by the supplier of the gas and must not be adjusted and has a lead seal to prevent tampering.

The main consumer's emergency control valve must be located in a readily accessible position and have a securely attached lever arm. When the valve is located in a vertical pipeline, it should be closed when the lever has been moved as far as possible downwards and in a horizontal plane. A label must be included to show the open and closed positions of the valve (see Gas Safety Regulation No 9).

Internal shafts

Where a continuous shaft is incorporated to accommodate multi-storey properties, the shaft must be constructed adjacent to, or within 2 m of, an external wall and be ventilated at high and low level to the external environment, thus preventing any build-up of gas in the event of a leak. The minimum free air opening depends upon the size of the duct and for both high and low openings this should be as follows:

Cross sectional area of duct (m²)	Free area of each opening (m²)
Less than 0.01	0
0.01–0.05	Cross sectional area of duct
0.05–7.5	0.05
Greater than 7.5	1/150th of cross sectional area

The individual branches to each apartment are taken off the riser, to include a service valve, and taken up vertically, to incorporate an expansion-type coupling, before turning horizontally through the shaft wall into the room, thus allowing for movement without fracture.

The horizontal pipe passing into the room must be fire-stopped as necessary, to prevent the passage of gas or smoke. The vertical shaft may be either one long continuous shaft or fire-stopped at each level, depending upon the building design and local Regulations, but in all cases the shaft must be ventilated in the way described above.

4 Gas Supplies

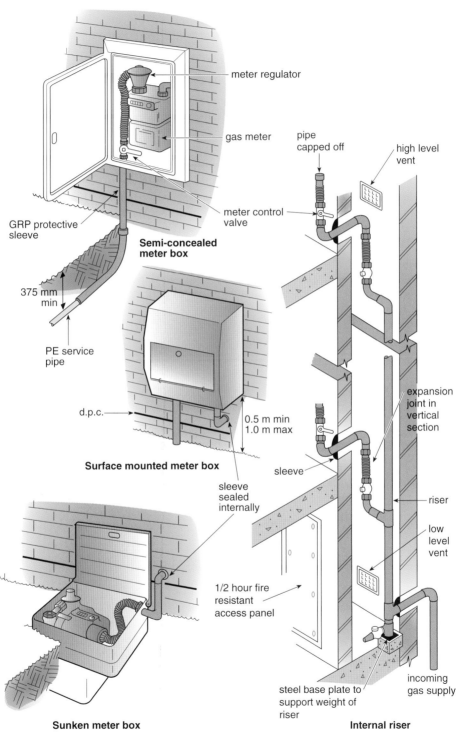

meter regulator

gas meter

pipe capped off

high level vent

meter control valve

GRP protective sleeve

Semi-concealed meter box

375 mm min

PE service pipe

d.p.c.

0.5 m min 1.0 m max

Surface mounted meter box

sleeve sealed internally

sleeve

expansion joint in vertical section

riser

low level vent

1/2 hour fire resistant access panel

steel base plate to support weight of riser

incoming gas supply

Sunken meter box

Internal riser

Gas supply to the consumer

Sizing of Domestic Gas Pipework

Relevant British Standard
BS 6891

If hot or cold water pipework is undersized, the worst that can happen is that an appliance becomes starved of water flow, which results in inconvenience and inefficient usage. There is no danger to the user or occupier of the building. With gas supply pipework however, undersized pipework may result in a dangerous situation. If the pressure is insufficient at the burner, not enough primary air will be drawn in to achieve complete combustion, which may result in the production of carbon monoxide. In very bad cases a flash-back into the pipework may occur which could lead to an explosion. The maximum pressure drop between the meter and the furthest appliance, under maximum flow conditions, must not exceed 1 mbar.

There are various methods that can be employed to pipe size a gas installation. The method shown here allows the system to be sized in stages and also proves useful to confirm if an existing system is suitably sized.

Shown opposite is a completed example of a gas carcass which has been sized in stages as follows:

Column 1   This is the section of pipework which is being sized; note that the gas carcass has been broken down into various sections.

Column 2   The actual length of the pipe is measured and inserted here.

Column 3   An allowance must be made for frictional resistance through fittings. Each time the gas passes round an elbow or tee fitting, 0.5 m should be allowed, and 0.3 m in the case of pulled 90° bends.

Example: In section A, three elbows are used and one tee fitting; therefore: $4 \times 0.5 = 2$ m.

Column 4   This is the sum total of columns 2 and 3 (thus in section A, 4.5 m + 2 m = 6.5 m), this being the effective length resulting from the actual length and additional length due to frictional resistance.

Column 5   This is found by adding the total kW rating for all appliances being supplied and undertaking the following calculation (note that the calorific value is 38.5 MJ/m^3):

$$\text{Gas flow rate} = \text{kW} \times 3.6 \div \text{calorific value}$$

Example: Section A = boiler: 17.0 kW

cooker: 14.6 kW

fire: 5.0 kW

Total   36.6 kW

Therefore gas flow rate = $36.6 \times 3.6 \div 38.5 = 3.42$ m^3/h.

Gas carcass serving boiler: cooker and fire

1 Section	2 Measured length (m)	3 Additional length due to fittings (m)	4 Effective length (m)	5 Required gas flow rate (m³/h)	6 Suggested pipe size (mm)	7 Maximum length allowed (m)	8 Pressure loss (mbar)	9 Progressive pressure loss (mbar)	Notes
A	4.5	2.0	6.5	3.42	22	12	0.5	0.5	✓
B	3.6	1.0	4.6	0.47	15	30	0.2	0.7	✓
C	4.0	0.5	4.5	2.95	22	15	0.3	0.8	✓
D	2.0	1.0	3.0	1.37	15	9	0.3	1.1	Pipe undersized
D	2.0	1.0	3.0	1.37	22	30	0.1	0.9	✓
E	0.6	0.5	1.1	1.59	15	6	0.2	1.0	✓

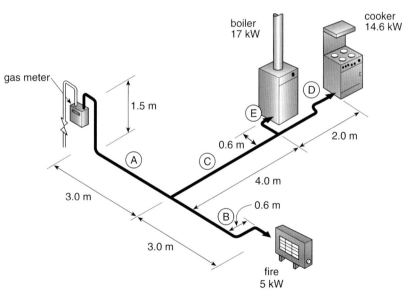

Sizing of gas pipework

4 Gas Supplies

Column 6   At this stage, a pipe size is suggested which the table will confirm (or not) for possible use.

Column 7   By referral to Table 1, opposite, the pipe diameter is aligned against the column showing the length of pipe through which the required gas flow will pass.

Example: In section A, where a 22 mm pipe was suggested and a flow of 3.42 m³/h required, the maximum length of pipe run is indicated to be 12 m, through which 3.9 m³/h of gas could pass, whereas inspection of the next column shows that only 3.4 m³/h of gas could pass through 15 m of pipe.

Column 8   The pressure loss for the section in question is found by dividing column 4 by column 7; thus, in section A, 6.5 ÷ 12 = 0.5 mbar.

Column 9   The progressive pressure loss is the sum total of the pressure losses for each section preceding the section in question. In section A, the progressive loss is the same as the pressure loss, but in section E, for example, the progressive pressure loss will be the sum total of sections A, C and E; therefore: 0.5 + 0.3 + 0.2 = 1.0 mbar which, in this example, will be of sufficient size.

In conclusion, one estimates a suggested pipe diameter and completes the table for the section to prove its suitability for use. If it proves undersized one simply goes back to column 6 to choose a larger pipe diameter. It may be that one has to increase the diameter of the first section, because in our example 0.5 mbar was generated before the first tee connection. Sometimes one opts for larger pipework to allow for possible extensions to the gas pipework in future years, e.g. a larger boiler or extra gas fire.

Meter sizing

To complete the work on pipe sizing one can find the meter size required; our example suggested we needed a supply of 3.42 m³/h; therefore, the smallest domestic meter would have been sufficient, which supplies up to 6 m³/h.

Table 1 Flow discharge of natural gas in m^3/h from copper tube with a 1.0 mbar pressure differential between each end

Pipe diameter (mm)	Length of pipe in metres							
	3	6	9	12	15	20	25	30
10	0.86	0.57	0.5	0.37	0.3	0.22	0.18	0.15
12	1.5	1.0	0.85	0.82	0.69	0.52	0.41	0.34
15	2.9	1.9	1.5	1.3	1.1	0.95	0.92	0.88
22	8.7	5.8	4.6	3.9	3.4	2.9	2.5	2.3
28	18.0	12.0	9.4	8.0	7.0	5.9	5.2	4.7
	Discharge of gas flow rate (m^3/h)							

Table 2 Flow discharge of natural gas in m^3/h from medium gauge steel tube with a 1.0 mbar pressure differential between each end

Pipe diameter (BSPT: in)	Length of pipe in metres							
	3	6	9	12	15	20	25	30
$\frac{1}{4}$	0.8	0.53	0.49	0.36	0.29	0.22	0.17	0.14
$\frac{3}{8}$	2.1	1.4	1.1	0.93	0.81	0.7	0.69	0.57
$\frac{1}{2}$	4.3	2.9	2.3	2.0	1.7	1.5	1.4	1.3
$\frac{3}{4}$	9.7	6.6	5.3	4.5	3.9	3.3	2.9	2.6
1	18.0	12.0	10.0	8.5	7.5	6.3	5.6	5.0
	Discharge of gas flow rate (m^3/h)							

(Tables 1 and 2 reproduced from BS 6891:1998 with the permission of BSI under licence number 2000 SK/0071.)

4 Gas Supplies

Domestic Tightness Testing and Purging

Relevant British Standard
BS 6891

Pressure tests are carried out on gas installations to check for leaks. Two types of test may be carried out, these being for new or existing installations. Should the installation not hold the pressure required, a leak may be present and should be found using a leak detection fluid (LDF). As the liquid is washed around the joints it will bubble up, provided a small pressure is maintained within the pipeline.

New installations prior to connection of meter

(1) All open pipe ends should be capped off except one onto which is fitted a testing tee, comprising a small plug cock and a pressure test nipple.
(2) A manometer, filled with water to register zero, is connected to the test nipple and air is pumped or blown into the system to register 20 mbar.
(3) The plug cock is shut off and there is a wait of 1 minute for the air to stabilise.
(4) After the initial 1 minute a further 2 minutes are waited in which there should be no further pressure drop at the manometer.

Existing installations (using the gas pressure from the service main itself)

The test method would depend upon whether a low or medium pressure regulator has been installed at the meter. The method explained here is the test to undertake where the pipework is no bigger than 22 mm and a low pressure regulator is found or where a test lever has been fitted downstream of a medium pressure regulator. Where a medium pressure regulator and no test valve is found it must be understood that a different test method is adopted, in which case further study would need to be sought from the British Standard or gas-related textbook.

(1) Turn off the main gas inlet control valve and all gas appliances and pilot lights.
(2) Remove the screw from the test nipple located on the outlet side of the gas meter and connect a manometer to it; adjust the gauge to zero.
(3) Slowly open the main gas cock to allow 10 mbar through, then re-shut the valve (and observe the gauge for 1 minute). There should now be no slow rise in the pressure reading at the gauge. If there is, it indicates that the main gas cock is letting gas through when closed; in these circumstances it is impossible to continue the test to achieve a true reading and the gas authority should be notified.
(4) Assuming the main gas cock does not let-by, **slowly** reopen the valve to allow gas to enter the pipework and come up to 20 mbar.
(5) Turn off the gas and wait 1 min for stabilization. Reset the gas to 20 mbar if necessary, then, over the following 2 min, the pressure registered on the manometer should show no further drop. Where appliances are connected and not isolated, a permissible pressure drop is allowed but must not exceed 4 mbar, using a U6 or G4 meter with a capacity up to 6 m^3/h and 8 mbar where an electronic E6 meter has been fitted.
(6) Upon the test proving satisfactory the test nipple is replaced and the gas turned on. Leak detection spray is now applied to the test nipple connection and any pipework preceding the meter regulator. Provided there is no smell of gas the system can be regarded as having no leaks.

Purging

Prior to gas entering the pipework, air which is trapped needs to be expelled. This is referred to as purging. Purging also removes debris, such as wire wool, from the pipeline. During any purging operations one must ensure good ventilation, prohibit naked flames and avoid the operation of electric switches. For small installations with pipework less than 22 mm, the procedure is carried out as follows:

(1) Turn off the emergency control valve at the meter and disconnect the supply pipe at the furthest appliance.
(2) Open the control valve and allow not less than $0.01\,m^3$ of gas through the meter, or $0.35\,ft^3$ where an old U6 meter is installed. A smell of gas must also be evident at the disconnected joint.
(3) Reconnect the appliance and check the broken joint with LDF to ensure it does not leak. Any further legs of pipe will only need to be opened until gas is smelt and is freely discharging from the open ends.

Gas meter with low pressure regulator

Gas meter with medium pressure regulator and test valve

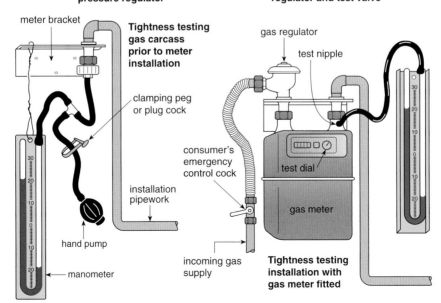

Tightness testing and purging

Pressure and Flow

Relevant British Standard
BS 6400

Standing pressure   This is the pressure available for use when all appliances are switched off. The standing pressure at the meter should be 20–30 mbar: where it is not within this range the supplier should be notified. The standing pressure can be taken anywhere throughout the system, from a suitable test point, on the gas carcass side of any regulating or governing devices.

Working pressure at meter   This is the measured pressure at the meter when gas is flowing. Only the supplier of the gas may adjust this pressure at the meter regulator. However, the installer should check for a reading of 21 mbar ± 2 mbar with one appliance operating at its full rate or, in the case of a cooker, three gas rings. Notify the supplier where incorrect.

Burner or operating pressure at appliances   This is the pressure recommended by the manufacturer and is adjusted while the appliance is burning fuel. Prior to lighting any burners, a manometer is positioned on the test nipple on the appliance side of the governor. The burner is ignited and left to run for some 5 to 10 minutes, thus allowing the pressure to stabilise. The pressure-adjusting screw on the governor is then turned as necessary to increase or decrease the pressure: it is turned clockwise to increase the pressure.

Working pressure drop of the system   This test is one which ensures that the maximum pressure drop across any gas supply installation does not exceed 1 mbar (as identified under pipe sizing, see page 192). The test is carried out as follows:

One manometer is positioned at the meter and a second one at the furthest appliance. All burners on the system are ignited and after a suitable stabilisation period (5–10 min) a reading is taken from each manometer, the difference between the two being the system working pressure drop. Where only one manometer is available, the test is done in two stages.

Heat input (gas rate or gas consumption)

The amount of gas supplying a burner is not only dependent upon the pressure but also on the injector size used; for example, the gas injector for LPG is much smaller than that used for natural gas. To ensure that the correct volume of gas is being consumed as recommended, i.e. to ensure that the correct injector is fitted or that there is no blockage in the injector, one must check the gas input as follows:

U6 gas meter

(1) Turn on the burner and wait for the gas to stabilise.
(2) Observe the test dial on the meter and record the time, in seconds, it takes to complete one complete revolution of the test dial (i.e. 1 ft^3).
(3) Divide the number of seconds in 1 hour (3600) by this figure to give the gas consumption in ft^3.

(4) Multiply by the calorific value of gas ($1035\,\text{Btu/ft}^3$) to give the Btu/h heat input
(5) Divide by 3412 to convert to kW.

$$\text{Hence } 3600 \div \text{time in seconds} \times 1035 \div 3412 = \text{kW}.$$

Example: Assume it takes 57 seconds for the gas passing to the burner to consume $1\,\text{ft}^3$ of gas.

$$\text{Therefore, } 3600 \div 57 \times 1035 \div 3412 = 19.16\,\text{kW}.$$

Note: Because all the numbers in the calculation except the time are constant, it is possible to compute the constant numbers $3600 \times 1035 \div 3412$ to give 1092. This gives a revised calculation of :

$$1092 \div \text{time} = \text{kW}$$

$$\text{Thus, } 1092 \div 57 = 19.16\,\text{kW}$$

E6 and G4 gas meters
(1) Turn on the burner and wait for the gas to stabilise.
(2) Note the LCD reading on the gas meter and immediately start a stop watch.
(3) After 2 minutes take a second reading of the LCD on the gas meter.
(4) Subtract the first reading from the second to give the gas flow in m^3 over the test period.
(5) Multiply the flow by 30 to give the gas consumption per hour.
(6) Multiply by the calorific value ($38.5\,\text{MJ/m}^3$) and divide by 3.6 to give the heat input in kW.
(7) Multiply by 3412 to convert to Btu/h if desired.

Hence, m^3 over 2 minutes $\times 30 \times 38.5 \div 3.6 = \text{kW}.$

Example: If the first LCD reading is 00396.326 and the second reading is 00396.401, the gas flow over 2 minutes would be

$$00396.401 - 00396.326 = 0.075\,\text{m}^3$$

$$\text{Therefore, } 0.075 \times 30 \times 38.5 \div 3.6 = 24.1\,\text{kW}.$$

Note: As with the previous U6 meter calculation, all the numbers in the calculation, with the exception of the meter reading, remain constant. Therefore the calculation simplifies to:

$$\text{m}^3 \text{ over 2 minutes} \times 320.38 = \text{kW}$$

$$\text{Thus, } 0.075 \times 320.38 = 24.1\,\text{kW}.$$

Note: The heat input calculated is the 'GROSS' input. Therefore, where the manufacturer specifies the 'NET' input on the data badge (found on modern appliances) the gross figure calculated will appear too high and will therefore need to be adjusted by dividing the heat input calculated by 1.11 in the case of natural gas, 1.09 for propane and 1.08 for butane:

Example: 9 kW gross $\therefore$ $9 \div 1.11 = 8.1$ kW net.

4 Gas Supplies

Gas Controls 1

Gas regulators (gas governors)

These are devices which are fitted into the pipeline to ensure that the gas arriving at the appliance does so at a constant pressure, as recommended by the manufacturer. Modern regulators use a spring to act upon a diaphragm, exerting a force; when the gas exerts a force on the underside of the diaphragm in excess of that of the spring, it causes the diaphragm to lift and carry with it the valve, thus reducing the gap through which the gas can pass to the appliance.

Should the gas pressure drop, the spring forces the diaphragm down again, increasing the gap and allowing more gas through. Conversely, should the pressure increase, the diaphragm is lifted higher, closing the valve gap even more. This continuous up and down movement of the valve ensures that a balance is maintained between the inlet and outlet to the regulator. The pressure acting on the spring can be adjusted by means of a screw on top of the regulator.

Sometimes regulators are found to be defective and fail to operate as they were intended. This usually results from one of the following faults:

- **No gas flow:** Often due to the valve being held in the locked up position, resulting from an appliance not being used for some considerable period. The valve has simply stuck onto the seating.
- **Low gas pressure:** Usually a blocked gas line or filter, a broken or fatigued spring, or a damaged diaphragm.
- **Gas pressure too high:** Often the result of let-by through the regulator, resulting from a badly seated valve, possibly due to dirt in the gas line.
- **Gas escaping through the breather hole:** The result of a punctured diaphragm.
- **Chattering noise:** The result of an oversized breather hole or the dust cap above the pressure adjusting screw having been removed.

Multi-functional valve

The multi-functional valve is really a series of different component valves incorporated within the one unit consisting of the following:

- regulator
- thermoelectric valve
- main burner solenoid
- pilot regulation screw
- working and burner test points
- filter.

Having all the components installed in one unit such as this saves on the space needed to incorporate the various controls.

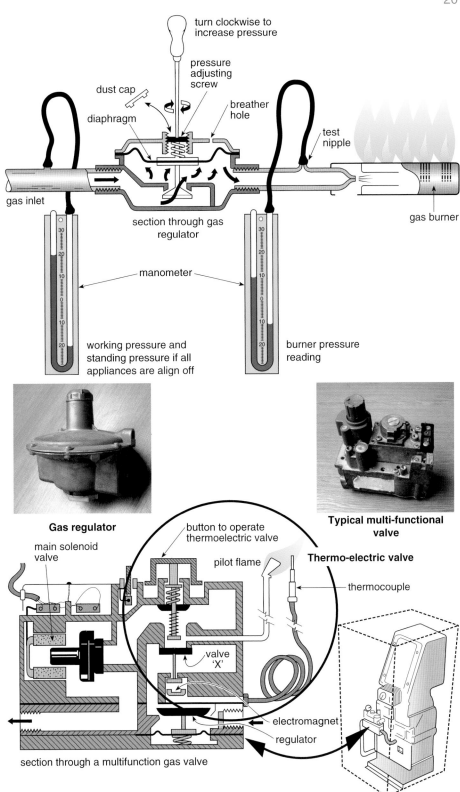

turn clockwise to increase pressure

pressure adjusting screw

dust cap

breather hole

diaphragm

test nipple

gas inlet

gas burner

section through gas regulator

manometer

working pressure and standing pressure if all appliances are align off

burner pressure reading

Gas regulator

Typical multi-functional valve

button to operate thermoelectric valve

main solenoid valve

pilot flame

Thermo-electric valve

thermocouple

valve 'X'

electromagnet

regulator

section through a multifunction gas valve

Gas controls 1

Gas Controls 2

Gas relay valve   A valve rarely found which is used to automatically control the gas supply to a burner. An outlet tube (the weep pipe) leads from the top chamber of this valve and if any valves fitted along its length are closed (e.g. rod thermostat or gas cock) the pressure will build up above the diaphragm to equalise that below it, resulting in the valve falling onto the seating under its own weight and closing off the supply. Conversely, should the weep line be open, the pressure in the top chamber is reduced faster than the bypass orifice can deliver. The weep pipe is generally terminated in the vicinity of the main burner. Sometimes a bypass adjusting screw is fitted to prevent the valve completely closing, allowing for a high/low flame control.

Low pressure cut-off device   A control somewhat similar to a relay valve but the diaphragm is now on the outlet side of the valve seating. This valve is used in situations where the supply may be isolated or cut off, such as schools where previously isolated gas taps may be opened by students or in LPG installations where the supply has run out and there is a possibility that an appliance has been left on. Should someone reinstate the gas supply unaware of the open gas line, unburnt gas will escape into the atmosphere potentially resulting in an explosion. Should the gas pressure be isolated and pressure lost from the outlet, the valve will close. Upon reinstating the pressure, the valve will remain closed due to there being insufficient pressure to overcome that acting upon the surface area of the valve itself. When all downstream valves have been closed and the reset plunger depressed, gas is allowed to enter the outlet and act under the diaphragm. With the pressure now acting upon this greater surface area, the pressure overcomes the force pushing downwards and the valve lifts open.

Solenoid valve   A solenoid valve is an electrically operated valve used to control the on/off flow of gas or liquid through a pipeline. The solenoid consists of a coil of wire; when an electric current is allowed to flow through the wire it creates a magnetic field and draws a plunger, usually of iron, into the solenoid, opening the valve. When the power is switched off, the return spring pushes the valve back onto its seating closing the valve.

Oxygen depletion sensor (ODS) atmosphere sensing device (ASD)   A device fitted to open flued appliances designed to shut off the gas supply in the event of the environment becoming starved of oxygen. The ODS utilises a controlled flame to heat up a thermocouple. In the event of oxygen levels decreasing, the pilot flame is made to search for oxygen which prevents it from playing onto the tip of the thermocouple, thus allowing it to cool and consequently closing off the gas supply. The ASD on the other hand detects spillage. A probe filled with a volatile fluid, or a thermistor, is positioned around the draught diverter. Should continued hot gases pass over this probe due to spillage, the probe heats up and causes the electrical contacts to open, breaking the electrical power supply to the gas valve or interrupting the current flow from the thermocouple, causing it to shut down.

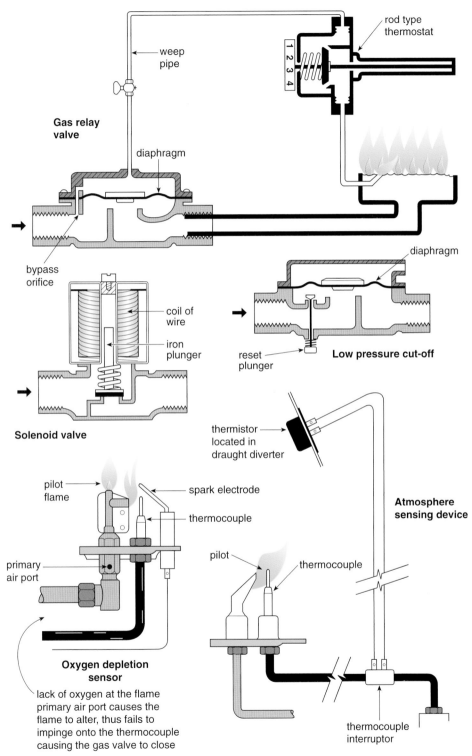

rod type
thermostat

weep
pipe

**Gas relay
valve**

diaphragm

bypass
orifice

coil of
wire

iron
plunger

diaphragm

reset
plunger

Low pressure cut-off

Solenoid valve

thermistor
located in
draught diverter

**Atmosphere
sensing device**

pilot
flame

spark electrode

thermocouple

pilot

thermocouple

primary
air port

thermocouple
interruptor

**Oxygen depletion
sensor**

lack of oxygen at the flame
primary air port causes the
flame to alter, thus fails to
impinge onto the thermocouple
causing the gas valve to close

Gas controls 2

Flame Supervision Devices
(Flame Failure Devices)

A device incorporated in the gas control valve of a burner designed to prevent gas discharging from the main burner until the pilot flame has become established. There are several methods of flame detection, including:

Bi-metallic strip An older method which consists of two different metals bonded together, each having different expansion rates, one high and one low. When heat is applied to the bi-metallic strip it is forced to bend in a particular direction because of the different expansion rates and opens the valve to the main burner.

Liquid vapour device One example of this device is found in the domestic gas oven. Gas is allowed to pass to the oven burner at a by-pass rate where a small pilot flame is allowed to play onto a small phial. Inside this phial and its connecting capillary tube is a volatile fluid which expands rapidly. This expanding liquid is forced into the bellows which in turn overcomes the power exerted by the spring and allows the valve to open. Should the flame go out, the liquid cools and the valve is made to close back down to the bypass rate.

Thermoelectric valve This valve is usually incorporated within a multi-functional gas valve. When heat is allowed to impinge onto a thermocouple, an electric current is generated which is used to operate the solenoid valve. The valve works as follows:

(1) When the reset button is depressed and held down, gas is permitted to flow to the pilot tube (sketch B).
(2) The pilot light is ignited and the flame impinges on the thermocouple, thus generating an electrical charge.
(3) After some 10–20 seconds the reset button can be released. The solenoid will now hold the valve open. As the reset button returns to its original position, gas is then allowed to pass onto the main burner pipe (sketch C).
(4) Should the flame no longer impinge onto it, the thermocouple will cool and will not produce an electrical current, thus the valve will close completely as in sketch A.

Flame rectification A method of flame supervision which utilises the fact that flames conduct electricity. An alternating current is passed between the electrode and the burner head and if the flame is present, the current will flow keeping the circuit energised, and indirectly holding a gas solenoid open. Should the gas go out for any reason, the electrical circuit is broken and the solenoid closes. Earlier systems of this design used a 'd.c.' supply. This, however, had the disadvantage that any short circuit within the system had the same effect as that of a flame, allowing gas to flow undetected. Flame rectification pushes alternating current through the electrode but only direct current is received at the burner head because the flow of ions through the flame is only detected in the direction of the large burner head, rather than the comparatively small electrode terminal. Should a short circuit result, 'a.c.' output would be detected and the device would fail to a safe condition.

See note on page 232 regarding testing of flame supervision devices.

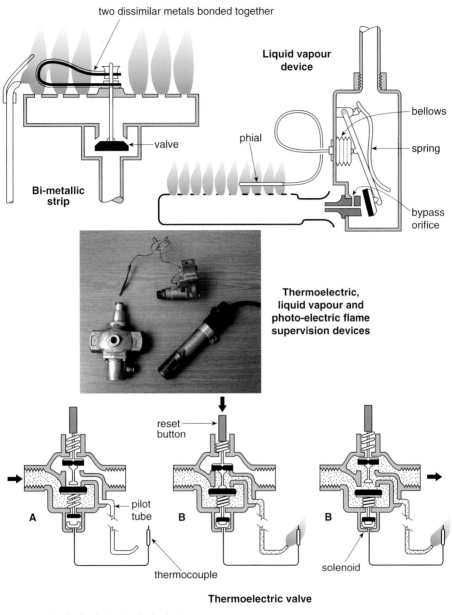

two dissimilar metals bonded together

Liquid vapour device

bellows

spring

phial

valve

bypass orifice

Bi-metallic strip

Thermoelectric, liquid vapour and photo-electric flame supervision devices

reset button

pilot tube

A

B

B

thermocouple

solenoid

Thermoelectric valve

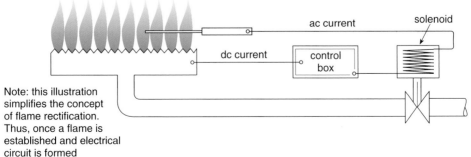

ac current

solenoid

dc current

control box

Note: this illustration simplifies the concept of flame rectification. Thus, once a flame is established and electrical circuit is formed

Flame supervision devices

Open Flued Appliances (Conventional Flue)

Relevant British Standard
BS 5440

When installing an open flued appliance, one must consider the location of the terminal and the route, length and size of the flue. Failure to do so may result in excessive condensation problems or, more seriously, the inadequate removal of flue gases which would create an inefficient and dangerous flueing system, possibly leading to the death of the building's occupants from carbon monoxide poisoning. The route that the flue takes on its way up through the building should be such that the flue gases are expelled as quickly as possible, thus ensuring that the gases are not cooled to the dew point of water (approximately 55°C) where condensation will occur.

Generally, increasing the height of the flue increases the flue draught because a high column of heated combustion products has a greater energy potential. However, surface friction also slows the flow, and therefore the height cannot be indefinite; also, increased height increases heat losses. A typical height would be 6–8 m. Internal flues are far less likely to suffer the problem of cooling gases; where exposure is likely, insulate the flue or use twin-walled flue pipe. The size of the flue or its cross-sectional area should not be too small as spillage of gases will occur, and if it is too large greater heat loss will occur.

The driving force to remove the flue gases comes either from thermal convection currents (natural draught), or from an electric fan (forced draught). Systems which incorporate a fan have several advantages over natural draught systems, including the increased positive removal of flue products, and greater flexibility in terminal location and siting of appliances. Alternatively, potential disadvantages include possible noises from the fan, its tendency to depressurise the room containing the appliance, and the need for additional safety devices to ensure that the gas supply is shut down if there is a fan failure.

A draught diverter is incorporated with the appliance to prevent excessive down- or updraught due to adverse weather conditions; it also serves the purpose of diluting the flue gases. If the diverter does not form an integral part of the appliance and is fitted separately, it must be installed within the same room or enclosure as the appliance. One should aim to have as much vertical flue as possible directly above the draught diverter (a minimum of 600 mm is recommended before any change in direction); this helps avoid the initial spillage of combustion products into the room.

The European Standard: BS EN 1443:2003 specifies the general requirements and basic performance criteria for chimneys (including connecting flue pipes and fittings) and identifies the markings to be used and their potential use.

Typical identification marking: BS EN 1443 – T250 P1 O W 1 R22 C25

For a full description of the ID marking see EN 1443. In short, the above would denote that the flue material used is suitable for a gas installation operating under a positive pressure and below 250°C. It can be installed with its outer surface no closer than 25 mm to any combustible material.

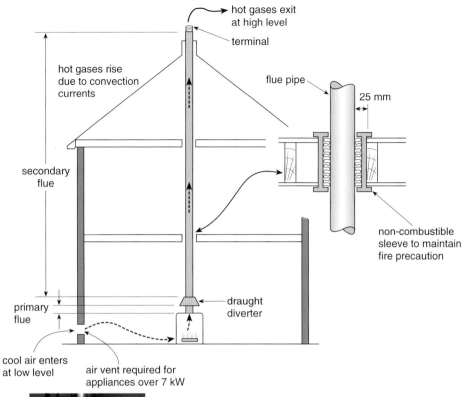

hot gases exit at high level

terminal

hot gases rise due to convection currents

flue pipe

25 mm

secondary flue

non-combustible sleeve to maintain fire precaution

draught diverter

primary flue

cool air enters at low level

air vent required for appliances over 7 kW

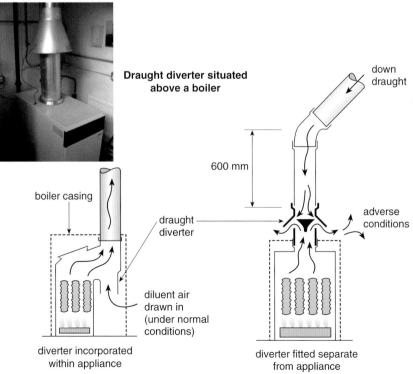

Draught diverter situated above a boiler

down draught

600 mm

boiler casing

draught diverter

adverse conditions

diluent air drawn in (under normal conditions)

diverter incorporated within appliance

diverter fitted separate from appliance

Open flued appliances

Terminal Location for Open Flues

Relevant British Standard
BS 5440

Positive and negative pressure zones When wind blows towards a structure it creates a positive force (+) on one side of the building and a negative force (–) on the other. Where the wind has to travel over high structures, the direction of the wind force changes, blowing round in circles, as in a vortex. In each case, these positive and negative pressure zones cause a blowing or sucking effect within their region. In the siting of any open flue terminal where the gases are being expelled by natural convection currents, it is essential to discharge the products above these zones, otherwise excessive up or down draughts will occur, affecting the correct operation of the appliance.

Natural draught systems The terminal should provide the extraction of flue gases under virtually all wind conditions. The ideal position is above the highest point of the roof (e.g. ridge) and not shielded by any other structures or objects which might cause a pressure zone. Providing the location of the terminal is not within 1.5 m of a vertical surface such as a chimney stack or dormer window, the flue should be at a height that maintains 1.5 m, measured horizontally from the underside of the flue terminal to the roofline. See illustration opposite. In the case of flat roofs the height is generally 600 mm from the roof decking to the underside of the terminal; however it can be reduced to 250 mm if the pipe passes directly through the roof and there is no parapet or structure within a distance of ten times the structure height.

It is possible to calculate the actual flue height through a pitched roof by undertaking the following calculation using a calculator with the scientific function keys **tan pitch × 1.5.**

Example: Calculate the terminal height where the roof pitch is 30°.

Answer: Tan $30 \times 1.5 = 0.866$ m

(Note: with older scientific calculators you would need to enter the angle into the calculator before pressing the tan button. Thus in the example given you would enter the following into the calculator: 30 tan × 1.5 = 0.866 m.)

Where the terminal is within 1.5 m of a vertical surface, the height of the flue needs to be at least 600 mm above the top of the structure. Note carefully from the example in the figure that the height indicated is from the highest point of the roof–flue structure to the base of the terminal.

Note: the sizes indicated here and in the illustrations opposite are applicable to boilers ≤ 70 kW (i.e. domestic).

Forced draught systems With fan-assisted systems, the flue route and final siting of the terminal are not critical to the performance of the appliance. However, positions where the combustion products may cause a nuisance should be avoided. Where the flue pipe terminates with a grille built into the wall, the effective free area opening of the outlet should be at least 70% of the area of the flue pipe.

wind direction

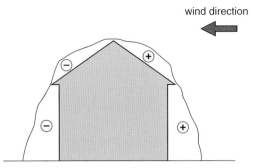

Positive and negative pressure zones
to be avoided for siting of flue terminals
relying on natural flue draught

high rise building

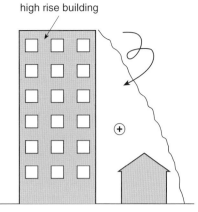

Unless there is no wind this example
will always be in a positive pressure
zone irrespective of wind direction

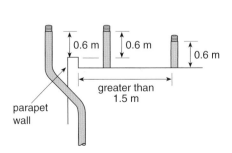

Typical gas flue terminals

600 mm

250 mm

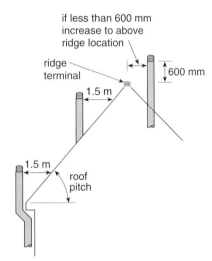

if less than 600 mm
increase to above
ridge location

ridge
terminal

600 mm

1.5 m

1.5 m

roof
pitch

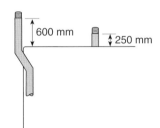

0.6 m   0.6 m

0.6 m

greater than
1.5 m

parapet
wall

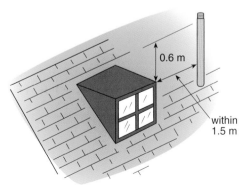

0.6 m

within
1.5 m

Terminal location for open flues

Materials and Construction of Open Flues

Relevant British Standard
BS 5440

Brick chimneys   Modern chimneys are lined during their construction and should therefore require no treatment. However, traditional $225 \times 225\,mm^2$ brick chimneys, designed for open fires, tend to be unnecessarily large for domestic gas burning appliances and as a consequence are prone to condensation problems; this inevitably leads to moisture passing into and through the brickwork. Generally speaking, gas fires and gas circulators do not require the lining of the flue-way whereas boilers, having a greater heat input, do (see BS 5440).

Flexible stainless steel liners   The size of the liner used should always be the same as, or larger than, the appliance flue outlet and is generally installed as follows:

(1) The existing chimney pot and its mortar flaunching are removed.
(2) A draw cord is lowered through the chimney and the liner attached to a dome-shaped bung and pulled up, or down if easier, through the flue-way.
(3) The liner is fixed at the top, using a clamping plate, and at the base, again using a clamping plate or directly on to the appliance.
(4) The flue terminal is fitted and the flaunching made good, thus discharging the water away from the terminal. Note that for older buildings it is possible to purchase terracotta terminals, suitable for gas and in keeping with the building design.
(5) The space between the base of the chimney and flue liner should be sealed (e.g. using glass fibre quilt) to create a void of trapped air and prevent flue gases passing up between the liner and the chimney.

Note that where the appliance is fitted externally to the chimney, a flue pipe should be used to run up into the chimney void and the liner fixed to a socket at this point. This joint should be made available for future inspections via an access plate.

Precast-concrete flue blocks   A specially designed flue system which is built into and forms part of the building structure, allowing for a larger living space. The system should incorporate a starter block at its base and, where possible, run vertically to its point of termination. Sometimes the system is only made up to the roof space and the final section run using twin-walled flue pipe terminating with a ridge terminal. Alternatively a brick chimney stack could be constructed through the roof section.

Flue pipes   Large diameter pipes, available in a range of sizes and made from vitreous enamelled mild steel, stainless steel, high grade aluminium and fireproof materials. Single- or double-walled pipe is available, the latter consisting of two concentric pipes separated by an air space or filled with an insulation material. When single-walled pipe is used with fire cement type joints, the socket should sit below the spigot, thus ensuring that any water resulting from condensation will not cause the cement to weaken and fall out of the joint.

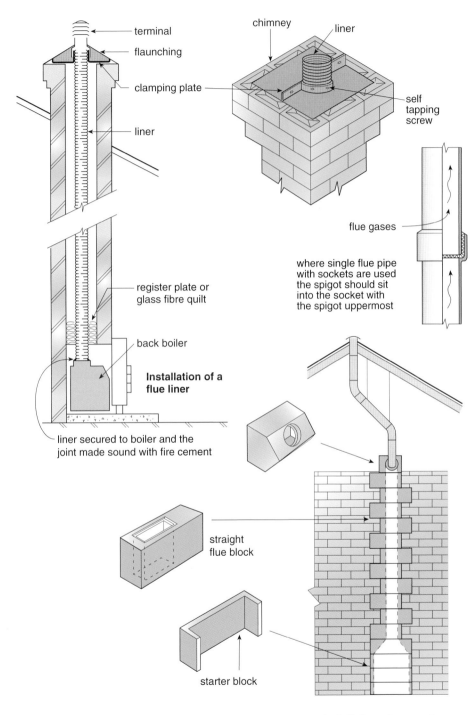

terminal

flaunching

clamping plate

liner

chimney

liner

self tapping screw

flue gases

where single flue pipe with sockets are used the spigot should sit into the socket with the spigot uppermost

register plate or glass fibre quilt

back boiler

Installation of a flue liner

liner secured to boiler and the joint made sound with fire cement

straight flue block

starter block

Installation of pre-cast concrete flue blocks

Materials and construction of open flues

Installation of Gas Fires

Relevant British Standard
BS 5871

Gas fires fall into three categories:

- *Traditional gas fires* These sit in front of the fireplace opening, making use of a closure plate.
- *Inset live fuel effect fires* These are either partially or fully inserted into the fireplace opening; the opening seal is usually incorporated in the fire.
- *Decorative fuel effect fires* These do not connect directly to the flue; the fire only sits in the grate within the fire opening. Such fires provide the room with very little radiant heat; they are primarily designed for decorative purposes (i.e. look like a real fire).

Connections to existing chimneys   When a gas fire is to be positioned into an existing masonry chimney which was previously used to burn some other fuel, the chimney must be swept thoroughly and all debris removed. The flue-way should be inspected for its sound condition and any dampers or restrictor plates removed or fixed permanently in the open position. A flue liner is not generally required, except where a large or long flue-way is encountered. It is not normally necessary to remove the existing chimney pot and fit a flue terminal.

When preliminary work on the chimney has been completed, it should be checked for a suitable up-draught (i.e. pull) using a suitable smoke pellet. The chimney is first warmed with a blowlamp, then the pellet inserted into the fire opening; all the smoke should be drawn up the flue-way.

No permanent ventilation is required for open-flued gas fires, providing they do not exceed 7 kW input. However, in the case of the decorative fuel effect (DFE) fire, purpose-provided ventilation of at least 100 cm^2 will be required on all appliances up to 20 kW input. Where more than one appliance is installed in the same room as a DFE, additional ventilation may be required; see BS 5871 for guidance.

Debris collection space   When the gas fire is installed against the fireplace opening it should be noted that a 250 mm minimum depth, or 12 dm^3 void, is to be maintained below the spigot entry point, thus allowing debris which may fall down the chimney to collect. When servicing any fire this space must be inspected.

Use of closure plate   The traditional gas fire requires a closure plate (see figure) to prevent the entry of excess air into the fire whilst maintaining the correct gap to allow for suitable air relief resulting from excessive down-draughts. The closure plate is secured in position using suitable adhesive tape, i.e. tape capable of maintaining its seal up to 100°C.

Fire precautions   Floor standing fires are usually mounted onto a non-combustible hearth having a minimum thickness of 12 mm. The hearth should extend 300 mm forwards from the back edge of the fire and 150 mm beyond each edge of the naked flame or radiant. Wall-mounted fires should be installed so that the flame or radiant strips are at least 225 mm above any carpet or floor covering.

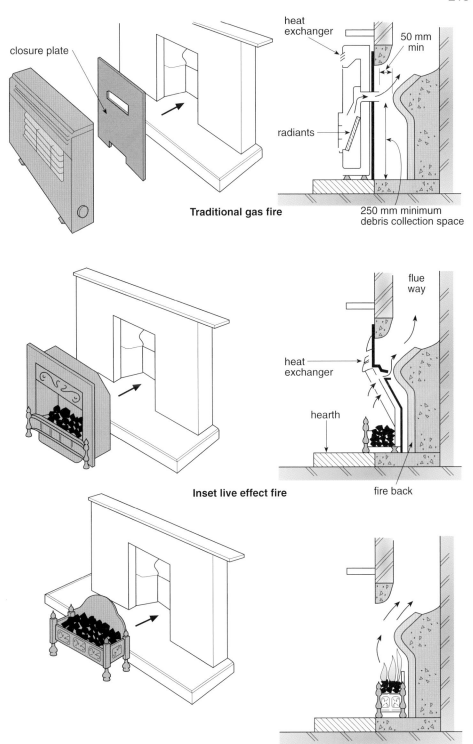

Traditional gas fire

closure plate

heat exchanger

50 mm min

radiants

250 mm minimum debris collection space

Inset live effect fire

flue way

heat exchanger

hearth

fire back

Decorative fuel effect fire

Installation of gas fires

Installation of Gas Cookers

Relevant British Standard
BS 6172

The domestic gas cooker consists of three basic components: the oven, hob and grill.

Oven   The oven is basically an insulated chamber with an access door to the front. Air enters at the bottom of this door for combustion purposes and the products are discharged to the top rear via a discharge flue. Two distinct designs of oven can be found. One is the direct oven in which the flame is located at the base. This design typically has different heat zones, the hottest at the top. In the second design of oven, the indirect oven, the combustion process occurs in a separate compartment from that used for cooking. With this design the oven temperature is more evenly distributed. Indirect ovens are used more in the USA and mainland Europe than in the UK.

Hob   Unlike the oven in which the burner is enclosed, the hob or hotplate is an exposed flame used for boiling and simmering. The majority of cookers have four burners, usually of varying heat inputs. To give a pleasing appearance, to hide the pipework and to catch liquid which spills from the pan, a spillage tray is incorporated into which sit the pan supports.

The gas supply to an oven and hotplate is made via a flexible connection at the rear to a pipe which connects to the float rail (the pipework manifold distributing gas to the individual gas taps). The float rail is located directly behind the hotplate tap knobs where there is often found a pressure test point.

Should a drop down lid be manufactured with the hob, a safety control will be incorporated to shut off the supply of gas should the lid be closed. The lid must therefore be raised during the tightness testing procedure.

Grills   The grill is used for toasting or browning of foods. Two types are found: the conventional and the surface combustion grill. With the conventional grill the gas flames play onto a grill fret which in turn glows red hot. With the surface combustion grill on the other hand, the whole grill fret becomes the burner, leading to a quick rise in temperature and even distribution of heat. The grill can be manufactured with an oven and be either high level or below the hotplate. Alternatively, it can be independently wall mounted.

When siting a cooker, its proximity to combustible surfaces should always be considered and it should be placed in accordance with the manufacturer's instructions. A stability bracket or safety chain should also be incorporated in all free-standing cookers to prevent them tilting forward when a heavy load is applied to the door.

Ventilation   The size of the room will determine whether additional ventilation is required or not. However, in most cases, a window direct to outside is required. See page 220 for the specified size requirement.

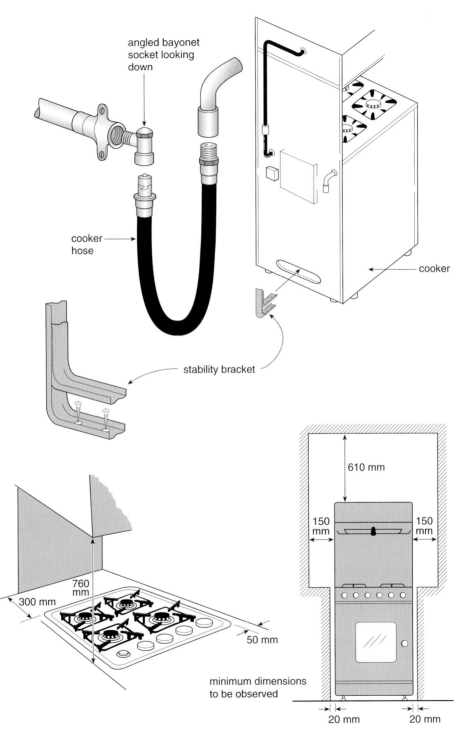

angled bayonet socket looking down

cooker hose

cooker

stability bracket

610 mm

150 mm

150 mm

760 mm

300 mm

50 mm

minimum dimensions to be observed

20 mm

20 mm

Gas cooker installation

Room Sealed Appliances 1 (Balanced Flue)

Relevant British Standard
BS 5440

Room sealed appliances are ones which use a flue system that draws its air for combustion purposes from outside the building; these are unlike open-flued appliances which draw air from the room. Because of their design concept, room sealed appliances should, whenever possible, be recommended, as they generally present no danger to the occupants of the building from carbon monoxide poisoning (the result of vitiated air (air lacking oxygen) circulating between the room and appliance).

Natural draught systems In most cases appliances which expel the flue gases by convection currents (natural air flow) need to be located adjacent to an external wall. Natural draught appliances are easily identified by a typical large square terminal located on the external wall face.

Fan assisted systems These appliances have advantages over natural draught systems in that they can be located some distance from the external wall. Because such appliances are more positive in the extraction of flue gases, smaller outlet terminals are used; also, such systems are much less affected by adverse wind conditions which means that the terminal need not be located in a clear expanse of wall. The fan can be located on the air intake or the flue extract, but in all cases the design of the burner should be such that, if the fan malfunctions, the burner will be prevented from firing.

Terminal location The location of the terminal should be such that the selected position ensures safe and efficient performance of the appliance. Unsatisfactory locations may result in:

- Flue gases entering the building through openings
- Possible fire hazards
- Staining to walls by the products of combustion
- Inefficient combustion of fuel.

Additional points to consider include:

- The provision of a terminal guard where the flue outlet is less than 2.1 m above ground, balcony or flat roof level, where people have access
- The provision of a suitable shield, at least 1 m long, to protect surfaces when a natural draught terminal is located within a distance of 1 m below a plastic gutter or within 0.5 m below any paintwork
- An assurance that the combustion products do not cause a nuisance to passers-by or people on adjoining property.

Manufacturers' instructions should be sought for the correct location of the terminal; however, the table opposite can be taken as a general guide to suitable positions.

Suitable room-sealed terminal locations for gas appliances

Terminal position	Minimum distance (mm)	
	Natural draught	Fan assisted
Vertically from a terminal on the same wall	1500	1500
From an opening in a car port (e.g. window)	1200	1200
From a terminal facing a terminal	600	1200
From a surface facing a terminal	600	600
From an internal or external corner	600	300
Below balconies or projected roofs	600	200
Horizontally from a terminal on the same wall	300	300
Above ground or roof level	300	300
Below the eaves	300	200
Below gutters or discharge pipework	300	75
Horizontally from vertical discharge pipe	300	150
Below opening into the building 0–7 kW Net input	300	300
7–14 kW	600	300
14–32 kW	1500	300
Over 32 kW	2000	300
Above opening into the building 0–32 kW Net input	300	300
Over 32 kW	600	300
Horizontally from opening to building 0–7 kW Net input	300	300
7–14 kW	400	300
Over 14 kW	600	300

4 Gas Supplies

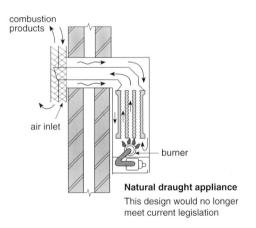

combustion products

air inlet

burner

Natural draught appliance
This design would no longer
meet current legislation

Room sealed boiler

fan

heat
exchanger

Fan assisted appliance

Natural draught terminal
Note: a guard is required
where the terminal is
accessible (i.e. < 2.1 m from
an accessible area)

Fan assisted terminal

Room sealed appliances 1

Room Sealed Appliances 2

Relevant British Standard
BS 5440

Vertical room sealed appliance

This design of flue system is often used where it is difficult to locate an appliance with either back or side outlet. Two systems can be found, either using a concentric flue and air duct or a separate parallel flue and air duct.

Typically, systems which use concentric flue and air ducts allow for a maximum flue length of up to 4–5 m, whereas systems using a separate parallel flue/air duct run to greater lengths (usually up to 30 m in equivalent length) depending on the flue route taken. The equivalent length is the total length of the air duct, plus the flue duct, plus any allowance for fittings used. See manufacturers' instructions for further advice.

Balanced compartments

An arrangement in which an open flued appliance is installed in a small room and the flueing and ventilation provisions are such that they, in effect, convert the operation of the appliance to that of a room sealed appliance. For the system to work it is essential that the air intake and flue gas extract are located within the same positive or negative pressure zones. This is achieved by ensuring that the extract and intake terminals are within 150 mm of each other. For this system to function correctly it must not be influenced by pressures within the building; therefore, a self-closing, tight-fitting door must be provided with a suitable draught sealing strip included. Should the door be opened, the electrical supply to the appliance must be broken, thus ensuring it will not work. A notice should be fixed to the door saying it must be kept closed. **Note:** These compartments must not open into a bathroom. No openings should be allowed into the compartment other than those intended for air supply and flue gas extract (see figure). The air intake size is dependent on whether a high or low vent is provided in the compartment.

Ducted to high level: $2\frac{1}{2}$ times 5 cm² for every kilowatt of the appliance
Ducted to low level: $1\frac{1}{2}$ times 5 cm² for every kilowatt of the appliance.

In each case this allows for cooling as well as for supplying air for combustion purposes.

These systems are particularly suitable for larger installations where room sealed appliances are unavailable, or as an alternative to installing long external flue runs such as adjacent to a tall building.

In most instances the room would be large enough for the commissioning engineer to stand inside to undertake the necessary tests. However, where this is not the case it is permissible to temporarily bypass the door switch.

4 Gas Supplies

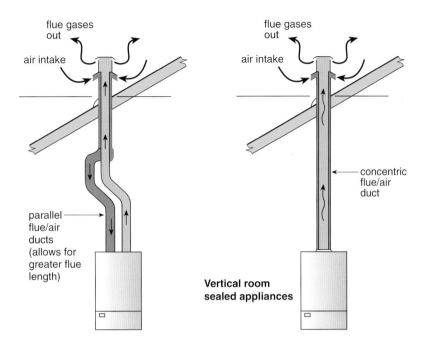

flue gases out

air intake

parallel flue/air ducts (allows for greater flue length)

flue gases out

air intake

concentric flue/air duct

Vertical room sealed appliances

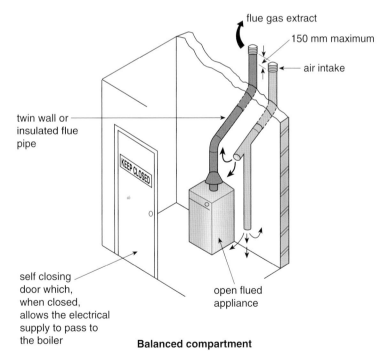

flue gas extract

150 mm maximum

air intake

twin wall or insulated flue pipe

KEEP CLOSED

self closing door which, when closed, allows the electrical supply to pass to the boiler

open flued appliance

Balanced compartment

Room sealed appliances 2

Ventilation Requirements up to 70 kW Net Input

Relevant British Standard
BS 5440

Effective free area   When installing an air grille it is essential to understand that a permanent, unrestricted air flow is required; a closing device must not be incorporated with the grille and no fly screens attached, which may consequently become blocked. The effective free area is that through which the air may pass. The air vent should preferably connect directly to the outside air. Where this is not possible, venting to an adjacent room is acceptable if the other room is provided with the correct size air vent.

Ventilation grilles to the outside environment may be located in any position (i.e. high or low), although when communicating between internal walls the grille should be located at low level (i.e. below 450 mm) to reduce the spread of smoke in the event of fire. The air supply requirements for an appliance will depend upon its location and heat input and can be calculated as identified below. Note that an air grille must not be positioned next to a flue terminal. In these cases, the British Standard should be consulted for the minimum distance to be observed.

Room sealed appliances   These do not require any air for combustion purposes.

Open flued appliances   Appliances with an input below 7 kW do not require any additional air requirements, but where the input to the appliance is in excess of 7 kW the room in which the appliance is installed must have an air grille with a minimum effective free area of 5 cm² for every kilowatt in excess of 7 kW net input.

Example: A 22 kW open flued boiler requires a grille size of:

$$22 - 7 = 15; \text{ therefore } 15 \times 5 = 75 \text{ cm}^2$$

Flueless appliances   Appliances such as cookers may require an air grille; it depends upon the size of the room in which the appliance is installed. Generally, an openable window, etc., direct to outside is required (see Table 1); however, the BS makes provision for internal rooms with no windows.

Table 1 Ventilation grille requirements for flueless appliances

Type of appliance	Maximum input (kW)	Room volume (m³)			
		<5	5–10	10–20	>20
Domestic oven, hotplate or grill	N/A	100 cm²	50 cm²*	None	None
Instantaneous water heater	11	Not allowed	100 cm²	50 cm²	None

*Not required if a door opens directly to the outside.

Decorative fuel effect (DFE) fires   (those which look like real fires) Unless manufacturers' instructions state otherwise, these require an air grille of 100 cm² (see page 212, Installation of Gas Fires).

Multi-appliance installations   With the exception of DFE fires (see BS 5871), where a room contains more than one appliance the air vent requirement is based on the following:

(1) the total rated input for all open flued fires, boilers and air heaters, or
(2) the total rated input for all flueless fires, boilers and air heaters, or
(3) any larger individual requirements for any other type of appliance.

Air for cooling purposes Where a room sealed or open flued appliance is installed in a small cupboard, the compartment will require air vents for cooling purposes. Table 2 gives the size of grille required per kW appliance net input rating. Where the manufacturer quotes a gross input, simply divide this figure by 1.1 to convert it to a net input.

Table 2 Air vent requirements for cooling (mm²/kW)

Type of appliance	Vented to:	High level grille (cm²)	Low level grille (cm²)
Room sealed	An adjacent room	10	10
Room sealed	The outside air	5	5
Open flued	An adjacent room	10	20
Open flued	The outside air	5	10

Some operatives/manufacturers refer to mm² in preference to cm². To convert one to the other, simply move the decimal point two places to the right or left accordingly, e.g. 5 cm² = 500 mm².

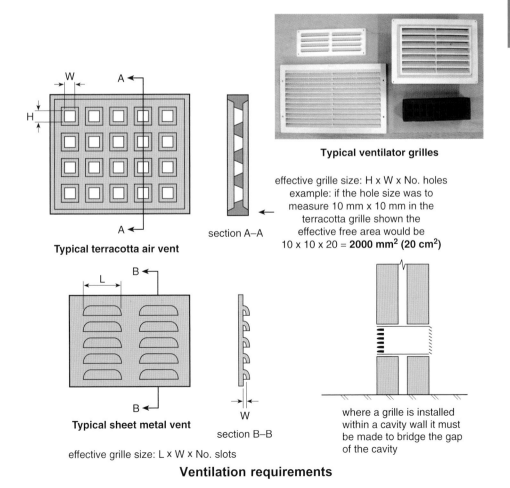

Typical ventilator grilles

effective grille size: H x W x No. holes
example: if the hole size was to measure 10 mm x 10 mm in the terracotta grille shown the effective free area would be
10 x 10 x 20 = **2000 mm² (20 cm²)**

section A–A

Typical terracotta air vent

Typical sheet metal vent

section B–B

effective grille size: L x W x No. slots

where a grille is installed within a cavity wall it must be made to bridge the gap of the cavity

Ventilation requirements

Other Flueing Systems

Relevant British Standard
BS 5440

'Vertex' or 'Solver' flue systems

These are trade names associated with a flue system which takes combustion air from the roof/loft space and delivers it to the appliance via a concentric flue and air pipe. At the point where the air is drawn into the system, a draught break is located; thus, in effect, the appliance is sealed from the room into which it has been installed. The products of combustion are discharged from the building via the secondary flue pipe, run to a suitably located terminal. Ventilation of the roof space is required according to the needs of the appliance as identified by the manufacturer.

Shared flue systems

Open flued appliances in the same room These are permitted to have their flue pipes joined together, provided each appliance has its own draught diverter and flame failure device. Note that the air vent requirements are based upon the total heat input of all appliances fitted within the room.

Open flued appliances in separate rooms These can be connected into a branched flue system only if all the appliances are of the same type and the air supply for ventilation purposes is taken from an identical aspect. The subsidiary branches are designed to provide each appliance with its own pull, ensuring that the products are removed up into the main duct. The height of each branch should therefore be at least 3 m for gas fires and 1.2 m for other appliances. The main flue should not serve more than ten consecutive stories.

Room sealed appliances These can be connected to a vertical shaft (see figure). The combustion products rise as a result of convection currents, cool air entering at the base of the shaft via horizontal ducts, or (in the case of buildings supported on columns) via air intake points referred to as the Se duct. In the case of a U duct system, air enters via a second vertical shaft run down from roof level.

Note: the room sealed appliance installed into these systems are of a special design.

Fan assisted open flues

Occasionally, due to construction difficulties, a fan located in the flue-way is required to assist the extraction of flue gases. Where this is the case, several automatic control measures must be in place, such as incorporating a proving unit to ensure that in the event of fan failure the appliance does not work. Where a fan cycles on/off in response to a thermostat, a pre-purge facility must be incorporated. Fans must be capable of handling the total products of combustion and excess air, and be positioned where they are accessible for maintenance purposes.

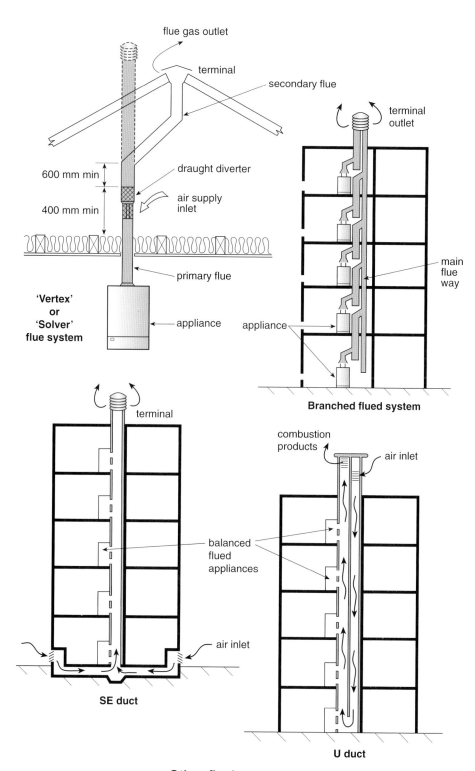

flue gas outlet

terminal

secondary flue

600 mm min

400 mm min

draught diverter

air supply inlet

primary flue

'Vertex' or 'Solver' flue system

appliance

terminal outlet

main flue way

appliance

Branched flued system

terminal

balanced flued appliances

combustion products

air inlet

air inlet

SE duct

U duct

Other flueing systems

Liquefied Petroleum Gas Installations 1

Relevant British Standards
BS EN 1949 and BS 5482

Many of the requirements for LPG are the same as those for natural gas; for example, the same ventilation and flue requirements must be met, as laid down in BS 5440. However, because LPG is not supplied by the national supplier from pipes in the road, and is delivered instead in pressurised liquid form via a bulk tank or cylinder, some special provisions need to be made.

The fundamental requirements for ignition to occur include a 1.9–8.5% gas-in-air mixture for butane and 2–11% gas-in-air mixture for propane, with an ignition temperature of 480–540°C. The relative densities of the fuels are 2.0 and 1.5, respectively; in both cases, therefore, the gases are heavier than air. LPG injectors are somewhat smaller than natural gas injectors.

Gas supply considerations Where cylinders are used to store LPG they should be located outside the building and afford easy access. Butane is sometimes fitted inside domestic premises. It is essential that all cylinders are stored upright with the valve at the top and it is essential that the stored supply is above ground level and never within 1 m of open drains or basements where gas resulting from a leak might accumulate. Bulk storage tanks may also be fitted in accordance with the requirements of the Home Office and the gas supplier's specification. LPG installations are divided primarily into high- and low-pressure stages, the high pressure being that in the cylinder and pipework up to the pressure regulator. Whenever possible, reserve cylinders should be connected to a manifold and fitted with non-return valves; this will permit one cylinder to be removed without the need to shut down the whole system.

Cylinder sizing It is essential to have cylinders sufficiently sized to ensure an adequate gas flow. Where large volumes of gas are to be drawn off, small cylinders may have adequate capacity of liquid LPG *but* are insufficiently large to allow the liquid to be converted into gas quickly enough. The size of cylinder needed is found by calculating the required kW input rating for all appliances and referring to the table below.

Cylinder sizes – recommended off-take

Propane (kg)	Off-take		Butane (kg)	Off-take	
	kW	Btu/h		kW	Btu/h
3.9	1.4	4776	4.5	4.7	16036
6	11.1	37873	7	6.8	23201
13	15.0	51180	15	9.5	32414
19	18.7	63804	Butane not recommended for		
47	33.5	114302	installations over 8.5 kW		

Example: The off-take for a propane installation consisting of a 14 kW cooker, 15 kW boiler and 4 kW fire is: 14 + 15 + 4 = 33 kW.

Therefore, either 1 × 47 kg *or* 2 × 19 kg cylinders are required.

OPSO and UPSO valves   These are special devices designed to close the supply if pressures are experienced outside the normal operating limits of the system. The two valves are:

Over-pressure shut off (OPSO) – causes valve to close at an elevated pressure, typically 75 mbar

Under-pressure shut off (UPSO) – prevents the valve opening unless pressure reaches a certain level, typically 25 mbar and will therefore identify any open ends. See low pressure cut-off device on page 202.

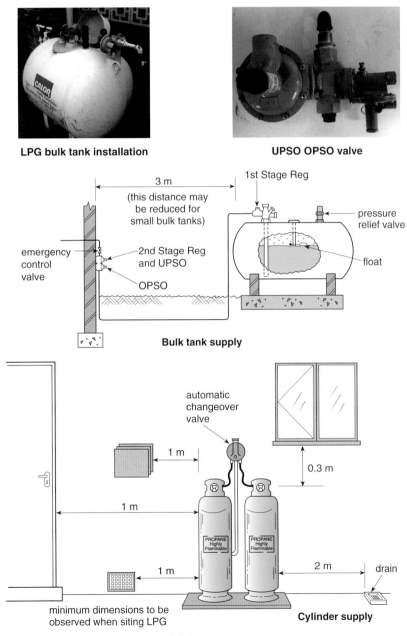

LPG bulk tank installation

UPSO OPSO valve

4 Gas Supplies

Bulk tank supply

Cylinder supply

minimum dimensions to be observed when siting LPG

LPG supply

Liquefied Petroleum Gas Installations 2

Relevant British Standards
BS EN 1949 and BS 5482

Testing the low pressure stage   For new installations, prior to gas connection, the system is charged with air to a pressure of 45 mbar; after a five minute temperature stabilisation period there should be no pressure drop over a further two minute test period (see page 196 for guidance on the procedure). Once the gas has been connected to the installation, the test pressure will be dependent on the gas being supplied, which should be 28 mbar for butane and 37 mbar for propane. Therefore, charge the system with the required pressure from the cylinder, purging the supply to ensure the whole system is filled with gas. Turn off the supply and burn off the gas until the pressure drops down to 5 mbar; close the appliance. Wait five minutes for temperature stabilisation, then observe the gauge over a two minute test period in which there should be no pressure rise, thus confirming the main supply valve is not letting by. Re-open the supply valve to recharge the system and re-close. Wait five minutes for temperature stabilisation then burn off the gas, turning off the appliance when the pressure at the appliance reads 30 mbar for propane, and 20 mbar for butane. Wait a two minute test period in which there should be no pressure drop. Any drop on the system must be found and repaired; however, if it is determined that an appliance is the cause of the drop due to appliance let by, a small loss is permissible as indicated in the British Standard.

For example, for a permanent dwelling without a meter this pressure loss would be up to 1 mbar where undertaking a 30 mbar test, and 0.5 mbar where undertaking a 20 mbar test. Note, where an inline valve is located downstream of the supply regulator, the test pressures are at the system operating pressures, thus: 37 mbar for propane and 28 mbar for butane.

The maximum working pressure drop across the system when all burners are operating should not exceed 2.5 mbar (see notes on page 198, Pressure and Flow). Because there is generally no meter fitted to LPG installations, it is not possible to calculate the heat input to the burner as with natural gas; it is therefore essential that the operating pressure is checked, and a check should also be made on the injector size and the combustion efficiency of the appliance (see page 230, Flue Efficiency); indication will be given of its correct input in accordance with the manufacturer's data.

Regulators   Several designs of regulators can be found, each having its own specific function. These can be broken down into two distinct types, single stage regulators or two stage regulators.

Single stage regulators   can withstand cylinder pressure and are designed to reduce it to the design operating pressure in one single go (i.e. 37 mbar for propane and 28 mbar for butane).

With two stage regulators, the high storage cylinder or bulk tank pressure is reduced to a level desirable to transfer the gas to the location where it is to be used. Then the gas is passed through a second regulator to reduce its pressure further still to that of the design system operating pressure.

Automatic changeover valves   These are single stage regulators which allow continuation of gas flow at all times and enable cylinders to be changed without the need to turn off the gas supply. Basically they are spring loaded regulators which can receive gas from two separate supply pipes, a non-return valve sealing off the pipe not in use.

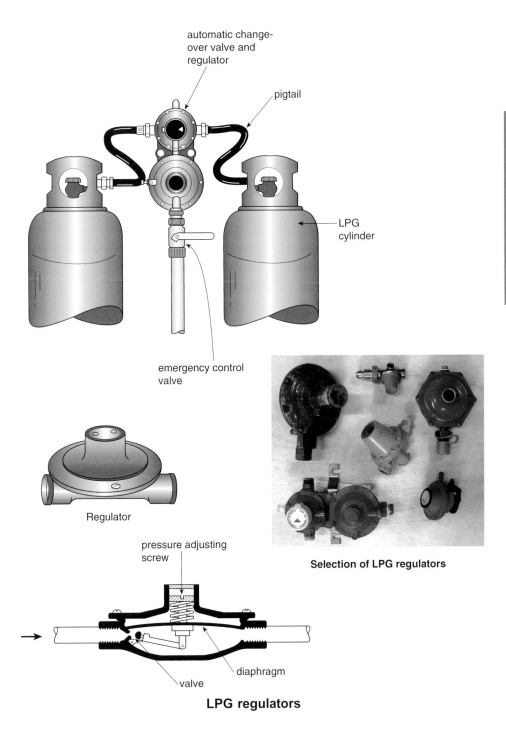

automatic change-over valve and regulator

pigtail

LPG cylinder

emergency control valve

Regulator

pressure adjusting screw

Selection of LPG regulators

valve

diaphragm

LPG regulators

Combustion Analysis

Relevant British Standard
BS 1756, 7927 7969 & BSEN 50379

The use of electronic combustion analysers is taking a much bigger role in the commissioning and servicing of appliances. This use may be to determine the efficiency of the appliance or its safe functioning. The analyser can also be used for safety checking of a gas appliance to determine whether a full service is required or not.

When using an analyser it is essential that the operative knows how to use it correctly and that the instrument is functioning correctly and caries a valid certificate of calibration.

Appliance sampling for commissioning purposes

Two specific uses of the analyser are put to use during the commissioning or servicing stage of a appliance to include:

1. A determination of the CO/CO_2 ratio to ensure safety is not compromised. This also provides an indication of the appliance efficiency.
2. Use to set the gas pressure/air ratio as applicable to the manufacturer's instructions.

To complete these tests you would need to observe manufacturer's instructions

Appliance sampling for safety purposes

Taking a reading of the combustion performance levels of a gas appliance gives an indication as to possible action levels to ensure safety. The readings need to be taken from 200 mm within the flue system or, in the case of a flueless appliance, at the point where the products discharge into the room. Whilst taking these samples the appliance must have been alight for at least 10 minutes and the pressure should have been adjusted to its highest working level. For the maximum performance CO/CO_2 levels, see the table opposite.

Where open flued or room sealed appliance fail to reach the minimum level the appliance needs to be treated as 'At Risk' In the case of a flueless appliance that fails to achieve the minimum level the appliance must be treated as 'Immediately Dangerous'.

Appliance sampling for the purpose of a safety check

Taking a sample of the flue gases will give some indication as to whether a full service should be undertaken to a boiler. If the CO/CO_2 reading is less than 0.004 there is no requirement to dismantle the appliance and complete a full service. However you must still be aware of the requirements of the Gas Safety Regulations in relation to the ventilation, fluing, operating pressure/heat input and general safe operation of the appliance.

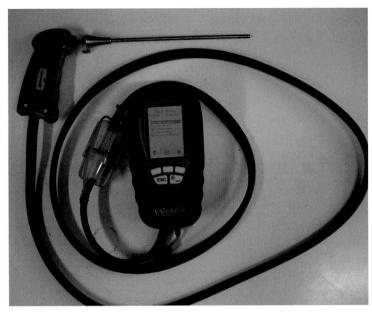

Combustion analysis

Maximum combustion performance levels

Appliance Type		Maximum CO/CO$_2$ Ratio
Freestanding boiler or warm air heater		0.008
Back boiler used with gas fire		0.02
Circulator		0.01
Open-flued convector fire or space heater		0.02
Flueless space heater		0.001
Flueless cabinet space heater		0.004
Water heater – Flued or Flueless		0.02
Cooker	Oven	0.008
	Hob	Visual inspection only
	Grill *(CE marked)*	0.01
	Grill *(not CE marked)*	0.02
Flued range oven		0.02
Tumble dryer	Flued	0.01
	Flueless	0.001

1. *Manufacturer's data may override the values given*
2. *Ensure new appliances/components have had sufficient time to burn off any oils etc.*

Flue Efficiency in Gas Appliances

Relevant British Standard
BS 1756

Test for spillage

This test is carried out on open-flued appliances to ensure that no combustion products are entering the room via the draught diverter. The spillage test helps confirm the adequate provision of a suitable air supply and ensures a suitable updraught within the flue. The test is carried out as follows:

(1) Close all windows and doors to the room and light the appliance. Wait for between 5 and 10 minutes for the flue to warm up.
(2) Ignite a smoke match and hold it about 3 mm up inside the lower edge of the draught diverter, along its whole length or perimeter. Spillage is indicated by the smoke discharging into the room.

Momentary spillage can be ignored, but if persistent, the situation must be rectified. Sometimes, especially in cold weather, one may need to wait longer than the initial 10 minutes to allow the flue to warm up sufficiently.

If an extractor fan is present in the same room as the appliance, or an adjacent room, the test should be carried out as above with any interconnecting door open and with the fan running. **Note:** Extractor fans include those on cooker hoods and tumble driers, etc. If spillage occurs the fault will need to be rectified. Generally an extra $50\,cm^2$ of free air ventilation will be sufficient in most cases to cure the problem.

Flue/appliance efficiency

The efficiency of the flue can be found by comparing the $CO_2\%$ with the flue gas temperature. This is done automatically by an electronic analyser; alternatively it can be calculated by using the following formula:

$$(0.343 \div CO_2\% + 0.009) \times (\text{flue gas temp} - \text{room temp})°C + 9.78 = \text{flue loss }\%$$

If one subtracts the energy due to flue loss from 100%, one can see how efficiently an appliance is operating.

Example: Assuming the CO_2 to be 5%, the flue gas temperature to be 180°C and the room temperature to be 21°C, calculate the efficiency of the appliance. Thus:

$$(0.343 \div 5 + 0.009) \times (180\ \ 21) + 9.78 = \text{flue loss }\%$$

$$\text{Therefore } 0.0776 \times 159 + 9.78 = 22.1\% \text{ flue loss}$$

$$\text{Therefore flue is } 100 - 22.1 = 77.9\% \text{ efficient}$$

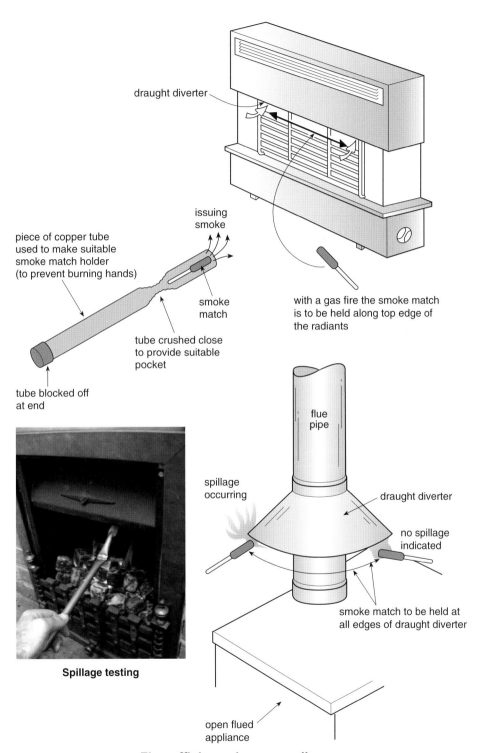

draught diverter

issuing smoke

piece of copper tube used to make suitable smoke match holder (to prevent burning hands)

smoke match

with a gas fire the smoke match is to be held along top edge of the radiants

tube crushed close to provide suitable pocket

tube blocked off at end

flue pipe

spillage occurring

draught diverter

no spillage indicated

smoke match to be held at all edges of draught diverter

Spillage testing

open flued appliance

Flue efficiency in gas appliances

Commissioning of Gas Installations

Upon the completion of any gas installation, the system should be commissioned in stages to ensure its safe and efficient use. Most of the tests described below have been identified elsewhere in this book; reference should be made to the appropriate pages if necessary. Some of the tests will not be applicable, e.g. spillage is not carried out on room-sealed appliances. In order to carry out these tests, a checklist, as shown on page 235, is used to ensure nothing is missed.

Tests prior to commissioning

(1) Complete a general visual check of the system and flue routes for obvious defects
(2) Carry out a tightness test (see page 196)
(3) Purge the gas carcass of air (see page 197)
(4) Carry out any pull tests to flue system as necessary (see page 212)
(5) Identify the standing pressure at the meter (see page 198)
(6) Check the correct provision of a suitable air supply to the appliance (see page 220)
(7) Where competent to do so, confirm the electrical connections are correct, i.e. earth continuity, polarity and insulation resistance, and ensure that the correct fuse has been fitted (see page 290)
(8) If applicable, confirm the boiler is filled with water.

Commissioning the appliance

(1) Check the standing pressure at the appliance (see page 198)
(2) Check the operation of the pilot flame and make adjustments if necessary; the flame's length should be about 25 mm and it should be playing on to the thermocouple
(3) Check and adjust the working pressure of the appliance (see page 198)
(4) Identify and confirm the heat input to the appliance (see page 198)
(5) With the appliance alight, spray the final section of pipework to the burner from the inlet control valve with leak detection fluid to confirm its tightness
(6) Check the operation of the flame supervision device. *See note below*
(7) Observe the general flame picture (a yellow flame indicates incomplete combustion of the fuel)
(8) Carry out a test for spillage (see page 230)
(9) Check the operation of the thermostat, including the high limit stat where applicable
(10) Carry out any required flue gas analysis (see page 228).

Testing flame supervision devices (FSDs)

With thermo-electric and flame rectification FSDs, it is possible to check that the valve is working correctly. One method uses a manometer connected to the inlet pressure test nipple, before the FSD and appliance governor. With the appliance alight, turn off the isolation valve, and power supply in the case of flame rectification. This will extinguish the gas. Wait a suitable time for the valve to click out, then turn back on the supply to re-charge the manometer with gas pressure and turn off again. If the solenoid is working correctly, the pressure will be maintained, and shown on the manometer, thus proving that the valve is working correctly and that the pipework downstream of the isolation valve is sound.

Final system checks

Check the system working pressure to ensure that the maximum pressure drop across the system does not exceed 1 mbar (see page 198).

Where a gas boiler is connected to a c.h. system, etc., the circulatory pipework will also need to be commissioned (see the notes for commissioning c.h. systems on page 182).

Handing over The working of the appliance should be demonstrated to the user and the best method of economic and efficient usage explained. All documentation supplied should be left with the owner/occupier who should be made aware of the benefits of a regular service contract to ensure that the equipment is maintained in an efficient and safe operating condition. Finally, the certification body needs to be notified (i.e. works notification through the Gas Safe Register) to ensure the appropriate certificate is issued to the customer in accordance with the law.

Commissioning of gas installations

Maintenance and Servicing

When servicing a gas appliance such as a fire or boiler the procedure is identical to that followed when commissioning gas appliances except that, in addition, the appliance should be checked for carbon or soot build up. For example, the burners will need to be exposed and brushed or blown through with air to rid them of deposits. The injector should also be inspected for blockage. The heat exchanger should be brushed and vacuumed clean as necessary and the air intake checked for such things as cat fur and dust build up, often referred to as lint.

With open-flue appliances which connect to a chimney, the appliance must be removed and the flue inspected; any debris from the collection space at the base of the chimney should be cleaned as necessary. When servicing it is always advisable to lay down a dust sheet; and a schedule should be completed to identify the tasks to be carried out. Should a boiler be connected to a c.h. system, this may form part of the service contract (see page 182).

When one goes to inspect, install, commission or service any gas appliance located within premises, an assumption is made that the gas service engineer last visiting the job is responsible for the whole installation. The fact that such an assumption is made worries many people and its implications are open to debate.

Provided that the previous service engineer has made no adjustments to the gas rate (i.e. has not adjusted any governing devices) and has not therefore changed the supply flow of gas to other appliances, there is little to worry about as to their operation. In such a case, by carrying out a visual inspection, one would not be deemed to have validated the appliance's safe and efficient use. Where adjustment to the gas pressure at an appliance has been made, or a new appliance installed, the rate of gas flow to other appliances may now be starved; in these circumstances, a check on the system working pressure must be made (see page 198). If necessary, a test nipple will need to be inserted at the furthest appliance.

The schedule shown opposite allows the installer or service engineer to complete a checklist which, upon completion, can be given to the owner/occupier; it informs them of the work completed and highlights any shortcomings.

Where an unsafe appliance has been located, the appliance must be rectified or shut down and isolated; the schedule will highlight this. Where an urgent problem such as an immediately dangerous appliance has been identified, if necessary the gas supplier must be notified directly by the gas engineer, as described on page 238. This will give the engineer some protection if the owner simply turns on the supply to the unsafe appliance as soon as the gas fitter leaves the premises. Schedules of this nature inform the client of possible faults and invariably give the installer more work checking other appliances brought to the attention of the client. The chart opposite should also be used when one is commissioning a new appliance (see page 232).

Certificate/Record of Gas Inspection

Page 1 of......

Gas Installer Details	**Client Details**	**Work Carried Out At:**
Name: Gas Safe Reg. No.: Address:	Name: Address:	Name: Address: Date:

Preliminary System Checks

20 mbar Soundness Test; to include Let By Check PASS ☐ FAIL ☐ N/A ☐	Standing Pressure of System (≤30 mbars) PASS ☐ FAIL ☐ N/A ☐

System/Appliance Details	Location 1: Type: Model/Make	Location 2: Type: Model/Make	Location 3: Type: Model/Make	Location 4: Type: Model/Make
Preliminary Electrical Checks				
Conductors Secure	PASS ☐ FAIL ☐ N/A ☐	PASS ☐ FAIL ☐ N/A ☐	PASS ☐ FAIL ☐ N/A ☐	PASS ☐ FAIL ☐ N/A ☐
Earth Continuity Maintained	PASS ☐ FAIL ☐ N/A ☐	PASS ☐ FAIL ☐ N/A ☐	PASS ☐ FAIL ☐ N/A ☐	PASS ☐ FAIL ☐ N/A ☐
Polarity Correct	PASS ☐ FAIL ☐ N/A ☐	PASS ☐ FAIL ☐ N/A ☐	PASS ☐ FAIL ☐ N/A ☐	PASS ☐ FAIL ☐ N/A ☐
Insulation Resistance >0.5 M Ohm	PASS ☐ FAIL ☐ N/A ☐	PASS ☐ FAIL ☐ N/A ☐	PASS ☐ FAIL ☐ N/A ☐	PASS ☐ FAIL ☐ N/A ☐
Bonding Maintained	PASS ☐ FAIL ☐ N/A ☐	PASS ☐ FAIL ☐ N/A ☐	PASS ☐ FAIL ☐ N/A ☐	PASS ☐ FAIL ☐ N/A ☐
Fuse Rating	Amps ☐ FAIL ☐ N/A ☐	Amps ☐ FAIL ☐ N/A ☐	Amps ☐ FAIL ☐ N/A ☐	Amps ☐ FAIL ☐ N/A ☐
Gas Utilisation Checks				
Pilot Flame Correct	PASS ☐ FAIL ☐ N/A ☐	PASS ☐ FAIL ☐ N/A ☐	PASS ☐ FAIL ☐ N/A ☐	PASS ☐ FAIL ☐ N/A ☐
Burner Pressure	mbar ☐ FAIL ☐ N/A ☐	mbar ☐ FAIL ☐ N/A ☐	mbar ☐ FAIL ☐ N/A ☐	mbar ☐ FAIL ☐ N/A ☐
Flame Picture Good	PASS ☐ FAIL ☐ N/A ☐	PASS ☐ FAIL ☐ N/A ☐	PASS ☐ FAIL ☐ N/A ☐	PASS ☐ FAIL ☐ N/A ☐
Heat Input	kW ☐ FAIL ☐ N/A ☐	kW ☐ FAIL ☐ N/A ☐	kW ☐ FAIL ☐ N/A ☐	kW ☐ FAIL ☐ N/A ☐
Operating Thermostat Correct	PASS ☐ FAIL ☐ N/A ☐	PASS ☐ FAIL ☐ N/A ☐	PASS ☐ FAIL ☐ N/A ☐	PASS ☐ FAIL ☐ N/A ☐
High Limit Thermostat Correct	PASS ☐ FAIL ☐ N/A ☐	PASS ☐ FAIL ☐ N/A ☐	PASS ☐ FAIL ☐ N/A ☐	PASS ☐ FAIL ☐ N/A ☐
Flame Failure Device Affective	PASS ☐ FAIL ☐ N/A ☐	PASS ☐ FAIL ☐ N/A ☐	PASS ☐ FAIL ☐ N/A ☐	PASS ☐ FAIL ☐ N/A ☐
Appliance Soundness Check	PASS ☐ FAIL ☐ N/A ☐	PASS ☐ FAIL ☐ N/A ☐	PASS ☐ FAIL ☐ N/A ☐	PASS ☐ FAIL ☐ N/A ☐
Door/frame seals Effective	PASS ☐ FAIL ☐ N/A ☐	PASS ☐ FAIL ☐ N/A ☐	PASS ☐ FAIL ☐ N/A ☐	PASS ☐ FAIL ☐ N/A ☐
Flue and Ventilation Checks				
Ventillation Grill Size	cm^2 ☐ FAIL ☐ N/A ☐	cm^2 ☐ FAIL ☐ N/A ☐	cm^2 ☐ FAIL ☐ N/A ☐	cm^2 ☐ FAIL ☐ N/A ☐
Flue Guard Fitted to Low Level Terminals	PASS ☐ FAIL ☐ N/A ☐	PASS ☐ FAIL ☐ N/A ☐	PASS ☐ FAIL ☐ N/A ☐	PASS ☐ FAIL ☐ N/A ☐
Flue Flow Performance (pull test)	PASS ☐ FAIL ☐ N/A ☐	PASS ☐ FAIL ☐ N/A ☐	PASS ☐ FAIL ☐ N/A ☐	PASS ☐ FAIL ☐ N/A ☐
Spillage Tests	PASS ☐ FAIL ☐ N/A ☐	PASS ☐ FAIL ☐ N/A ☐	PASS ☐ FAIL ☐ N/A ☐	PASS ☐ FAIL ☐ N/A ☐
CO_2 %	%	%	%	%
CO %	%	%	%	%
CO/CO_2 Ratio				
Flue Temperature	°C	°C	°C	°C
Appliance Efficiency	%	%	%	%
Appliance Safe to Use	YES ☐ NO ☐	YES ☐ NO ☐	YES ☐ NO ☐	YES ☐ NO ☐

Post System Checks

Meter Working Pressure (21 mbar ± 2 mbar)　　PASS ☐ FAIL ☐ N/A ☐	Working Pressure Drop Across System (max. 1 mbar)　　mbar　PASS ☐ FAIL ☐ N/A ☐

In accordance with Regualtion 34 of the Gas Safety Installation and Use) Regulations 1998 I am duty bound by law to inform you that the gas appliances itemised:–
........................; found in the locations identified above, have not been inspected by the gas installer hereon and that their safe functioning cannot be guaranteed, therefore operation and safe usage with the owner/occupier of the said installation as signed on this schedule.

Recommendations and/or Urgent Notification

Gas Installers Signature ....

Owner/Occupier Signature ....

Carbon Monoxide Detection

Relevant British Standard
BS 7860 & BSEN 50291

Carbon monoxide (CO) gas is produced as the result of incomplete combustion of a fuel. Where this gas is to be found within an environment in sufficient quantities it can prove fatal within a very short period of time. The following table identifies the symptoms.

Typical causes & effects of CO poisoning

<0.01%	Slight headache after a couple of hours
0.01–0.02%	Headache and feeling dizzy after a couple of hours
0.02–0.05%	Strong headache; sickness & palpitations after 1–2 hours
0.05–0.15%	Severe headache & sickness after ½ hour
0.15–0.3%	Severe headache & sickness after 10 minutes, leading to convulsions and possible death after 15 minutes
0.3–0.6%	Severe symptoms as above within 1–2 minutes
>0.6%	Immediate symptoms & death within 1–3 minutes

Carbon monoxide detectors

Different types of CO detector can be found, although the 'Spot' type detectors should be regarded with suspicion as they fail to work effectively after a short period of time and are affected by humidity and various halogens and nitrous gases within the atmosphere. The spot type indicator works by the orange spot on the surface of the card turning grey or black in the presence of CO. The best kinds of detector and the only type to be recommended are those operated by electrical power, such as through a battery or via connection to the supply. These conform to the above BS and have the combined function of both audible and visual alarms, identified via a flashing light. These detectors should operate within 3 minutes where the CO detected is in excess of 300 ppm (0.03%) and at worst, such as where the CO is as low as 30 ppm (0.003%), just after 2 hours.

The location of an alarm should be as indicated by the manufacturer of the unit, but generally should be above any door or window and at least 1.5 m away from the appliance, whilst also being 150 mm below the ceiling.

Maximum combustion CO/CO$_2$ performance levels

When an appliance is installed within a room it is sometimes necessary to check the combustion performance levels to ensure that they do not increase the level of CO within a room to a point that could be regarded as dangerous. This would be essential where a flueless gas fire/heater is installed. Determining the CO/CO$_2$ level has already been identified on page 228.

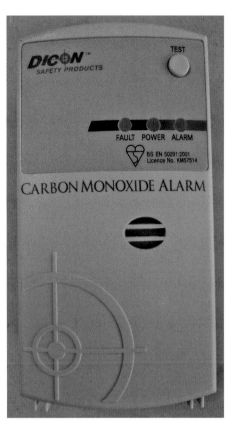

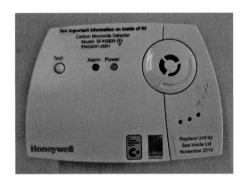

Suitable CO detectors

Unsuitable CO detectors

Unsafe Gas Installations

Due to changes in specifications, lack of maintenance or bad installation practices, installations invariably fall to a standard below that acceptable. Where this is the case the installer has to make a judgement as to the severity of the case and categorise the installation as *Not to Current Standard, At Risk* or *Immediately Dangerous*. Gas installers have a duty to take appropriate action where a defect has been identified and would be in contravention of the Gas Regulations if they failed to do so.

Not to Current Standard installations These pose no immediate threat to life or property and are usually the result of changes in Regulations and Standards. Where an installer identifies such an installation they *must* advise the person responsible for the property (e.g. owner/occupier) of the defects found, with recommendations that the faults be corrected, ideally keeping a record of the faults identified on worksheets for future reference. The gas installer may leave the gas appliance in operation. Note, however, that should two or more faults be identified it may be more appropriate to treat the installation as At Risk.

At Risk installations An At Risk installation or appliance is one which, if used, may be a risk to life or property. For example, an open-flued appliance installed in a bathroom, or pipework of inappropriate materials. Note, should two or more ventilation and/or flue faults be identified, it may be more appropriate to treat the installation as Immediately Dangerous.

Immediately Dangerous installation An installation which, if the system or appliance is left connected to the gas supply, poses an immediate danger to life or property. Examples include gas escapes and appliances that fail a spillage test.

Where an *At Risk* or *Immediately Dangerous* situation has been identified, with the home owner's agreement, the gas installer should try to rectify the fault, making the appliance safe. Where this is not possible, the following should be undertaken:

(1) Turn off the appliance and inform the customer of its dangers, warning not to use.
(2) Attach a WARNING! DO NOT USE label to the installation in a prominent position.
(3) Complete a warning notice, signed by the customer, and keep it for future reference.

Specific to At Risk: Inform the user/responsible person not to permit the appliance to be used, as in your opinion it is unsafe.

Specific to Immediately Dangerous: Inform the user/responsible person that in your opinion the appliance/installation is dangerous and must be disconnected immediately from the supply until the fault has been rectified. If permission to disconnect is not granted, make immediate contact with the gas supplier to explain the concern you have.

The gas installer may be required to report At Risk or Immediately Dangerous appliances to the HSE using a RIDDOR form, particularly if the situation is the result of unsatisfactory fittings or workmanship.

Dealing with gas installations which do not comply with current Regulations and Standards

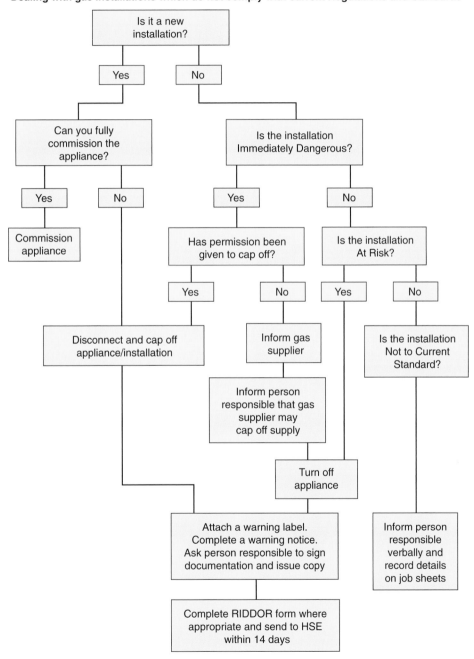

Part 5
Oil Supplies

Plumbing, 4th Edition. R. D. Treloar.
© 2012 Blackwell Publishing Ltd. Published 2012 by Blackwell Publishing Ltd.

Properties and Combustion of Fuel Oils

Relevant British Standard
BS 5410

Liquid fuels in the United Kingdom are available in two grades: distillate and residual. Residual grades are very thick and are only used in heavy industry. Two types of distillate grade fuels are used for domestic burners:

- Class C2 28 second fuel or kerosine
- Class D 35 second fuel or gas oil.

Specification data for Class C2 and D distillate grade fuels

Property	Class C2	Class D
Colour	Clear	Red
Calorific value	46.3 kW/imp. gal	48.1 kW/imp. gal
	(10.18 kW/litre)	(10.58 kW/litre)
Relative density	0.79	0.83
Viscosity (Redwood No. 1 scale)	28 s	35 s
Solidification temperature	−40°C	−10°C
Flash point	38°C	55°C
Sulphur content	0.2	1.0

Note: 1 imperial gallon = 4.546 litres.

5 Oil Supplies

Note: Class D oil has a greater calorific value than Class C oil; thus Class D oil gives out more heat per unit volume and it is slightly cheaper. However, it is prone to solidification (waxing up) in cold weather and, because it has a high sulphur content, produces sulphur dioxide (SO_2) during combustion which, when combined with water, produces sulphurous acid (H_2SO_3). This tends to corrode the heat exchanger more quickly, thus shortening the life of the boiler. As a result, not many manufacturers recommend the use of Class D fuel; where it is used, a preheater is often required to warm the fuel at the burner, thus reducing its viscosity.

The *viscosity* of a fuel governs its ability to flow easily. Treacle flows more slowly than and is more viscous than water. The Redwood No. 1 viscometer is a device which measures the length of time, in seconds, taken by $50 \, cm^3$ of a liquid at 37°C (100°F) to flow through a hole of defined size.

Liquid fuel oils, like natural gas (or any other fuel), are hydrocarbons, being composed primarily of hydrogen and carbon atoms. Both Class C and D oils comprise about 84% carbon and 16% hydrogen. Although the principles of the combustion process are the same as those for gas, it will be found that where insufficient oxygen is available to consume the fuel, vast amounts of unburnt carbon (i.e. smoke) will be discharged from the flame; this unburnt carbon is deposited as soot in the heat exchanger and flue-way.

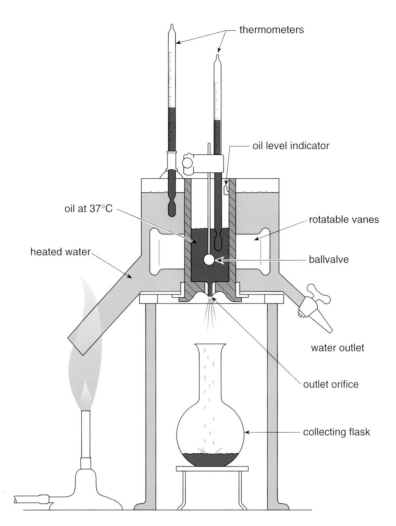

thermometers

oil level indicator

oil at 37°C

rotatable vanes

heated water

ballvalve

water outlet

outlet orifice

collecting flask

section through a Redwood No. 1 Viscometer

Properties and combustion of fuel oils

Oil Storage

Relevant British Standard
BS 799 & BS 5410

Traditionally oil storage tanks were single-skinned vessels, but current legislation dictates that unless it can be proven that there is no environmental risk in the event of an oil spillage, the vessel used must be positioned within a secondary containment area, i.e. bunded. Integrally bunded tanks are available with an inner tank that will hold the contents in the event of a leak or overfilling.

The tank may be constructed of steel or plastic (polyethylene). Where a steel tank is to be used and surrounded by a concrete/brickwork bund it should be positioned on the supporting wall so as to provide a slope of 20 mm/m towards the drain-off valve, thereby allowing the removal of any sludge etc. In the case of a bunded plastic vessel de-sludging is achieved by inserting a suction hose into the top fill point. Plastic tanks must be fully supported throughout their entirety.

Storage Capacities

Total Appliance Heat output	Nominal Capacity	Steel Tank Size Length × Width × Depth	Plastic Tank Size Length × Width × Depth
<15 kW	1100–1500 litres	1.52 × 0.61 × 1.22	2.03 × 0.69 × 1.36
15–25 kW	1500–2000 litres	1.83 × 0.61 × 1.22	1.37 × 1.06 × 1.25
25–35 kW	2000–2500 litres	1.83 × 1.22 × 1.22	2.02 × 1.36 × 1.36
35–40 kW	2500–3500 litres	2.44 × 1.22 × 1.22	1.84 dia × 1.84 dia × 1.75

Fill Connection: The use of open lids to fill the vessel is no longer to be recommended; therefore a suitably capped 2" BSPT parallel male iron threaded connection should be provided to enable the hose connection from the fuel tanker. Where the storage tank is located at an elevated position, or some distance from the point where the delivery can be made without compromising safety, you should run an extended fill point to terminate with a non-return and valved 2" BSPT connection located above a drip tray as shown.

Vent Pipe: A vent pipe should be provided with an internal diameter of no less than the diameter of the fill pipe. The vent should be provided with a return bend or mushroom-type vent to prevent the entry of rain. The termination height of this vent must be higher than the fill pipe but no higher than 500 mm to prevent any pressure restriction. Where the vent pipe cannot be seen from the fill point, an alarm should be fitted to warn of possible overflowing.

Isolation Valve: A valve suitable for use with oil should be fitted as close as possible to the tank in an accessible location. In addition to this valve a second valve should be fitted as close as possible to the appliance.

Contents Gauge: Traditionally on single-skin tanks sight tubes were used to give an indication as to the contents of the tank, and where these are used they need to be supplied via a spring-loaded isolation valve. Other gauges, such as those that use a float-operated mechanism, are less likely to give rise to possible leaks. Electronic gauges, where used, can be read both locally and remotely, displayed inside the dwelling. Note that where a bunded tank is used the sight gauge must also identify if there is any oil within the secondary containment chamber.

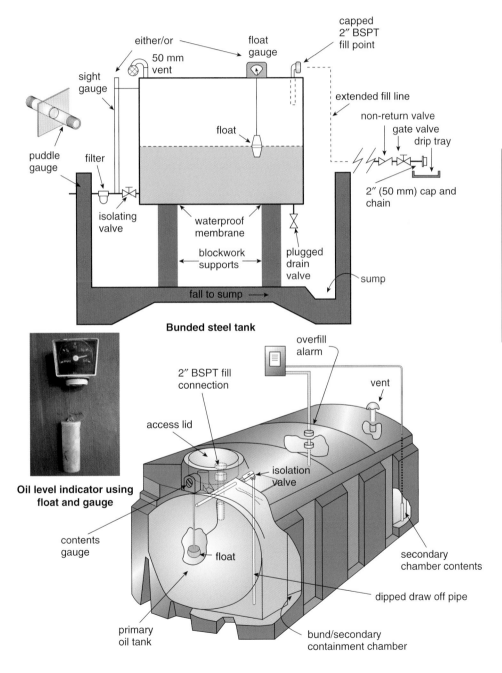

Bunded steel tank

Oil level indicator using float and gauge

Integrally bunded plastic oil tank

Oil storage

Oil Feed Pipework

Relevant British Standards
BS 799 and BS 5410

The pipe supplying oil to the burner should have a minimum diameter of 10 mm. It should incorporate two stop valves, one at the storage tank and one by the appliance. A filter and fire valve should also be located within its length. It is recommended that two filters be positioned for pressure jet burners, one at the tank and one prior to the burner. If just one filter is used, this must be located at the boiler. The oil line should be run to avoid trapping air which may restrict or stop the flow of fuel. Where 35-second Class D fuel is used it may be necessary to insulate the pipework in exposed positions to prevent the fuel waxing up (solidifying) in winter.

Gravity feed systems   Where the oil tank is fitted at a raised elevation above the burner, its head pressure should not exceed 3 m to the underside of the tank; equally, to ensure a good flow this head should be at least 0.3 m (see top figure). The actual size of the oil line will be dependent on the location and head pressure created (see table). Where high points cannot be avoided, they should not be above the outlet of the tank and they should be fitted with a means of manual venting.

Maximum pipe length

| | Head (m) | | | |
Pipe dia (mm)	0.5	1.0	1.5	2.0
10	10	20	40	60
12	20	40	80	100

Sub-gravity system   Where the bottom of the tank is less than 0.3 m above the burner it will be necessary to run a 10 mm return pipe (see middle figure). A spring-loaded non-return valve will be required on the supply pipe to prevent oil draining back to the tank. Alternatively, an oil deaerator, e.g. the Tiger-loop, may be used (see bottom figure); this removes the air from the oil feed on a single pipe lift. The burner pump is piped to the deaerator as shown. The deaerator must be installed outside the dwelling unless the design is intended to be installed internally, which allows the trapped air to be vented to the outside or reabsorbed into the fuel line. Also note that where the design of the tank provides exit via a top oil supply pipe, an anti-siphon device needs to be included to prevent the oil from being siphoned from the tank should a leak materialise in the oil supply pipe. Where the suction head is excessive, an oil lifter can be used which draws the oil up to an elevated position from which it can flow by gravity to the burner.

The materials recommended for the oil supply line are: fully annealed copper tube (R220) with manipulative type compression fittings (soft soldered joints should not be used), or low carbon steel pipe (not galvanised). Tapered threads should be cut on the pipe; running joints such as long screws, etc., should not be used.

Petroleum resisting compounds and PTFE tapes which remain slightly plastic make the most satisfactory joints; hemp and hard-setting jointing compounds should be avoided.

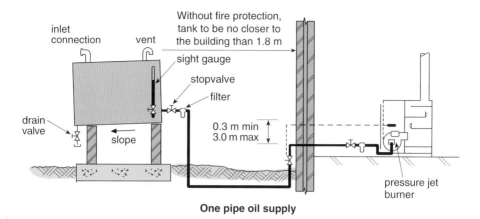

One pipe oil supply

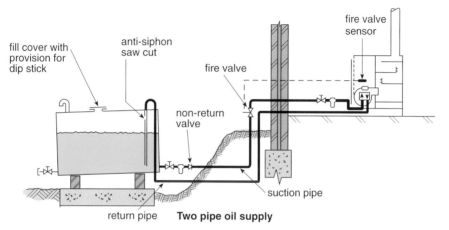

Two pipe oil supply

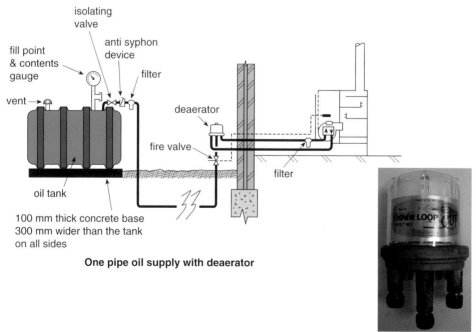

One pipe oil supply with deaerator

Deaerator

Oil feed pipework

Controls Used on Oil Feed Pipework

Relevant British Standard
BS 5410

Filter   This device is fitted in the pipeline from the oil tank to the burner and is designed to prevent the passage of particles that may block the burner jets of the appliance. Two types of filter will be found: those which contain a fine mesh which can be removed for cleaning purposes, and those which contain a paper element which is replaced periodically.

Fire valves   These are special valves designed to shut off the flow of oil to an oil-burning appliance automatically in the case of a fire. There are three basic types: the *fusible link type*, the *leaded handle* type and the *pressure* type.

With the fusible link type, the valve is held open by a tensioned wire and the arrangement is set up as shown in the figure. The fusible link has a low melting temperature and in the event of a fire will break; the valve will then close, assisted by a weight or attached spring. The fusible link type would normally only be used in a purpose-made boiler house.

For domestic oil fired installations, the pressure type is generally used. This consists of a bellows type valve which is connected to a heat-sensitive bulb. This bulb overheats in the event of a fire and causes an increased vapour pressure in the bellows, resulting in its expansion; thus the valve is closed.

The heat sensitive bulb should be located above the combustion chamber at a minimum height of 1 m, although where this is impracticable the sensor can be fitted as close as possible to the appliance, at a similar level; however, it must be in the same room as the appliance, with the actual valve outside.

Older supplies used a leaded handle fire valve, which consisted of a small spring-loaded stop valve in which the thread in the wheel head was made of a low melting solder. In the event of a fire, the solder melts and the valve springs shut. This type of valve should no longer be fitted as it does not conform to current practices.

When an external tank is being used, the fire valve should be fitted as close as possible to the point where the supply enters the building, ideally externally.

Constant oil level control   This device will only be found on the supply to a vaporising burner. It is designed to control the feed of oil to the appliance at a constant pressure and flow, irrespective of the oil level in the storage tank. If the oil were to flow directly to the burner from the oil tank, it would flood the appliance, as its operation is dependent on a maintained tray of oil at its base. The vapour from the oil rises and is ignited or, in the case of a wallflame burner, drawn up to be dispersed round the perimeter of the combustion chamber.

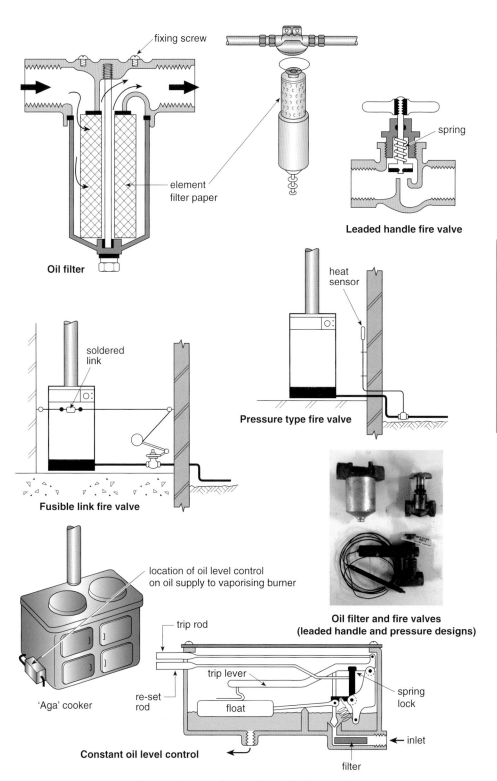

fixing screw

element
filter paper

Oil filter

spring

Leaded handle fire valve

heat
sensor

Pressure type fire valve

soldered
link

Fusible link fire valve

location of oil level control
on oil supply to vaporising burner

**Oil filter and fire valves
(leaded handle and pressure designs)**

trip rod

trip lever

spring
lock

re-set
rod

float

inlet

'Aga' cooker

Constant oil level control

filter

Controls used on oil feed pipework

Pressure Jet Burners 1

Relevant British Standards
BS 799 and BS 5410

All oil boilers sold today are of this design and consist of a self-contained package. The burner is reasonably quiet in operation and basically works as follows:

(1) Should the thermostat require heat, the motor runs, rotating a fan which allows air to be blown into the combustion chamber. This rids the appliance of any residual vapour.
(2) After a set time a solenoid opens to allow oil to be pumped through a small nozzle into the combustion chamber, emerging in a fine spray (a process referred to as atomising), where it mixes with the air. Simultaneously a spark is generated at two electrodes inside the chamber, located at the tip of the nozzle, to allow ignition to occur.
(3) When the flame is detected by a photoelectric cell, which distinguishes light rays from within the combustion chamber, the electrodes cease to spark.
(4) The motor continues to run, allowing fuel and air into the combustion chamber until the thermostat is satisfied, at which the solenoid closes, allowing the motor to stop.

The burner unit is located in a steel heat exchanger, which consists of a rectangular combustion chamber. The waterway surrounds the chamber with flow-and-return connections at high and low level. Oil burners have a large void through which the hot flue gases pass to prevent the build-up of soot; therefore, a series of baffle plates are located within the chamber to direct the flow of hot gases on to the walls of the heat exchanger.

Upon initial commissioning it is essential to check that these plates are in position in order to avoid damage to the unit. The hot gases pass up through the chamber and are expelled from either an open or balanced flue.

When installing burners with a two-pipe oil supply or a one pipe system incorporating a Tiger-loop, it is essential to install a by-pass screw into the pump, as supplied by the manufacturer, thus preventing damage.

Lighting up a new burner which has also had a new supply run may take several attempts. It is best to bleed the supply pipe of air by allowing the oil to discharge into a container. A combined air bleed manifold and pressure gauge (to register 0–20 bar) should be connected to the appropriate oil pump connection on the burner, and the thermostat turned up (assuming the electricity supply is on and calling for heat).

When the motor starts it may be necessary, on the one pipe system, to open the bleed screw on the test manifold to remove the air. If the burner locks out, wait about a minute or so and reset the lockout button on the control box. Once the burner fires up, the burner pressure should be checked and adjusted to the correct pressure as recommended by the manufacturer, usually in the region of 7 bar ($100\,lb/in^2$).

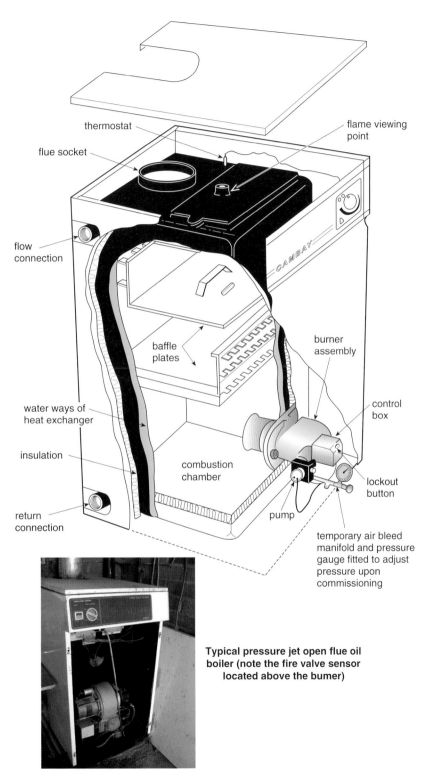

thermostat

flame viewing point

flue socket

flow connection

baffle plates

burner assembly

control box

water ways of heat exchanger

insulation

combustion chamber

lockout button

return connection

pump

temporary air bleed manifold and pressure gauge fitted to adjust pressure upon commissioning

Typical pressure jet open flue oil boiler (note the fire valve sensor located above the bumer)

Pressure jet burners 1

Pressure Jet Burners 2

The pressure jet burner consists of the following component parts:

Control box   An electrical device which controls the operation of the burner in a predetermined sequence. In the event of a boiler malfunction, the control box will cause the sequence of operations to stop and dislodge a trip switch (referred to as lockout) causing the complete shutdown of the appliance. To restart the appliance a reset button needs to be manually depressed. **Note:** Some burners will automatically carry out a second attempt at ignition.

Electric motor   Provided to rotate the air fan and fuel pump simultaneously.

Air fan   Designed to supply air, under pressure, into the combustion chamber. The amount of air supplied will be dependent upon the air shutter opening which is adjusted during commissioning.

Fuel pump   A device which receives the oil from the oil storage tank at atmospheric pressure and increases its pressure to around 7 bar (approximately $100 \, lb/in^2$) thus forcing the oil through the small hole in the nozzle.

Solenoid valve   (an electromagnetic valve) This device prevents oil from entering the combustion chamber until the appliance has been purged with air. It also allows the oil to come up to pressure before entry into the chamber, allowing the fine spray to be formed. When the appliance has reached temperature, the solenoid also causes quick shutdown of the fuel supply.

Air diffuser   A device which causes the air molecules to vibrate and swirl around, allowing for a better fuel/air mixture.

Electrodes   Two conductors positioned at the nozzle tip to allow electricity to flow between their ends; this causes a spark which is used to ignite the fuel.

Transformer   The device which increases the voltage of the power supply to some $10\,000 \, V$, which is sufficient to produce the required spark at the electrode tips.

High tension (HT) leads   The cable which carries the high voltage to the electrodes, similar to spark plug leads on a car. **Note:** Some burners, such as the Riello, do not use these leads; the electrodes simply plug into the transformer located in the control box.

Photocell   A device which detects light. It consists of a small resistor which, when exposed to light, makes or breaks an electrical circuit.

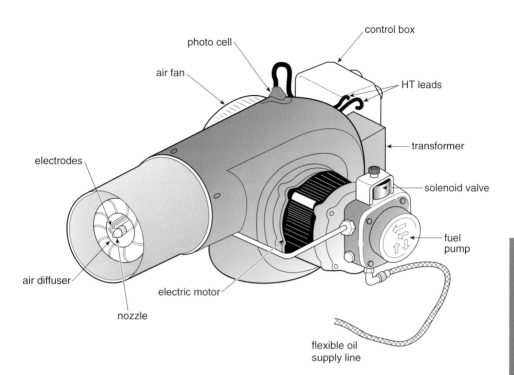

photo cell

control box

air fan

HT leads

transformer

solenoid valve

electrodes

fuel pump

air diffuser

electric motor

nozzle

flexible oil supply line

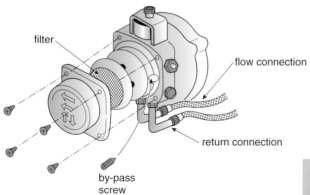

filter

flow connection

by-pass screw

return connection

Inserting a by-pass screw into the pump for use with a two pipe system

Pressure jet oil burner

Pressure jet burners 2

Pressure Jet Burners 3

Nozzle and ignition setting

The nozzle is the device which atomises the fuel and causes it to leave the burner in a swirling motion in readiness for ignition by the electrodes. Stamped on the body of the nozzle will be the spray angle in degrees and the flow rate in US gallons per hour, plus an indication of the type of spray formation. This information must always be checked upon initial set up and servicing to ensure compliance with the manufacturer's data and efficient and economic use. The spray angle for domestic boilers is usually 45°, 60° or 80°, the angle of spray being governed by the length of the combustion chamber. Generally a long chamber requires a small spray angle.

The flow rate is the volume of oil consumed per hour (the US gallon is used because the pressure jet burner was developed in the United States). Where data are unavailable, the following formula can be used to suggest a suitable flow rate:

$$\frac{\text{Boiler output}}{\text{Calorific value of oil}} \times \frac{1.2}{0.8} = \text{Flow rate (US gallons/hour)}$$

Note: 0.8 is used because it is assumed that the boiler is 80% efficient; the 1.2 changes imperial gallons to US gallons.

Example: Suggest a nozzle size for a pressure jet burner when the output is $17\frac{1}{2}$ kW and the fuel used is Class C2, 28-second oil, having a calorific value of 46.3 kW/gal.

$$\text{Flow rate} = \frac{17.5}{46.3} \times \frac{1.2}{0.8} = 0.57 \text{ US gallons/hour}$$

for which the nearest nozzle flow size is 0.6 US gallons/hour

The electrode setting needs to be checked. Where a problem is encountered with ignition, bear in mind that the ignition spark will take the shortest distance when jumping a gap, so ensure that the narrowest gap is between the electrode tips and not between one electrode and, say, the nozzle. Where possible, set the electrodes alongside the flat of a nozzle rather than its corner, thus giving a bigger gap clearance.

Also check for carbon build-up on the electrode tips. As a general rule, the electrodes are positioned 3 mm apart and 15 mm away from the centre line of the nozzle outlet, usually slightly forward of the nozzle itself.

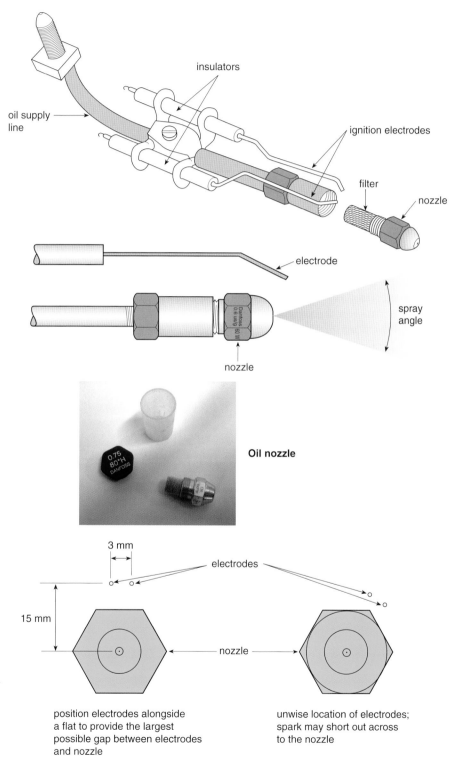

5 Oil Supplies

insulators

oil supply line

ignition electrodes

filter

nozzle

electrode

spray angle

nozzle

Oil nozzle

3 mm

electrodes

15 mm

nozzle

position electrodes alongside a flat to provide the largest possible gap between electrodes and nozzle

unwise location of electrodes; spark may short out across to the nozzle

Pressure jet burners 3

Open Flued Appliances

Relevant British Standard
BS 5410

Where a traditional brick chimney is encountered it should be suitably lined with a stainless steel liner. It may be necessary to install loose-fill insulation material in the void between the liner and the exposed chimney to avoid condensation problems. Any liner or flue pipe used should be of the same diameter as the outlet from the appliance. Chimneys and flue pipes should be as straight as possible and free from bends exceeding 135° (45°).

The terminal location for an open flued system should be above any excessive positive or negative pressure zones: generally a distance of 1 m above the roof line is sufficient. The flue should not exceed 6 m in height, as this may cause excessive flue draught problems. The flue draught for pressure jet burners should be around 0.09 mbar (0.035 in water gauge (wg)). However, if the draught exceeds 0.4 mbar (0.15 in wg) a draught stabiliser should be fitted.

The *draught stabiliser* consists of a hinged flap, usually fitted to the base of a chimney. It is fitted in the same room as the appliance and its purpose is to ensure that the chimney maintains the required draught condition. Should the draught within the chimney be too high (because of high winds, for example) the flap will open, permitting air to enter directly into the chimney and thus reducing the draught through the boiler. The draught stabiliser is also forced open should the pressure inside the chimney become excessive.

Air supply requirements

Open flued oil burning appliances with an input below 5 kW do not require any additional ventilation. Where the input is in excess of 5 kW, the room in which the appliance is installed must have an air grille with a minimum effective free area of 5.5 cm² for every kW in excess of 5 kW.

Example: A 22 kW open flued oil burner requires an effective grille size of:

Input above 5 kW = 22 − 5 = 17 kW; thus grille size 17 × 5.5 = 93.5 cm².

Where a draught stabiliser is fitted, this grille size should be doubled.

Should a boiler be located in a cupboard or small confined space, suitable air grilles should be provided for cooling purposes. These would consist of grilles at high and low level to the outside air of 5.5 cm² per kW input; this figure is doubled where the grille connects to an internal room, which itself needs to be suitably vented. The effective free area of a grille is defined under ventilation for gas appliances (see page 220).

Where the flue gas temperature does not exceed 260°C the flue pipe or brick chimney, etc., should be installed in the same way as for gas installations. However, if this temperature is in excess of 260°C, Part J of the Building Regulations should be observed and installation of pipes, etc., should be as for solid fuel appliances.

Note: Domestic oil boilers need to meet the requirements of the Building Regulations and be of the high efficiency/condensing type.

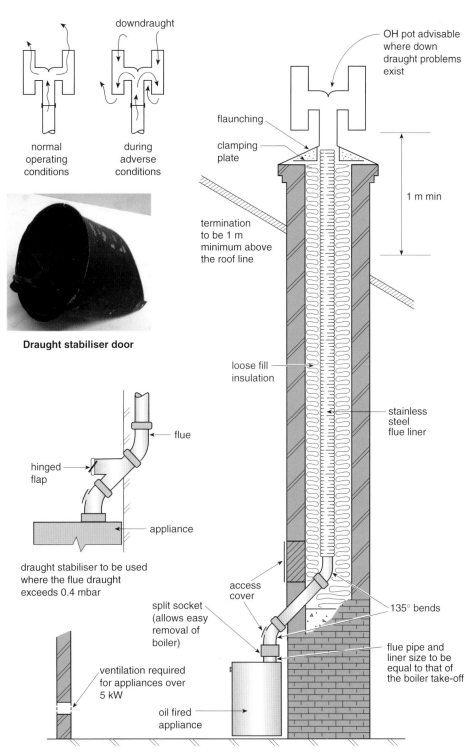

downdraught

normal operating conditions

during adverse conditions

Draught stabiliser door

flue

hinged flap

appliance

draught stabiliser to be used where the flue draught exceeds 0.4 mbar

split socket (allows easy removal of boiler)

ventilation required for appliances over 5 kW

oil fired appliance

OH pot advisable where down draught problems exist

flaunching

clamping plate

termination to be 1 m minimum above the roof line

1 m min

loose fill insulation

stainless steel flue liner

access cover

135° bends

flue pipe and liner size to be equal to that of the boiler take-off

Open flued appliances

Room Sealed Appliances

Relevant British Standard
BS 5410

With the forced draught of air into the combustion chamber in the pressure jet burner, the extraction of flue gases is that little bit easier. As a result, the installation of the balanced flued appliance is often favoured. This allows for a faster installation without the need to worry about flue draughts and permanent ventilation, as required with open flued systems. The air required for combustion purposes is drawn directly from the outside environment.

Where a balanced flued boiler is to be installed it is essential that the casing is securely sealed; otherwise, should the pressure outside the building be in excess of that within the room, incomplete combustion will occur, resulting in smoke and carbon emissions from the flue terminal. This will generally stain the brickwork. The principle of the balanced flue relies on the pressure within the boiler casing and the pressure within the combustion chamber to be the same as that at the point of flue gas discharge.

Note that some boilers need to have an air blanking plate inserted in the base to seal the unit, as the boiler manufacturer often uses the same casings for both open and balanced flued appliances.

Boilers operating with Class D 35-second gas oil are not permitted to have their products discharged at low level. Where the terminal is located it is essential that a terminal guard be fitted for the protection of any person who might otherwise come into contact with the terminal itself. The location of terminal positions is indicated in the following table:

Suitable room sealed terminal locations for oil appliances with pressure jet burners

Terminal position	Minimum distance (mm)
From a surface facing a terminal	600
Vertically from a terminal on the same wall	1500
Horizontally from a terminal on the same wall	750
From an internal corner	300
From an external corner	300
Horizontally from any opening, e.g. window or door	600
Directly below any opening, e.g. window or door	600
Below the eaves or discharge pipework (without protection)	600
Below the eaves or discharge pipework (with protection)	75

It is possible to have a room sealed arrangement with a vertical balanced flue where it may not be possible to discharge through an external wall. Where a vertical balanced flue is used it must terminate at least 750 mm from any wall face.

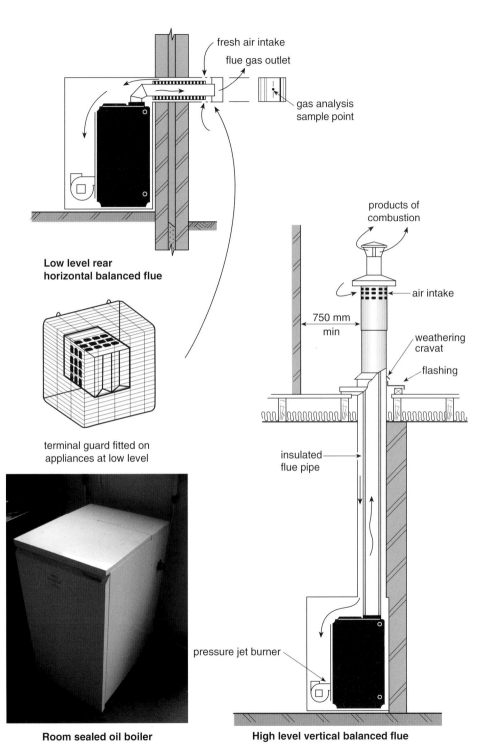

**Low level rear
horizontal balanced flue**

terminal guard fitted on
appliances at low level

Room sealed oil boiler

High level vertical balanced flue

Room sealed appliances

5 Oil Supplies

fresh air intake

flue gas outlet

gas analysis
sample point

products of
combustion

air intake

750 mm
min

weathering
cravat

flashing

insulated
flue pipe

pressure jet burner

Vaporising Burners

Relevant British Standards
BS 799 and BS 5410

The vaporising burner is no longer installed to serve as a boiler, its use being generally restricted to Aga type cookers. The burner warms the oil, changing it into its gaseous state where it can readily mix with the oxygen in the air which is required to support combustion; in this form it can be ignited easily. There are three designs of vaporising burner:

- The natural draught vaporising burner (natural draught pot burner)
- The forced draught vaporising burner (fan assisted pot burner)
- The wallflame or rotary vaporising burner.

To ensure that not too much oil is fed to the burner at any time and at any pressure, a constant oil level control is fitted at the base of the appliance. This contains a float which allows a small quantity of oil into its reservoir (see page 248).

The natural draught 'pot' burner is the simplest type of oil burner and consists of a circular container (the pot) with a series of holes through which air can pass to the combustion chamber. It works as follows:

(1) When the appliance requires heat, the igniter is activated, which glows red hot; then a solenoid valve opens which allows oil to trickle into the burner from a constant oil level control. (Note that some burners are ignited manually with a lighted match or taper).

(2) The oil spreads as a thin film over the base of the burner and the vapour from the oil is ignited (it is the vapour that burns, not the liquid).

(3) As the fuel burns, it generates heat which speeds up the vaporising process; thus the flame burns more fiercely and rises up the chamber to obtain sufficient air for combustion.

(4) Eventually the flame burns out at the top of the pot where the waterways of a heat exchanger may be located.

(5) When the thermostat is satisfied the fuel is reduced to the burner. Otherwise, the appliance will continue to burn owing to the oil and vapour in the pot.

It is *most important that no attempt be made to relight a warm or hot burner,* as the ignition of the fuel vapour could blow back with explosive force.

The forced draught burner is identical to the natural draught burner, except that air is blown into the combustion chamber.

In the wallflame burner, the oil is drawn up from a well at the base of the boiler through two oil distribution tubes, which spin in the centre of the appliance when heat is required. The oil is thrown out to the rim of the combustion chamber and is ignited by a sparking high tension ignition electrode. As the droplets fall they are quickly vaporised by the previously heated rim heater and rise to mix with the oxygen, which is also blown through the centre. The flame quickly becomes established and burns around the perimeter of the chamber in a wall of fire.

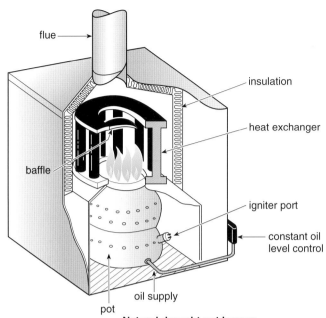

flue

insulation

heat exchanger

baffle

igniter port

constant oil
level control

oil supply

pot

Natural draught pot burner

Typical pot burner
('Aga' type cooker)

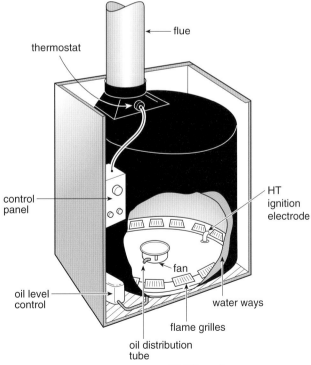

flue

thermostat

control
panel

HT
ignition
electrode

fan

oil level
control

water ways

flame grilles

oil distribution
tube

Wallflame burner

Vaporising burners

Combustion Efficiency Testing

Relevant British Standard
BS 1756

When commissioning an oil burning appliance it is important to ensure that it works at maximum efficiency and that no undue smoke is produced. This can be achieved by taking the following measurements in the section of flue directly above the appliance, ensuring that any hole drilled is plugged on completion. In the case of the balanced flue appliance, measurements should be taken at the terminal location. The equipment used to administer these tests may be an electronic analyzer, as identified on page 228, or more traditional, as described below. In all cases, allow the appliance an initial 10–15 minute warm-up period.

Flue draught   Not applicable to balanced flue appliances. A probe is inserted into the flue to give a draught reading which should be as recommended by the manufacturer. A typical draught is 0.09 mbar (0.35 in wg).

Smoke reading   A test in which a special pump is used to draw a quantity of flue gas through a filter paper. A clean, dry piece of filter paper is inserted into the pump and the probe inserted into the test hole. Ten pulls are given to the pump handle and the paper inspected for signs of discoloration. Where any signs are detected, the sample is held against the smoke gauge supplied. A reading of 0–1 should be achieved. To reduce the smoke content, the air supply to the combustion process will need to be increased by adjusting the air shutter on the burner.

CO_2 percentage   The carbon dioxide content within the flue gas is used in conjunction with the flue temperature to find the efficiency of the appliance.

When the smoke has been adjusted to allow the minimum amount of air possible, the CO_2 reading can be taken. Ideally one is looking for as high a CO_2% as possible, which indicates greater combustion. Closing the air shutter raises the CO_2 but increases the smoke emissions. If the air shutter is altered at this stage, the smoke reading must be retaken. The CO_2 indicator consists of a special container filled with a liquid capable of absorbing CO_2.

When the probe is inserted into the test hole, the hand pump is operated to purge the connecting hose. The plunger (see figure) is now connected to the container and held down, causing a non-return valve in the CO_2 indicator to open. The pump is depressed 18 times and during the 18th deflation the plunger is removed. The container is now inverted several times; this causes the CO_2 to be absorbed into the liquid. With the gas absorbed into the liquid a partial vacuum is formed in the container and a diaphragm at the base of the vessel flexes upwards, which raises the level of the fluid. The CO_2% can now be read from a scale at the side of the container.

Flue gas temperature   This is taken using a flue gas thermometer or electronic pyrometer probe inserted into the test hole.

Flue/appliance efficiency This is found by comparing the CO_2% with the flue gas temperature, either by using a sliding chart or by calculation using the following formula:

$$(0.477 \div CO_2\% + 0.0072) \times (\text{flue gas temp} - \text{room temp})°C + 6.2 = \text{flue loss }\%$$

Thus, if you subtract the energy lost in the flue from 100% you will have an indication of the efficiency of the appliance.

Example: With 11% CO_2 in the flue gas, a flue gas temperature of 195°C and a room temperature of 21°C, the appliance will be operating at:

$$\text{flue loss (\%)} = (0.477 \div 11\% + 0.0072) \times (195 - 21) + 6.2$$
$$= 0.0506 \times 174 + 6.2 = 15\%$$

Therefore the appliance is $100 - 15 = 85\%$ efficient.

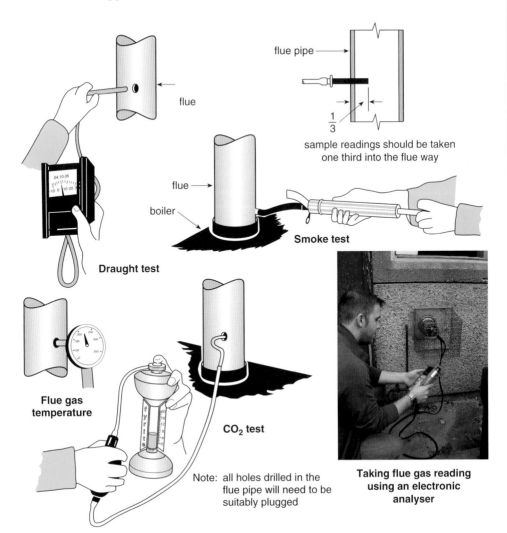

flue pipe

flue

$\frac{1}{3}$

sample readings should be taken
one third into the flue way

Smoke test

flue

boiler

flue

04.10.05

-10 0   10 20 3

Draught test

200   250
180   300
-100   350

**Flue gas
temperature**

CO₂ test

Note: all holes drilled in the
flue pipe will need to be
suitably plugged

**Taking flue gas reading
using an electronic
analyser**

Combustion efficiency testing

Commissioning and Fault Diagnosis

Relevant British Standard
BS 5410

When an appliance has been installed it should be commissioned in accordance with the manufacturer's instructions. However, the following checklist can help to avoid lengthy problems in setting the burners up to operate at maximum efficiency.

Tests prior to ignition of the appliance

(1) Complete a general visual check of the system and flue to identify obvious defects.
(2) Disconnect the oil supply as close as possible to the burner, bleed the supply pipe and drain 4–5 litres of oil into a clean container. Inspect the oil for water or any impurities; repeat as necessary. Do not reconnect until clear, as damage may occur to the unit. Clean all filters and de-sludge the oil tank if necessary.
(3) Look for leaks in the oil supply pipework, inspecting the tank also.
(4) Check that the baffles are correctly positioned inside the boiler combustion chamber.
(5) Drill a hole in the flue-way to serve as a test point for open flue appliances, where applicable.
(6) (a) With *balanced flue appliances*, is the casing sealed correctly from the room?
  (b) With *open flue appliances*, is there an adequate supply of air for combustion?
(7) Where competent to do so, confirm the electrical connections are correct, i.e. earth continuity, polarity and insulation resistance, and ensure that the correct fuse is fitted (see page 290).
(8) Check that the system is filled with water.
(9) Remove the burner as necessary to check that:
  (a) the by-pass screw is fitted where applicable (see page 253)
  (b) the correct nozzle is fitted (see manufacturer's instructions)
  (c) the electrode settings are correct (see manufacturer's instructions)

Fit the combined bleed manifold and replace the burner.

Commissioning the appliance

(1) Ignite the appliance and adjust the burner pressure (see page 250).
(2) Check the lockout function; either cover the photoelectric cell or remove the solenoid coil to simulate flame failure.
(3) Check the operation of any thermostats; where a high limit stat is fitted, remove the control thermostat phial to simulate its malfunction.
(4) Carry out combustion efficiency tests (see page 262).
(5) Check the manual operation of the fire valve to ensure its operation.
(6) Remove the bleed manifold and give a final check for leaks.

Fault finding: pressure jet burners

Symptoms	Possible causes
Burner will not run although no lockout light has come on	No power is reaching the burner (i) Power is off (ii) Timeclock/programmer is in the *off* position (iii) Fuse has blown (iv) Thermostat not calling for heat
Burner will not run and the lockout light is indicated	Motor not running (i) Motor defective (ii) Motor capacitor defective (iii) Fuel pump seized To confirm (iii) remove pump and restart burner; if motor runs, pump is at fault.
Burner runs but goes into lockout without firing	Starvation of fuel at the burner nozzle (i) Fuel tank empty (ii) Valves on fuel line shut (iii) Fuel pump drive sheared or worn; repair or renew Nozzle not atomising fuel – renew nozzle Spark not reaching fuel (i) Electrodes incorrectly set (ii) Electrodes earthing out (iii) Defective high tension leads (iv) Ignition transformer defective
Burner runs and fires but goes into lockout	Dirty or defective photocell Starvation of fuel (i) Partially blocked filter. **Note**: In winter the fuel may be waxing up (ii) Valves in fuel line shut (iii) Air in fuel, e.g. oil level low (iv) Water in fuel (v) Faulty solenoid valve Faulty control box Blockage in flue, possibly causing back pressure in the combustion chamber
Burner runs and fires but fails to switch off	Faulty thermostat (i) Defective (ii) Loose connections (iii) Not correctly housed in appliance

5 Oil Supplies

Part 6

Electrical Work

Basic Electrical Theory

Electricity is the flow of free electrons moving along a conductor. Free electrons are small electrical charges moving round an atom. Millions of atoms make up a solid conductor, such as copper, but the electrical charge is not felt because the electrons are moving in all directions. It is when all these electrons are made to travel in the same direction that the electrical *current* is produced. A flow of about 6 billion electrons per second produces a current of about 1 *ampere*.

Electricity will not flow through certain materials, such as wood or plastic, because there are no free electrons in these materials. These materials are therefore known as insulators. To make the free electrons flow in the same direction, an electromotive force (emf), expressed in *volts*, needs to be applied, such as that produced by a battery or generator. The flow of free electrons can only occur if the electrons have somewhere to go, which is achieved by allowing the electrons to flow in a circuit. Break the circuit and the electron flow (current) will stop.

Resistance   As the electrons flow through the cable they slowly lose their ability to move (emf). When the electrons are forced to pass through thin sections they all rub together in order to get through; thus they are restricted and once again there is a drop in emf. This voltage drop is due to the resistance of the system and its components; resistance is measured in *ohms* (Ω).

The relationship between current (amps), emf (volts) and resistance (ohms) is expressed in *Ohm's Law*:

volts ÷ amps = ohms; similarly volts ÷ ohms = amps, and amps × ohms = volts

Example: where a 12 volt battery is used to serve a circuit in which the total resistance is 4 ohms the current flow will be 12 ÷ 4 = 3 amps. **Note:** increasing the voltage increases the amperes.

Power   When the electrical energy produced is used up very quickly, such as in passing through a light bulb filament, heat energy is produced, due to resistance; this is referred to as power and is measured in *watts*. Again there is a relationship between current and emf:

volts × amps = watts, watts ÷ volts = amps, and watts ÷ amps = volts

So in the previous example, where the 12 volt battery had a current flow of 3 amps, the power would have been 12 × 3 = 36 watts.

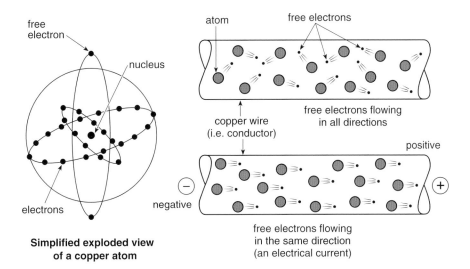

free
electron

nucleus

electrons

**Simplified exploded view
of a copper atom**

atom   free electrons

copper wire
(i.e. conductor)

free electrons flowing
in all directions

positive

negative

free electrons flowing
in the same direction
(an electrical current)

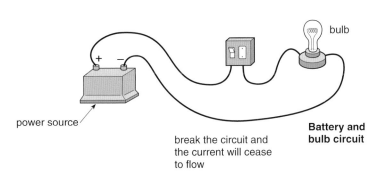

bulb

power source

break the circuit and
the current will cease
to flow

**Battery and
bulb circuit**

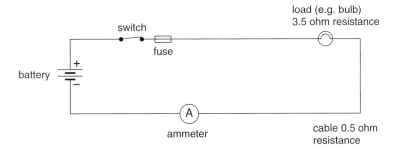

load (e.g. bulb)
3.5 ohm resistance

switch

fuse

battery

ammeter

cable 0.5 ohm
resistance

Equivalent circuit diagram with meter installed

Basic electrical theory

Electrical Current

Simple circuits

Series circuit   A system where the current is made to flow through each resistor (e.g. a bulb) in the circuit. Because of this, the resistance slowly diminishes as the current flows round the system and as a result the total voltage of all the bulbs must not exceed the total available voltage, otherwise the bulbs will not glow sufficiently.

Parallel circuit   In this system the same emf is applied to each element, so the voltage at all the bulbs needs to be equal to that of the battery. The bulbs will glow just as brightly whether one is removed or not, but the life of the battery is reduced when more bulbs are used, i.e. more power is used. With this system, when a bulb blows the others stay alight, whereas with a series system, when one bulb blows, it breaks the circuit and all the bulbs go out. For example, the lighting in a house is wired up in parallel and Christmas tree lights are wired up in series.

Direct current (d.c.)   In a d.c. electrical circuit, the electron flow is in the same direction all the time. One example would be a battery in which a metal such as copper slowly destroys another metal such as zinc (see page 44, Corrosion) and in so doing creates a small electric current; d.c. is also produced when a thermocouple is heated.

Alternating current (a.c.)   With a.c. the electron flow travels continually back and forth because of the way in which the current is produced. Alternating current is produced by moving a magnet in and out or around a coil of wire wound onto a soft iron core. The movement of the magnetic field around the wire causes electrons to flow, providing the circuit is made. As long as the magnet is moving, the current will flow.

The electric generator at the power station produces a.c. on this principle: one wire is connected to the overhead power cable, the other to 'earth'. Thus, when the cable in the home is connected to earth a circuit is formed and the current flows. From this we can deduce that in our home, which is supplied with 230 V a.c., the current flowing to an appliance is flowing from the phase (live) wire and then from the neutral wire (which is, in effect, connected to earth); the current direction changes 50 times a second (50 hertz).

Three phase supply   An a.c. supply where not only one circuit/wire is connected to the generator but three. This gives three batches of current in sequence (three phases). In industry, where large motors are used, a single phase supply of 230 V would not produce sufficient power to operate the plant. The three phases, when all circuits are wired to the machinery, give an approximate emf of 400 V.

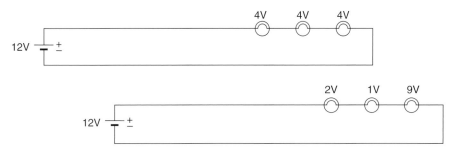

Electrical circuits in series

Electrical circuits in parallel

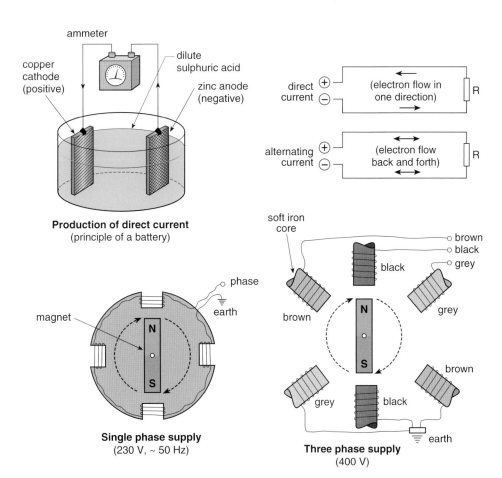

ammeter

copper
cathode
(positive)

dilute
sulphuric acid

zinc anode
(negative)

direct
current $\oplus$ $\ominus$ (electron flow in
one direction) R

alternating
current $\oplus$ $\ominus$ (electron flow
back and forth) R

Production of direct current
(principle of a battery)

soft iron
core

brown
black
grey

magnet

phase

earth

N

S

N

S

black

grey

brown

grey

brown

black

earth

Single phase supply
(230 V, ~ 50 Hz)

Three phase supply
(400 V)

Electrical current

Electrical Supply

Relevant British Standard
BS 7671

The electrical supply to the domestic dwelling is distributed by a network of above- and below-ground cables, via various substations; these run to the home via one of the following three systems:

1. *TT system*: Designed for older properties with overhead power supply; the earth is achieved by an earth stake.
2. *TN-S system*: An older system which brings in the supply from below ground level. A metallic armoured casing encloses the line wire (phase) and neutral cables. The earth is made by connection to the metal casing.
3. *TN-C-S system*: Also called the PME system. The system most commonly used today, it uses a special concentric cable, the line wire being surrounded by the neutral. The earth and neutral are the same outer wire of the cable; it is only at the house that the earth is separated from the neutral. **Remember**: Neutral and earth are to all intents and purposes the same thing.

The conductors terminate at the dwelling in an electricity authority sealed fuse unit. From here the earthing cable is run to the main terminal block; the line wire and neutral are fed through a meter and into the main consumer unit where a double pole switch is incorporated to isolate the system. A fuse or circuit-breaker is incorporated to provide automatic protection to the system. Three wires are run from the incoming supply to the various circuits, these being:

- *The line conductor*, commonly called the phase wire
- *The neutral conductor*
- *The circuit protective conductor* (cpc), commonly called the earth.

The circuit protective conductor is designed as a back-up. Should a fault develop within the system where a live cable touches some metallic part (e.g. the appliance itself): the electric current flows through the cpc and travels down to earth. If you were to touch the metal part you would be prevented from getting a shock because your body would offer a greater resistance to the electron flow than the cpc. All the metalwork in an installation that could allow a current flow to earth must ultimately be connected to the main earth terminal block.

Colour identification of cables   To standardise cable colour Europe-wide, the insulating cover has been changed as shown in the table opposite. **Note**: When both old and new colours are used in one building, a warning notice must be displayed at the meter.

In the latest edition of the British Standard it should be noted that in all new works all systems must be RCD-protected.

		Prior to 2005	Since 2005
230 V	Line	Red	Brown
	Neutral	Black	Blue
	CPC	Green & yellow	Green & yellow
400 V	Phase 1	Red	Brown
	Phase 2	Yellow	Black
	Phase 3	Blue	Grey
	Neutral	Black	Blue

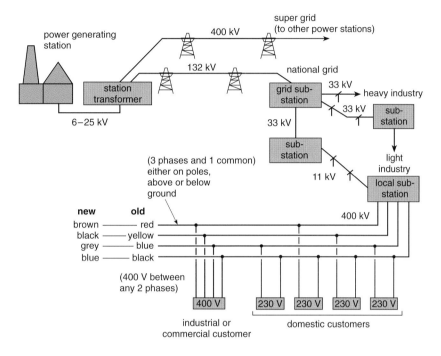

**Single and three phase connections
from generator to consumer**

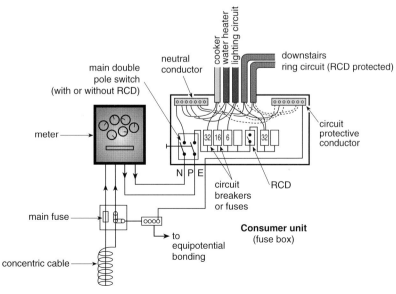

Electrical supply

Electrical Safety

Relevant British Standard
BS 7671

It is a requirement that no person should carry out any electrical work unless competent to do so. This brief introduction to electrical installation work is designed to give only a basic knowledge.

Circuit protection

When a current in excess of that required passes down a conductor (cable) due to an overload or a fault, causing a short circuit, it is essential that the supply is automatically cut off. The Wiring Regulations (BS 7671) stipulate that the current is disconnected automatically within 0.4 seconds where a 13 amp socket outlet is used, should a greater current flow through a conductor than that for which the circuit was designed. The time is increased to 5 seconds for circuits with fixed appliances, providing they are not located in a bathroom or shower-room. Several devices are used to detect such overloads, including the fuse, the miniature circuit-breaker (MCB) and the residual current device (RCD).

The **fuse** is simply a short length of wire which will heat up and melt should the flow of electrons exceed its design ampere rating. The most common fuse is the cartridge type, although it is still possible to find rewirable fuses. The fuse rating is calculated by applying the formula, watts ÷ volts = amps; this formula does, however, ignore the resistance of the supply cable.

It is worth noting that the cartridge fuse fitted into a 13 A plug feeding an appliance is often overrated. For example, a 100 W light bulb only requires a fuse of $100 \div 240 = 0.4$ A, to which a 1 A fuse should be fitted. However, 1 A is more than sufficient to kill at 230 V. Usually, for the sake of convenience, a 3 A fuse, which may be purchased at the local shop, is fitted. It may not blow immediately but the cable is protected, this being the prime function of the fuse. The earth is to protect the user! Equally a 13 A fuse should not be fitted because it is unnecessarily high. The **MCB** is a device which usually includes a temperature-sensitive element to operate an electromagnet, causing it to trip a switch which closes, breaking the flow of current. They are more expensive than fuses but have the advantage of being resetable – and they can be used as a switch to control the circuit they serve. The RCD is a device that detects the current flowing along the line and neutral conductors; they should be the same. However if there is a leak to earth, then this will be detected and within a few milliseconds the RCD will close off the supply.

Safe isolation

The Electricity at Work Regulations require operatives working on exposed electrical circuits to ensure correct and locked off isolation and that test equipment is working prior to and immediately after testing the supply as safe. Therefore before commencing electrical work on any circuit you must ensure that the circuit is completely dead and that it cannot be switched on without your knowledge, i.e. lock off the supply and/or carry the fuse in your pocket. Approved fused test lamps or multimeters should be used, which are designed with probes with a minimal amount of exposed metal and fused test leads to prevent any electric shock.

MCB, cartridge and
re-wirable fuse holders

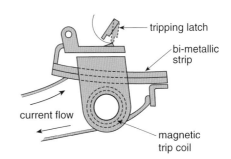

tripping latch

bi-metallic
strip

current flow

magnetic
trip coil

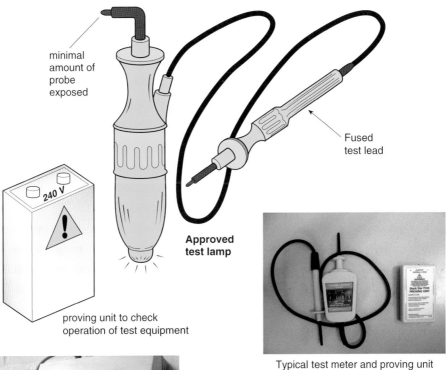

minimal
amount of
probe
exposed

240 V

Fused
test lead

**Approved
test lamp**

proving unit to check
operation of test equipment

Typical test meter and proving unit

Locked off electrical circuits

Electrical safety

Earth Continuity

Protective equipotential bonding   It is a requirement of the Wiring Regulations that all exposed metalwork entering a building be bonded together and connected to the main earthing block at the consumer unit. In the case of any metal pipes entering a building (e.g. water, gas or oil) the buried pipe itself may provide a path for any stray electrical currents; however, the current may cause corrosion to the pipework; also, should someone disconnect any part of the service pipework they might receive an electric shock. To prevent this, the supplies are bonded together at their point of entry to the building and, in the case of the gas supply, within 600 mm of the gas meter.

The main protective equipotential bonding conductor should have a cross-sectional area of not less than 10 mm² and be run to all extraneous conductive parts (i.e. any exposed metal which is not part of the electrical installation), including accessible structural steel members, main ventilating ductwork, lightning conductors and communal central heating pipes brought into the building where externally heated, as well as the main services.

Additional supplementary bonding   The main protective equipotential bonding conductor connects only to one point of the metalwork. To ensure that earth continuity is maintained throughout the system, additional cross-bonds may be required. The Wiring Regulations only specify supplementary bonding requirements within bathrooms. There is no requirement to bond anywhere else within a building. Often supplementary bonding is found connecting to appliances and kitchen sinks, etc., but this is not a legal requirement. The size of the bonding wire used may be quite small where protected; however where no protection is provided it should not be less than 4 mm².

Earth clips   When securing the bonding wire to pipework, a special clip should be used (see figure). Note that the cable is hooked onto the clip and tightened, leaving a section of insulation material intact. Where a conductor is to loop to another service pipe, the cable should not be cut. On making the connection, a good contact must be made with the metalwork.

Temporary bonding wire   At any time, should a section of metalwork be removed (e.g. a section of pipework), it is essential that the earth continuity is maintained before the disconnection occurs. This is achieved by bridging the section with a 10 mm² piece of cable. Thus, if there is a fault on the system and you touch live metal while a current is passing down to earth, you are prevented from getting an electric shock because, should disconnection with the earth route via the metalwork occur, the current will flow through you.

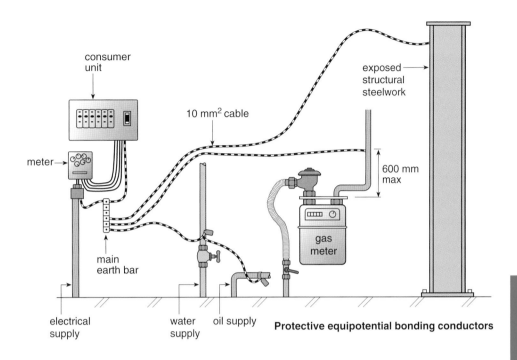

consumer unit

meter

10 mm² cable

exposed structural steelwork

600 mm max

main earth bar

gas meter

electrical supply

water supply

oil supply

Protective equipotential bonding conductors

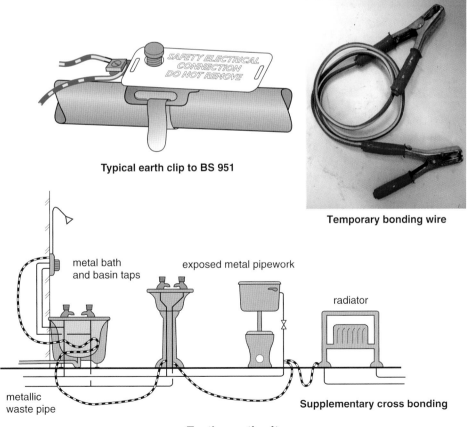

Typical earth clip to BS 951

Temporary bonding wire

metal bath and basin taps

exposed metal pipework

radiator

metallic waste pipe

Supplementary cross bonding

Earth continuity

Domestic House Wiring 1

Relevant British Standard
BS 7671

In general, three types of wiring circuits will be found in domestic installations: the supply to the 13 A socket-outlets, the lighting circuit, and the various radial circuits to fixed appliances.

13 A socket-outlets

13 A socket-outlets are either fed from a continuous ring circuit or from a radial circuit. The *ring circuit* is the most common system in which the line, neutral and circuit protective conductor (cpc) are connected to their respective terminals in the consumer unit. From here the cables circulate from one socket to the next, passing round to all the sockets in one big loop, and returning to the consumer unit.

There is no limit to the number of outlets served by the ring main, although the maximum floor area in a domestic situation served by a single 30 A fuse or MCB must not exceed 100 m². The *radial circuit* differs in that the final socket on the system does not feed back to the consumer unit and as a result the maximum floor area to be served must not exceed 50 m² (i.e. half that of a ring circuit) with a maximum fuse/MCB of 20 A. The size of the conductor used is usually 2.5 mm² twin and earthed PVC insulated cable.

Spur outlets These are used where it is inconvenient to incorporate a socket-outlet within the loop of the circuit, such as at an isolated location, or an addition, it is possible to run a single twin and earth wire to the socket. The spur is connected to the circuit, usually via a joint box or directly from the back of an existing socket. If the spur is fused, an unlimited number may be connected to the circuit; however, where not fused it may supply only a single- or double-socket outlet. The total number of spurs must not exceed the total number of socket-outlets connected directly to the circuit. The cable size for a spur should be equal to that of the main circuit.

Lighting circuit

The lighting system is a radial circuit which feeds each ceiling rose or wall light in turn as it passes round the system. To prevent the light being *on* continuously, the line conductor does not pass directly to the bulb but via a switch, usually mounted at the entrance to the room at about 1.2 m above floor level.

The circuit is protected with a 6 A fuse or MCB and the usual size of conductor is 1.5 mm² twin and earth PVC insulated cable. Generally a one-way switch is required to turn the light on and off; however, a two-way switch is used on stairways, etc., so that the light can be operated from more than one location. The two-way switch requires a special switch control and is wired up as shown. Because of the size of the fuse and to prevent nuisance tripping when many lights are on, the house is often divided into two circuits (e.g. one upstairs and one downstairs).

The latest edition of BS7671 now requires **all** circuits, including the lighting circuits, to be RCD-protected unless special precautions are made to ensure there is no possibility of getting an electric shock.

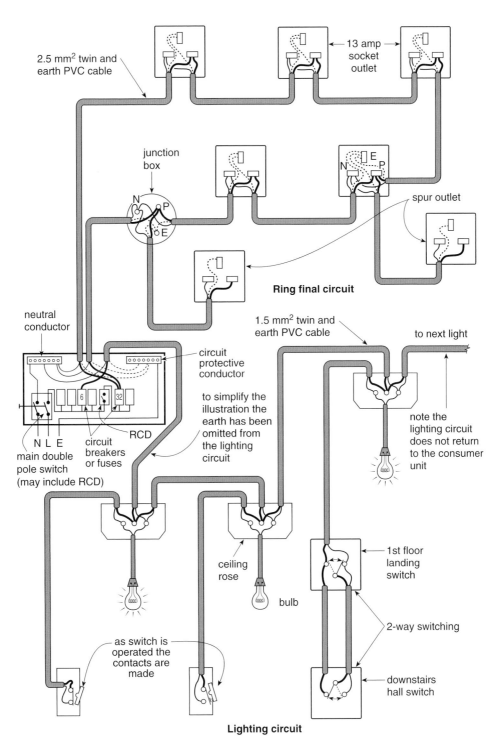

2.5 mm^2 twin and earth PVC cable

13 amp socket outlet

junction box

spur outlet

Ring final circuit

neutral conductor

circuit protective conductor

1.5 mm^2 twin and earth PVC cable

to next light

to simplify the illustration the earth has been omitted from the lighting circuit

note the lighting circuit does not return to the consumer unit

N L E

RCD

circuit breakers or fuses

main double pole switch (may include RCD)

ceiling rose

1st floor landing switch

bulb

2-way switching

as switch is operated the contacts are made

downstairs hall switch

Lighting circuit

Domestic house wiring 1

Domestic House Wiring 2

Relevant British Standard
BS 7671

Radial circuits to fixed appliances

Various radial circuits will be found in a home, including the supply to an immersion heater, a shower heater and a cooker control unit. The line, neutral and cpc are run to terminate close to the appliance with a double pole switch (i.e. both line and neutral are switched). From this point, the supply is run to the appliance as necessary.

Power supply to an immersion heater   This is a typical radial circuit in which 2.5 mm^2 twin and earth PVC-insulated cable is run from the consumer unit, the line supplied via a 16 A fuse or MCB, to terminate at a 21 A double pole switch with neon indicator. From this point, a 21 A heat-proof flex is run to the immersion heater element, the line being supplied via the thermostat as shown in the figure.

Fixed appliances such as a boiler must be wired up so that when the power is switched off and no longer feeds the appliance, both line and neutral conductors are disconnected. This is usually achieved using a double pole fused spur box, although it is possible to use an *unswitched* socket outlet; thus, to ensure that the power is off, the plug must be removed.

Switches in bath and shower rooms   No 13 A sockets are permitted in these rooms and where switches are required to operate lights, pumps and heaters, etc. these must be via a pull cord located at high level.

Termination of a 13 A 3-pin plug

The wires to be terminated into a plug need to be securely anchored in position. The earth must always be the conductor which would be the last to pull from its connection if the cable were to be pulled; thus the earth continuity is maintained for as long as possible. The wires into the plug are connected as indicated in the following chart.

Note: although the plug is referred to as 13 A, this does not mean that the fuse to be fitted should be of this size (see page 274, Electrical Safety).

Colour use for flexible cable

Conductor	Present colour code	Older appliances	Notes
Line	Brown	Red	To connect via the fuse
Neutral	Blue	Black	
Earth	Green & Yellow	Green	To the largest pin

6 Electrical Work

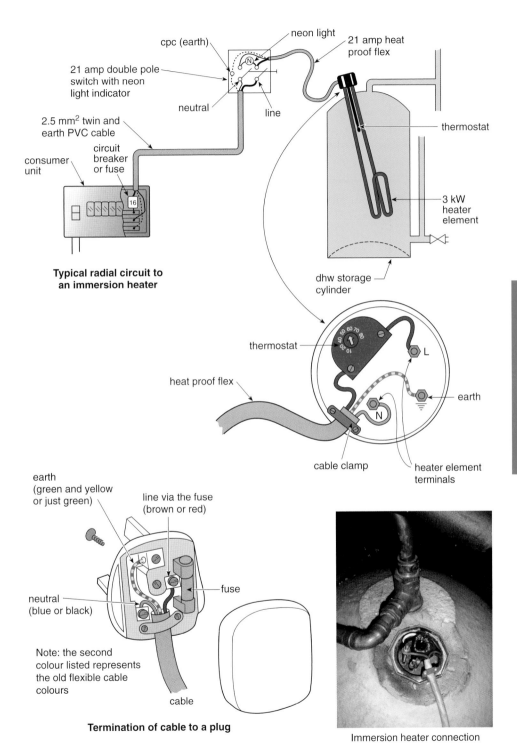

neon light

21 amp heat proof flex

cpc (earth)

21 amp double pole switch with neon light indicator

neutral

line

2.5 mm² twin and earth PVC cable

circuit breaker or fuse

consumer unit

thermostat

3 kW heater element

Typical radial circuit to an immersion heater

dhw storage cylinder

thermostat

L

heat proof flex

earth

N

cable clamp

heater element terminals

earth (green and yellow or just green)

line via the fuse (brown or red)

fuse

neutral (blue or black)

Note: the second colour listed represents the old flexible cable colours

cable

Termination of cable to a plug

Immersion heater connection

Domestic house wiring 2

Installation Practices

All electrical installation work must comply with the latest edition of the Wiring Regulations BS 7671. The Wiring Regulations are not legally enforceable, but in following their requirements you are deemed to be following the law. The electricity supplier will not provide electricity until the conductors and apparatus are of sufficient size, and, as far as practically possible, constructed, installed and protected against danger and damage. The following identify just a few of the requirements to be met:

- All cable used is to be suitably insulated; usually for domestic circuits, PVC twin and earth cable is used.
- The cable must be suitably supported and installed so that it is not liable to damage.
- Cables passing through holes in metalwork should be bushed using rubber grommets.
- Where the cable terminates it must be securely anchored; the cpc (earth) must be the last conductor to be disconnected should the supply cable suffer undue stress.
- Where the cpc is exposed it should be sleeved using green and yellow striped insulating material.
- A means of isolation to appliances must be provided in readily accessible locations.
- All circuits must be suitably protected with a fuse or circuit-breaker to prevent overloading the cable. In addition an RCD must be incorporated to close the supply if there is a leak to earth.
- Where the cable is concealed and installed in cement or plaster it should be protected against damage by covering it with a metal channel; alternatively, the cable may be run in metal conduit, suitably bushed at each end, or kept at least 50 mm below the surface.
- If the cable is to be run under wooden floors or above ceilings, the cable should be fixed to the side of the joists and where it is necessary to pass it through a joist, a hole must be drilled at least 50 mm from the surface to prevent damage caused from screws and nails securing any boards.
- Care should be taken to ensure that the cable is not allowed to come into contact with any metalwork such as gas or water pipes, and must be placed at least 25 mm from them.

Termination of cables Where a cable is run into an appliance or accessory, such as a 13 A socket, the cable must be securely anchored in position. It is, however, essential to ensure that the wire is not so taut as to prevent movement. The reason for this is that when the current flows through the conductor a certain amount of heat is generated, causing the conductor to expand and contract; this movement, if restricted, will cause the conductor to pull out from the point of termination.

Work notification/certification Since January 2005 it has been a requirement that specific electrical works, as identified opposite, undertaken in domestic dwellings are self-certificated by an approved operative, as being completed and tested to a satisfactory safe standard. An approved operative is someone who has demonstrated competence and obtained registration through an appropriate validation body.

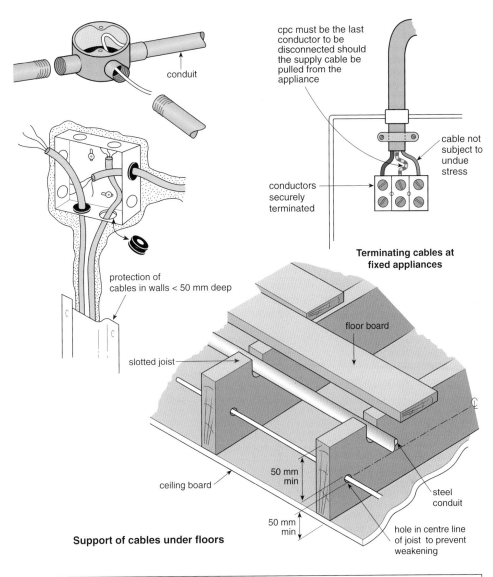

conduit

cpc must be the last
conductor to be
disconnected should
the supply cable be
pulled from the
appliance

cable not
subject to
undue
stress

conductors
securely
terminated

**Terminating cables at
fixed appliances**

protection of
cables in walls < 50 mm deep

slotted joist

floor board

ceiling board

50 mm
min

50 mm
min

steel
conduit

hole in centre line
of joist to prevent
weakening

Support of cables under floors

Notifiable electrical work	Outside a bathroom or kitchen area	Within special locatation or kitchen
Installing bonding	x	✓
Complete new/rewire	✓	✓
Changing the consumer unit	✓	✓
Installing an additional socket	x	✓
Installing a new shower circuit	✓	✓
Installing an additional light	x	✓
Replacing damaged equipment or component	x	x
Running a supply to an outbuilding	✓	N/a
Installing a socket in an outbuilding	✓	N/a
Installing a pond pump	✓	N/a
Installing a hot air sauna	✓	✓
Work on a caravan	x	x

Installation practices

Electrical Components 1

Relevant British Standard
BS 7671

Thermostats

Thermostats are devices which detect a change in temperature and make or break the contacts of a switch. Thermostats generally comprise one of the following:

Bi-metallic strip: Usually found in room thermostats. It consists of two metal alloys, such as brass and Invar, which have different expansion rates, secured together. As the environment heats up, the brass will expand more than the Invar; this causes the bi-metallic strip to bend, breaking the electrical contact.

Differential expansion: Typically found in an immersion heater. Again, it consists of two metal alloys with different expansion rates which are secured at one end (see figure). As the brass tube heats up it expands, pulling with it the Invar rod, onto which is attached the moving contact. On cooling, the brass contracts to remake the connection. Note that the brass tube does not actually touch the water but sits in a pocket in the heater element.

Liquid expansion: Most commonly found in boilers in which the sensor sits in a pocket at the top of the heat exchanger. A volatile fluid, such as mercury or alcohol, is enclosed in a sensor (phial) and capillary tube which is connected to a bellows. As the fluid heats up it expands into the bellows chamber which fills and becomes enlarged; this in turn acts upon the push rod which causes the electrical contact to disconnect.

Ignition systems

To light the fuel of gas- and oil-burning appliances, some form of ignition (often automatic) is required. The ignition of a flame, often a pilot flame in the case of gas appliances, can be achieved by either filament or spark ignition.

Filament ignitors: A small coil of thin resistance wire, usually platinum. When power is supplied to the coil, (at around 3 V) the wire heats up and glows red hot. The power may be supplied by a battery or a step-down mains transformer.

Spark ignitors: Several designs of spark ignitor can be found, including the piezo-electric device and the step-up transformer:

With the *piezoelectric device* two crystals, e.g. quartz, each about 6 mm in diameter and 12 mm long, are positioned in a metal holder and separated by a metal pressure pad. This pad is connected to the spark electrode and the holder suitably earthed. When pressure is applied to the crystals, usually by the operation of an impact spring device, an emf of about 6000 V is produced; this causes a spark to jump between the electrode tip and the metalwork of the appliance.

The *step-up transformer* is used to produce an emf of around 5000–15 000 V, which causes a series of electrical impulses to discharge a spark across the gap between two electrodes, or one electrode passing to earth. Systems of this nature are used to ignite non-permanent pilot lights and pressure jet oil burners.

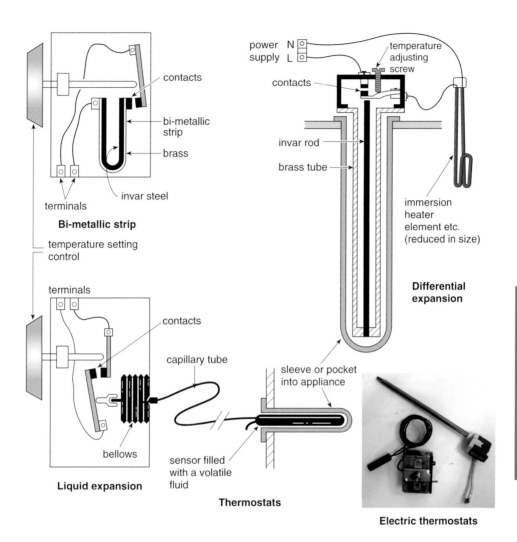

Bi-metallic strip

temperature setting control

Liquid expansion

Differential expansion

Thermostats

Electric thermostats

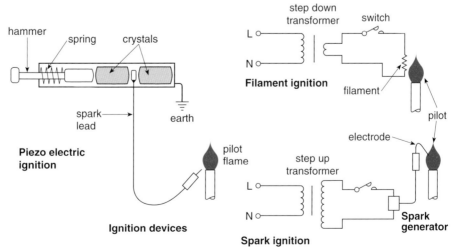

Piezo electric ignition

Ignition devices

Filament ignition

Spark ignition

Spark generator

Electrical components 1

Electrical Components 2

Relevant British Standard
BS 7671

Transformers Devices which step up or down the voltage from an a.c. supply. When a coil of wire is wound onto a soft iron core and an electric current passed through the wire, a magnetic field is generated. If the supply is a.c., the current will flow back and forth, and cause the lines of magnetic flux to change direction also. When a magnetic field is passed over a coil of wire, the current is induced in the wire conductor (see page 270 on electrical current referring to a.c.). Therefore, if a second wire is also wrapped onto the soft iron core the electric current will be induced from one coil (the primary coil) to the other (the secondary coil). The number of turns on the secondary coil determines whether the voltage will be stepped up or down.

Example: Where the secondary coil has half the number of turns as the primary coil, the voltage will be halved; conversely, where it has twice the number of turns, the voltage will be doubled. Where a 110 V transformer is used on site, a centre tapping has been made into the secondary coil and run to earth. Under normal operations 110 V will be supplied; however, in the event of a fault, only 55 V will travel down the conductor to the appliance and its user, and 55 V will run down to earth.

Solenoid valve (magnetic valve) A device used to hold a valve open, typically found in washing machines and gas valves. The solenoid consists of a coil of wire wrapped round a soft iron core. When a current of electricity is passed through the wire it creates a magnetic field which attracts a soft iron armature, drawing it up into the coil. Attached to the armature is a valve; thus, when energised, the valve will open. When the current flow is switched off, the valve, assisted by a spring, closes.

Relay The relay is a device which uses the current from one circuit (e.g. low voltage) to switch on the current to another circuit (e.g. mains voltage). The relay consists of a solenoid which, when energised, draws in the armature to cause two contracts to be made on a switch. Relays will be found in many situations such as when switching the mains voltage to a pump from a low voltage thermostat circuit.

Rectifier (a.c./d.c. converter) A device which converts a.c. to d.c. It is like a non-return valve allowing current to flow only in one direction. When a single diode rectifier is fitted in a conductor, the current will flow intermittently (i.e. stop and start) as the a.c. voltage flows back and forth between phase and neutral. However, when a bridge rectifier containing four single rectifiers is used, the current is allowed to flow continuously due to its design. Rectifiers are found in many appliances, such as transistor radios which work from a d.c. supply.

Capacitor (condenser) A device which stores a charge of electrical energy. The capacitor consists of two sheets of metallic foil separated by a thin sheet of insulation material, usually wrapped in a roll for convenience. When an a.c. voltage is fed to the capacitor, the electrons, when flowing in one direction, enrich one metal foil. When the current is flowing in the opposite direction, the second foil becomes enriched with electrons, whilst at the same time, the first foil discharges its previous negative charge. The process continues as long as the current flows. This, in effect, creates an out-of-phase differential which is like having a two-phase supply.

A capacitor is typically fitted to small motors, such as in c.h. pumps to give the initial impetus to get the motor going. It must be borne in mind that when the current stops flowing, the capacitor still has a charge of electricity. Therefore, when removing a capacitor, the terminal connections should be earthed to remove the charge.

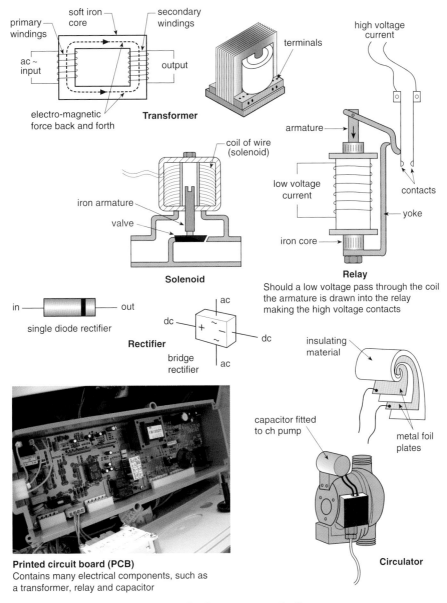

Electrical components 2

Central Heating Wiring Systems

When a c.h. system has been completed it will need to be wired up in accordance with the manufacturer's instructions. It is not necessary to know the installation function of the various elements or how the circuit operates to make the required electrical connections. Manufacturers go to great lengths to design their wiring systems in such a way as to ensure that they can be installed by anyone competent to do so. Basically all one has to do is follow a series of steps laid down by the manufacturer, which identify where to run the cables. Usually, there is a wiring diagram which shows the location of all cable connections. Note that in one diagram below, the Honeywell controls shown opposite are also used.

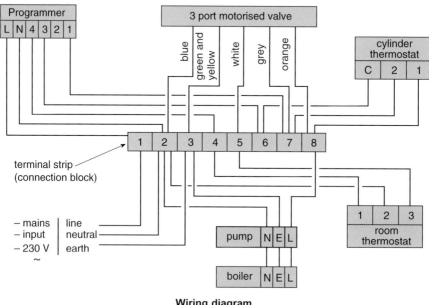

Wiring diagram

The boiler and pump differ from one make to another. I have chosen a basic boiler. The cables should be run, complying with Wiring Regulations, to the terminal strip and completed as shown in the wiring figure opposite. It is not possible to identify all the various wiring plans available; however, a manufacturer of these controls, such as Honeywell or Landis & Gyr, would be happy to provide a simple-to-follow guide.

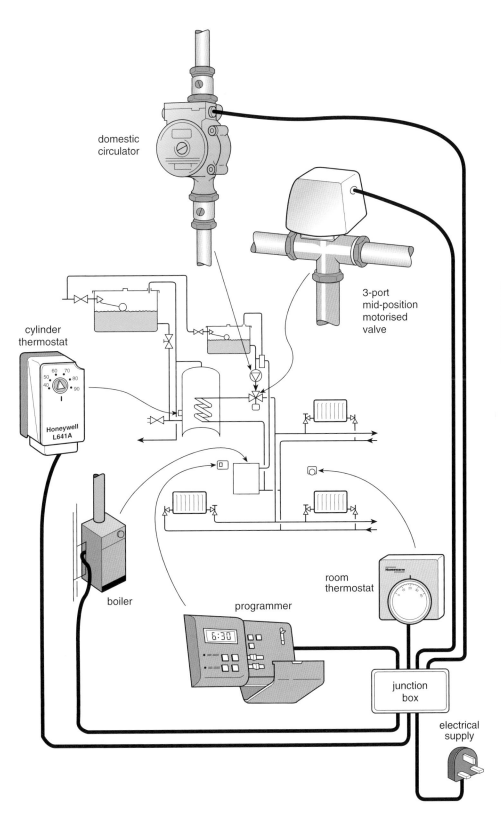

Central heating wiring systems

Inspection and Testing of Electrical Work

Relevant British Standard
BS 7671

Before carrying out any work on electrical circuits it is essential that the electrical supply is isolated. The main MCB or isolating switch should be capable of being locked open, breaking the circuit, or any fuses should be removed.

A visual inspection should be made of the system to ensure that all connections are properly secured into the terminals and that all components are securely fixed. The fuse should be checked to ensure it is not overrated (see page 274). The plumber is generally only interested in testing earth continuity and polarity of wires at an appliance; however, to fully ensure the system is safe an insulation resistance test should be undertaken. The testing identified here relates to the final connection at the appliance and not the circuit feeding it.

Testing for earth continuity

This test is carried out to ensure that any exposed metalwork (e.g. appliances such as boilers, exposed pipework, sink tops, etc.) is effectively connected to the main earth terminal block. To carry out this test, either an ohmmeter or multi-meter set to a low resistance (Ω) setting is used.

The procedure to follow is to position one probe onto the exposed metalwork and the other probe onto a known earth (e.g. the metal screw exposed on the surface of a 13A 3-pin socket-outlet). The reading on the dial should be *less* than one, which indicates a very low resistance to current flow and ensures a good earth. It may be necessary to use a cable to extend the length of the test leads; where this is the case, the resistance of the test lead should be taken, then subtracted from the final reading.

It should be noted that the method described here confirms a good earth continuity but does not say how good the earth connection is, i.e. it may only be one thin strand of copper. This connection should therefore be visually inspected.

Testing for correct polarity

This test is done to ensure the line conductor is connected to its correct location at the appliance and not crossed somewhere, which may result in the appliance casing becoming live. It is achieved by testing the continuity of the conductor from its source to the appliance, using a low resistance (Ω) setting on the meter. The test is similar to the earth continuity test. One probe of the multi-meter is positioned onto the line conductor of the fused spur box and one onto the connection of the appliance; the reading should be *less* than one. Keeping the probe in position at the spur, the second probe is now repositioned, onto the neutral and then the earth terminals, respectively; in both cases the readings should indicate one. Testing for polarity with the power off is certainly safe; however, it does not confirm that the spur box, or source, is itself wired up correctly. That may subsequently also need to be confirmed.

Testing for insulation resistance

This test is undertaken to ensure that the cable used is in good condition and that there is no possibility of a stray current shorting to earth. The test can be undertaken using a multi-meter set to its highest resistance (Ω) reading; however, this would not subject the wire to a suitable voltage, the multi-meter only supplying a 9 V charge. Instead an insulation resistance tester supplying 500 V should be selected. When one probe is positioned on the earth conductor and the other on either the neutral or line conductor, each in turn, the resistance should read greater than 1.0 M Ω when 500 V is supplied.

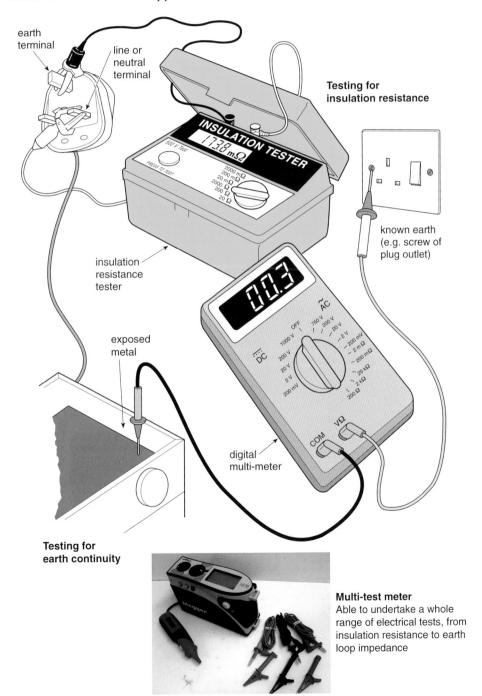

earth terminal

line or neutral terminal

Testing for insulation resistance

INSULATION TESTER

1738 mΩ

500 V Test

PRESS TO TEST

2000 mΩ
200 mΩ
20 mΩ
2000 Ω
200 Ω
20 Ω

known earth (e.g. screw of plug outlet)

insulation resistance tester

00.3

AC

OFF

750 V

1000 V

200 V

200 V

20 V

2 V

200 mV

2 mΩ

200 mΩ

20 kΩ

2 kΩ

200 Ω

exposed metal

DC

200 V

20 V

2 V

200 mV

digital multi-meter

COM

VΩ

Testing for earth continuity

Multi-test meter
Able to undertake a whole range of electrical tests, from insulation resistance to earth loop impedance

Inspection and testing of electrical work

6 Electrical Work

Part 7

Sanitation

Plumbing, 4th Edition. R. D. Treloar.
© 2012 Blackwell Publishing Ltd. Published 2012 by Blackwell Publishing Ltd.

Sanitary Accommodation

Relevant British Standard
BS 6465

Sanitary accommodation is the term used for a room containing a water closet (WC) or urinal; it may or may not contain other sanitary fittings. The room is used solely for ablutionary (washing) and/or excretory purposes. The sanitary accommodation should not be entered directly from either:

- a room used for the preparation of food
- an office or working area
- a habitable room, except a bedroom, where additional sanitary accommodation is provided elsewhere in the dwelling.

Part G of the Building Regulations identifies the requirements for sanitary accommodation. In general, every new dwelling should be provided with at least one WC, one bath or shower, one washbasin and one sink. The appliances should be suitably located, usually within the same bathroom with the WC in the same or adjoining room, thus allowing the WC compartment direct access to the washbasin. Where the dwelling is for more than five people it is recommended that a second WC compartment be installed. All sanitary appliances must be installed to a suitable drainage discharge system and connected to the drain via a trap. The requirements for buildings other than simple domestic dwellings are clearly identified in BS 6465 which should be consulted for further study.

Ventilation requirements These are listed in the Building Regulations under Part F in which all sanitary accommodation should have airflow rates as shown in the following table:

Extract ventilation rates (litres/second)

Room	Minimum intermittent rate	Minimum
Bathroom	15	8
Sanitary accommodation	6	6

Spatial requirements The internal layout and arrangement of the sanitary fittings and facilities should be determined by:

- the dimensions of the human body, allowing for an activity space, in which there is room to move
- the number of persons using the facility
- the size of each component, the associated services and their location in relation to the component. Also, provision should be made to gain access, especially to drainage, for servicing and maintenance purposes.

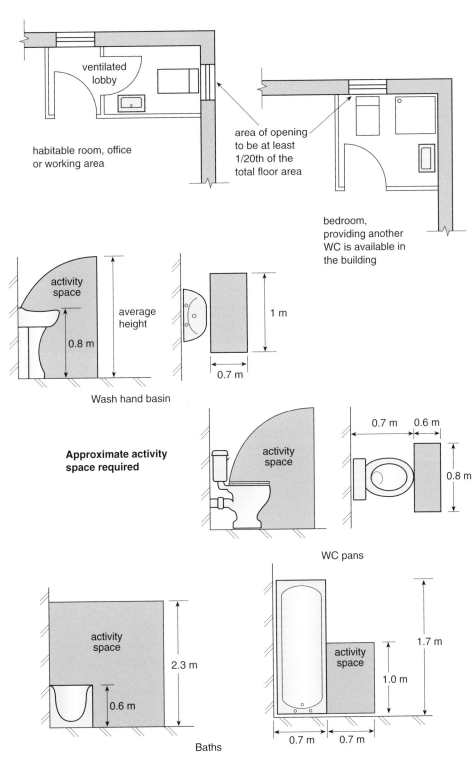

ventilated lobby

habitable room, office or working area

area of opening to be at least 1/20th of the total floor area

bedroom, providing another WC is available in the building

activity space

average height

0.8 m

1 m

0.7 m

Wash hand basin

Approximate activity space required

activity space

0.7 m   0.6 m

0.8 m

WC pans

activity space

2.3 m

0.6 m

Baths

activity space

1.7 m

1.0 m

0.7 m   0.7 m

Sanitary accommodation

7 Sanitation

Sanitary Appliances 1

Relevant British Standard
BS 6465

Sanitary appliances can be broadly divided into two groups: those which are used for washing purposes, i.e. waste or ablutionary appliances; and those used for the removal of human excreta, i.e. soil appliances.

Waste appliances

Sinks Several designs of sink may be found. Firstly, there are those fixed at a height not exceeding 900 mm with adjoining storage cupboards and used for washing items. These include stainless steel and enamelled pressed steel crockery sinks, and glazed fireclay butler sinks.

The most common design of butler sink is the 'Belfast' design, having an integral overflow; this kind of sink is mostly found in utility rooms. A second type of sink is the bucket or cleaner's sink which has a protective strip fitted to its front edge and is provided with a hinged metal grating on which to rest a bucket. These sinks are fitted at low level to facilitate the emptying of buckets.

Washbasins These are bowl-shaped fittings used for the purpose of washing the upper parts of the body; they are usually made from vitreous china. The fixing height of a basin is usually determined by a pedestal or vanity unit, although where brackets are used it should be fixed at a height of 760–800 mm unless specified differently by the client.

Bidets These are appliances designed for the purpose of washing the lower parts of the body. Bidets are made from the same materials as washbasins and are usually floor standing. Two designs of bidet will be found: those with an over-rim water supply and those with a submersible ascending spray (see page 118, Connections to Hot and Cold Pipework).

Baths Most baths installed today are made from acrylic materials, for which adequate support is essential. The bath is intended to allow complete bodily immersion in the reclining position, although it is possible to have a bath with a stepped bottom (sitz bath) where floor space is restricted or for the disabled person. A bath is generally fitted as low as possible, its height usually being determined by the height of the accompanying side panel.

Showers These are sometimes incorporated as part of a bath; however, it is possible to purchase shower trays to give maximum usage of floor space, the tray forming part of a shower cubical.

In general, all appliances should be securely fixed using the brackets supplied. The support of service and waste connections should not be relied upon. The appliance should be levelled in along its top edge; the fall is built in with its design. The sizes of the waste pipe connections are listed on page 304.

7 Sanitation

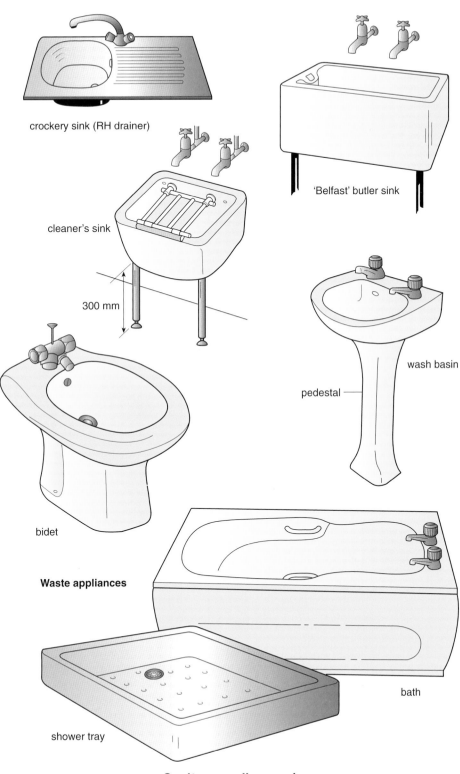

crockery sink (RH drainer)

'Belfast' butler sink

cleaner's sink

300 mm

wash basin

pedestal

bidet

Waste appliances

bath

shower tray

Sanitary appliances 1

Sanitary Appliances 2

Relevant British Standard
BS 6465

Soil appliances

WC pan An appliance used in conjunction with a flushing cistern to remove the contents of solid and liquid excreta from the building by the flush of water. The pan may be floor standing or mounted on the wall, leaving the floor area clear. Two designs of WC will be encountered: the *washdown* and the *siphonic* pan.

The washdown pan relies solely on the momentum of the flushing water entering the pan from above to remove the contents of the bowl whereas the contents of the siphonic WC pan are removed by siphonic action in addition to the momentum of the falling water. Two designs of siphonic pan are found: the single trap and the double trap; in each case, the principle is to create a negative pressure below the trap seal. This is achieved by restricting the outflow in the case of the single trap pan, or in the case of the double trap pan, drawing the air from the void between the two traps. With the double trap pan, the suction of air is achieved when the water from the flush passes over a pressure-reducing fitting. Due to the reduced flushing volume, siphonic WC pans are now rarely sold or found in the UK.

The slop hopper or slop sink An appliance, similar to a WC pan, with a flushing rim, supplied with water from a flushing cistern; it is used to discharge the contents of bed pans and urinal bottles. The difference between slop hoppers and slop sinks is simply that the hopper sits on the floor whereas the sink is installed at a height of about 1 m. In addition to the flushing cistern, these appliances are installed with hot and cold supplies fed to bibcocks located above the appliance.

Urinals Appliances found only in male toilets for the removal of urine. Four designs of urinal will be found: the bowl, trough, slab and stall. Bowl and trough urinals are installed at a height of 610 mm to the front lip, 510 mm for junior boys. The stall and slab are floor-standing appliances, with a channel positioned at ground level not exceeding 2.4 m to the discharge outlet.

Urinals are flushed periodically with water from an automatic flushing cistern which cleans the whole surface of the appliance. Bowl and stall urinals are fitted with a spreader at each location (see figure), whereas for trough and slab urinals a sparge pipe may be used consisting of a perforated pipe the same length as the appliance. Thus, a discharge of water along the whole length of the slab/trough back is facilitated. Alternatively, spreaders may be fixed at 600 mm centres. To provide privacy to the user, division may be included with the installation allowing 600 mm per unit.

It is now possible to use 'waterless' urinals that do not require flushing, saving a vast amount of water. There are two basic designs: one using a fluid treatment and the other a disposable cartridge. The principle is that the waste passes through a deodorising agent-impregnated pad on its way to the waste outlet. The cartridge will last for about 7000 uses. Discharge systems employing this design will require adequate ventilation.

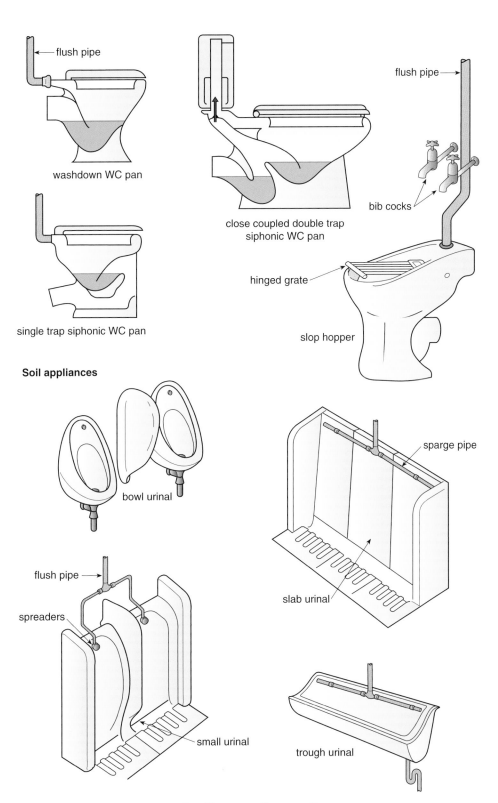

flush pipe

washdown WC pan

single trap siphonic WC pan

Soil appliances

close coupled double trap
siphonic WC pan

flush pipe

bib cocks

hinged grate

slop hopper

bowl urinal

sparge pipe

flush pipe

spreaders

small urinal

slab urinal

trough urinal

Sanitary appliances 2

Flushing Cisterns 1

Relevant British Standards
BS 1125 and BS 1876

In recent years, many changes have taken place regarding the volume of water contained in a WC flushing cistern. Prior to 1993, a 9 litre flush was employed; this was subsequently reduced to 7.5 litres and current Water Supply Regulations have further reduced this to a maximum 6 litres. Dual flushing cisterns have been reintroduced, allowing the user to select a full 6 litre flush or a reduced 4 litre flush. In order to meet these stringent new requirements, a valve style flushing mechanism is allowed.

Plunger type flushing cisterns Water enters the cistern at a high level; a float-operated valve (ballvalve) controls the water level in the cistern so that it contains the required volume. When a flush of water is required, the flushing arm is operated by pulling a chain, on high level cisterns, or by turning the handle on low level cisterns, causing the plunger in the cistern to lift. This forces a flow of water over the top of the U-shaped siphon bend; as the water falls down the flush pipe it carries with it the air from within the pipe, lowering its pressure.

The atmospheric pressure acting on the water in the cistern pushes down and forces all the water from the cistern by siphonic action, up and over the bend and down the flush pipe. When the water level drops to reach the base of the plunger, air can get in to equalise the pressure inside the tube and the siphonic action ceases. The cistern may be bolted directly to the WC pan; it may be part of a close-coupled suite or fitted with a flush pipe in which the cistern is fitted at either high or low level. An overflow will need to be run from the cistern to terminate in a visible location, usually outside the building.

Valve type flushing cistern Despite being commonplace in Europe and North America, the flushing valve type cistern has only been permitted in the UK since 2001. To flush the cistern the lever arm is operated and simply lifts a valve from the seating, allowing water to cascade directly into the flush pipe. The valve can be operated either by a manual or pneumatic push button control. There is also a touch-free system which is particularly hygienic and is becoming increasingly popular in commercial buildings. The system illustrated opposite uses one such system and operates as follows: when a hand is waved in front of the sensor, located in front of the cistern, it activates a three port diverter valve to divert the water flow, under supply pressure, to the underside of a piston. This causes the piston to lift which brings with it the valve from the base of the cistern, allowing water to flow into the flush pipe. As the valve lifts it also causes an arm to lift that pushes onto the float-operated valve arm. This prevents water entering the cistern and ensures the correct 6 litre flush is maintained. After an electronically timed flush period has elapsed the diverter valve turns to re-divert to water flow into the float-operated valve. The flushing valve simply drops to close, assisted by positive spring pressure.

7 Sanitation

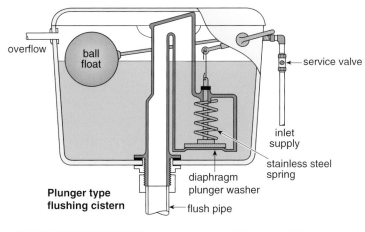

overflow

ball float

service valve

inlet supply

stainless steel spring

diaphragm plunger washer

Plunger type flushing cistern

flush pipe

Flushing siphon

Flushing valve

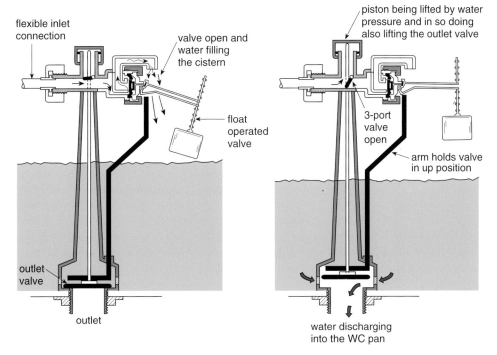

flexible inlet connection

valve open and water filling the cistern

float operated valve

outlet valve

outlet

piston being lifted by water pressure and in so doing also lifting the outlet valve

3-port valve open

arm holds valve in up position

water discharging into the WC pan

Operation of valve-type flushing cistern

Flushing Cisterns 2

Dual flush An arrangement in which a flushing cistern has two different volumes of water for flushing purposes. The user has a choice of flush, which is made depending upon the contents of the WC pan. When the flushing mechanism is operated in a piston type cistern, flushing commences and continues until air gets in to break the siphonic action. Should a reduced flush be selected, air is admitted half way down the siphon; however, where a full flush is required a seal is made that blocks the hole and therefore air cannot get in to stop the siphonic action until all the contents of the cistern have been emptied. In the figure shown this is achieved by the user holding the handle down until the full flush has been delivered. Other arrangements, as used with a valve type cistern, make use of timed control.

Automatic flushing cistern A flushing cistern designed to discharge its contents of water at regular intervals into a urinal. The rate at which the water will flush depends upon the rate at which the water is fed into the cistern and, for a single installation, this should not exceed 10 litres per hour.

Where several urinals are installed, the filling rate should not exceed 7.5 litres/h per bowl, stall or 700 mm-width slab. To prevent the wastage of water from these cisterns, during times when the cistern is not used (such as at weekends), an automatic flow cut-off device should be fitted.

The automatic cistern illustrated operates as follows:

(1) As the water rises on filling, the air is trapped inside the dome; as the air is compressed, enough pressure is created to force the water out of the U tube, reducing the air pressure inside the dome.
(2) The reduced air pressure immediately allows siphonic action to start, thus flushing the appliance.
(3) When the flush is finished, the water in the upper well is siphoned out through the siphon tube and refills the lower well and U tube.

Automatic flow cut-off to WC facility To prevent wasting water in automatic flushing cisterns during times when the facility is not in use, such as evenings and weekends, some form of automatic water flow cut-off device should be fitted. Several methods can be employed to detect when the toilet facility is not in use, including: infrared sensors to detect movement, pipeline sensors to detect the presence of urine within the waste pipe and a hydraulically-operated valve that senses no water is being drawn off from the basin taps; this last type is shown opposite.

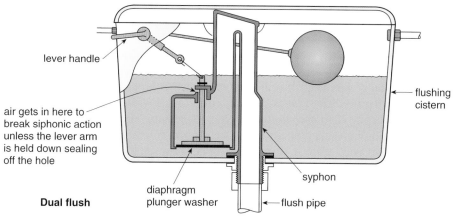

lever handle

air gets in here to
break siphonic action
unless the lever arm
is held down sealing
off the hole

flushing
cistern

syphon

diaphragm
plunger washer

flush pipe

Dual flush

Dual flush siphon
Photo shows hole through which air
can enter to break the siphonic action
where a reduced flush is required

Automatic flushing siphon

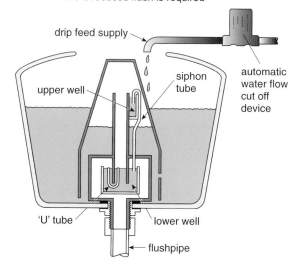

drip feed supply

upper well

siphon
tube

automatic
water flow
cut off
device

'U' tube

lower well

flushpipe

Automatic flushing cistern

**Automatic water flow
cut-off control**

Flushing cisterns

Waste Pipe Connections

Relevant British Standard
BS EN 12056

When an appliance has been installed it will need to be connected to the above-ground sanitary pipework. This is done by installing a waste fitting into the appliance and connecting it to the discharge pipe via a trap. The minimum size of the waste fitting and discharge pipe and trap are: 32 mm for washbasins, bidets, drinking fountains and bowl urinals; 40 mm for sinks, showers, baths, washing machines and trough urinals; and 65 mm for stall and slab urinals.

Appliances such as sinks and baths use a combination overflow and waste fitting, which does away with the need for a separate overflow pipe discharging from the building; note that the overflow connects into the waste above the trap seal.

Appliances such as the Belfast butler sink, washbasins and bidets have an integral (built in) overflow and therefore the waste fitting needs to have a slot to allow its connection. The urinal bowl, on the other hand, has no overflow at all; thus an unslotted waste fitting must be chosen. Most waste fittings connected to the appliance are best fitted using a rubber-type material to give a good seal; however, materials such as Plumbers' Mait (a non-setting putty) or silicone rubber can be used, providing the appliance is completely dry when it is applied.

Traps

A trap is a fitting or integral part of an appliance designed to retain a body of water, thus preventing the passage of foul air. There are many different designs, including those with a vertical outlet (S traps), those with a near-horizontal outlet (P traps), and those fitted in a pipe run, called running traps.

The depth of the trap seal would depend upon the circumstances and usage of the pipe but, in general, pipes of less than 50 mm internal bore should have a trap with a seal of not less than 75 mm; however, for baths and shower trays and for basins with spray taps, provided that a flush-grated waste is fitted (i.e. no plug), this depth may be reduced to 50 mm. If the appliance is discharging into an open gully, this depth may be reduced to 38 mm. For pipes with an internal bore larger than 50 mm, a trap with a seal of only 50 mm is required as trap seal loss is much less likely to occur in so large a pipe.

Traps up to 40 mm outlet size may be either tubular or bottle trap design (see figure). Tubular traps tend to be less prone to blockage. (See page 312 for an example of a special trap which is designed to maintain its water seal under adverse design conditions.)

Waste pipe valve

One manufacturer has designed a fitting that provides an adequate seal from the drain by way of a rubber type seal. Should water discharge from the water fitment, the seal opens to allow the water to flow. Conversely, when no water is flowing the rubber seals the outlet shut simply by the rubber sucking together. These valves can be fitted into horizontal or vertical pipework.

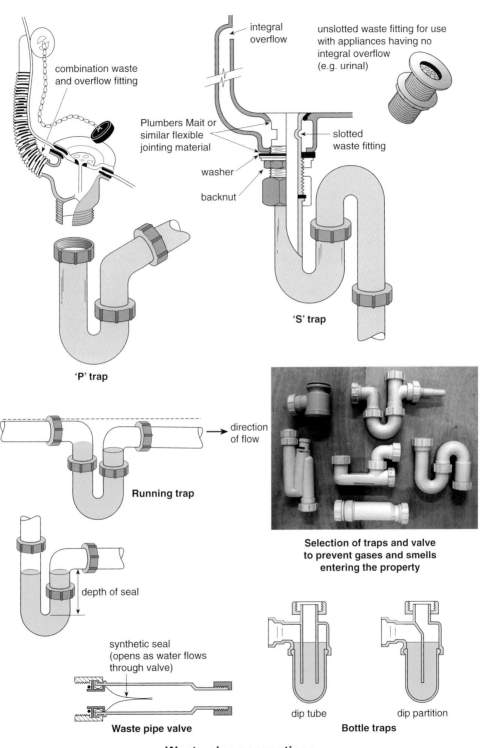

combination waste and overflow fitting

integral overflow

unslotted waste fitting for use with appliances having no integral overflow (e.g. urinal)

Plumbers Mait or similar flexible jointing material

washer

backnut

slotted waste fitting

'S' trap

'P' trap

Running trap

direction of flow

depth of seal

Selection of traps and valve to prevent gases and smells entering the property

synthetic seal (opens as water flows through valve)

Waste pipe valve

dip tube

dip partition

Bottle traps

Waste pipe connections

Sanitary Pipework

Relevant British Standard
BS EN 12056

A system of above-ground drainage designed to remove all the foul and waste water to the below-ground drainage system. Originally, the foul water from soil appliances was kept separate from the water from waste appliances and two separate discharge stacks were required; the water only joined at ground level in the below-ground drainage system. This type of system was known as the two-pipe system.

All sanitary pipework today is based on the one-pipe system in which one discharge stack is used to convey both foul and waste waters. There are three basic systems in use, these being the *ventilated discharge branch system*, the *secondary ventilated stack system* and the *primary ventilated stack system*.

With the **ventilated discharge branch system** a ventilating pipe is extended to connect to each of the individual branch pipes throughout the system; it is designed to safeguard against trap seal loss (see page 312). This system is generally adopted in situations where it is not possible to have close groupings of sanitary appliances, and long branch discharge pipes can be expected.

In the **secondary ventilated stack system** only the main discharge stack is ventilated, to overcome pressure fluctuations. With this system the branch discharge pipes connect directly into the main stack without the need for a branch ventilating pipe. This system, therefore, is only suitable for buildings in which the sanitary appliances are closely grouped to the main stack.

Finally there is the **primary ventilated stack system**, which is used in similar situations to the secondary ventilated stack system, the difference being that the stack ventilating pipe can be omitted if the discharge stack is large enough to limit pressure fluctuations. This system is often referred to as the single-stack system.

Whichever system is chosen, all work must comply with Part H of the Building Regulations.

Branch discharge pipe   This pipe should connect to the main discharge stack in such a way as not to cause any 'crossflow' into other pipes (see figure). Branch discharge pipes should be at least the same diameter as the appliance trap. Oversizing the pipe to avoid self-siphonage could prove uneconomic and lead to an increased rate of solid deposit accumulation. Bends should be avoided but where they are unavoidable long radius bends should be used. The gradient of branch discharge pipes should be between 1° and 1¼° (18–22 mm drop per metre run).

Branch ventilating pipe   No branch ventilating pipe should connect to the discharge stack below the spillover level of the highest fitting served. The minimum size for a branch ventilating pipe serving a single appliance should be 25 mm; however, where the branch run is longer than 15 m, or contains more than five bends, or serves more than one appliance, the minimum pipe size should be 32 mm.

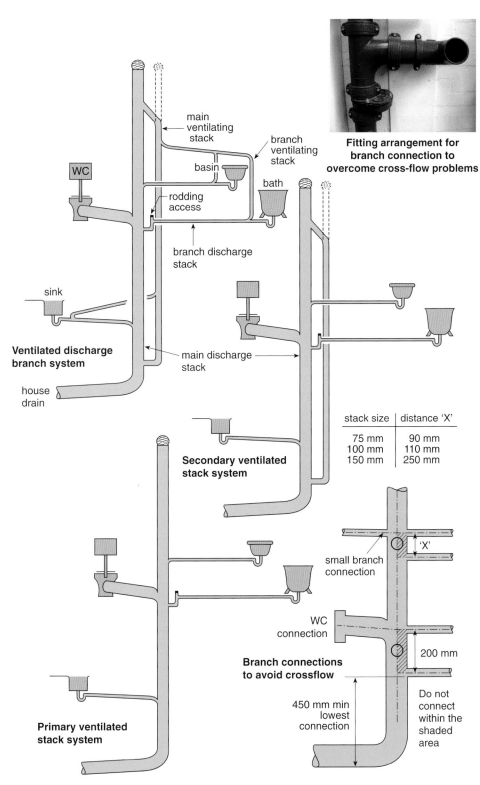

main
ventilating
stack

branch
ventilating
stack

basin

WC

bath

rodding
access

branch discharge
stack

sink

**Ventilated discharge
branch system**

main discharge
stack

house
drain

**Fitting arrangement for
branch connection to
overcome cross-flow problems**

**Secondary ventilated
stack system**

stack size	distance 'X'
75 mm	90 mm
100 mm	110 mm
150 mm	250 mm

small branch
connection

'X'

WC
connection

200 mm

**Branch connections
to avoid crossflow**

450 mm min
lowest
connection

Do not
connect
within the
shaded
area

**Primary ventilated
stack system**

Sanitary pipework

Primary Ventilated Stack System

Relevant British Standard
BS EN 12056

Most buildings are designed to meet the criteria of this system, therefore reducing the cost of installation. This system was referred to as a 'single stack' system prior to the current European standard. It is designed so that no separate ventilating pipes are required to prevent trap seal loss. This can only be achieved by observing the following guidelines:

(1) All sanitary appliances must be closely grouped to the discharge stack, within the limits shown.
(2) All appliances, as far as possible, should be fitted with a P trap or waste pipe valve with a discharge pipe diameter equal to that of the trap, and bends in branch pipes should be avoided – the gradient being kept to a minimum.
(3) The vertical discharge stack must be as straight as possible, with a long radius bend fitted at its base.
(4) The lowest connection to the discharge stack must be a minimum of 450 mm above the invert of the drain. If the building is over three storeys, this distance should be increased to 750 mm; for buildings over five storeys, no ground-floor appliances should connect into the stack. In buildings over 20 storeys, no first floor, or ground floor appliances should be connected into the stack.
(5) Where a range of appliances are installed, they should comply with the following table:

Unvented discharge pipes serving more than one appliance

Appliance	Max no. fitted	Min pipe size (mm)	Gradient (mm/m)
WC	8	100	9–90
Washbasin	4	50	18–90
Bowl urinal	5	50	18–90
Stall urinal	7	65	18–90

(6) The main discharge stack must be large enough to limit pressure fluctuations without the need for a ventilating stack; as a general guide, 100 mm diameter is required for buildings with up to five storeys and 150 mm diameter for those with up to 20 storeys, with two groups of appliances on each floor. (**Note:** a *group* of appliances consists of one WC, bath, basin and sink.)
(7) Branch connections must join the main vertical discharge stack at an angle of 45° *or* at a radius of 25 mm for pipes up to 75 mm in diameter and a radius of 50 mm for pipes over 75 mm in diameter.

Where the discharge pipe exceeds the criteria listed, it should be vented by a branch ventilating pipe located at the highest point, extended to the atmosphere or connected to a ventilating stack, in which case the system is generally referred to as a ventilated discharge branch system. Sometimes air admittance valves, or resealing traps, are used which allow air into the system in order to prevent trap seal loss.

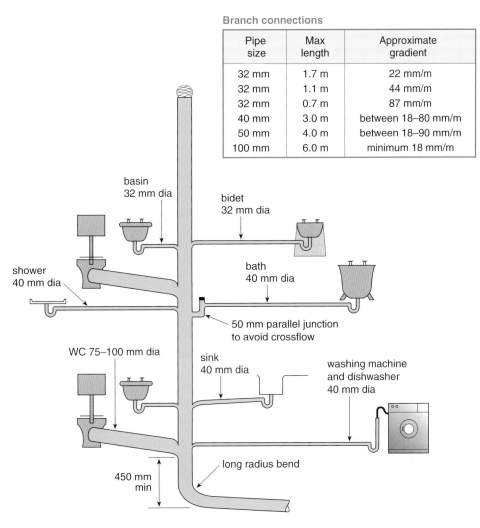

Branch connections

Pipe size	Max length	Approximate gradient
32 mm	1.7 m	22 mm/m
32 mm	1.1 m	44 mm/m
32 mm	0.7 m	87 mm/m
40 mm	3.0 m	between 18–80 mm/m
50 mm	4.0 m	between 18–90 mm/m
100 mm	6.0 m	minimum 18 mm/m

basin 32 mm dia

bidet 32 mm dia

shower 40 mm dia

bath 40 mm dia

50 mm parallel junction to avoid crossflow

WC 75–100 mm dia

sink 40 mm dia

washing machine and dishwasher 40 mm dia

450 mm min

long radius bend

Typical primary ventilated stack system

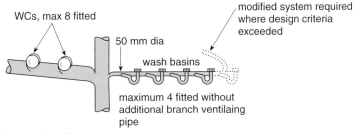

WCs, max 8 fitted

modified system required where design criteria exceeded

50 mm dia

wash basins

maximum 4 fitted without additional branch ventilaing pipe

Range of appliances
(only to be used where discharge stack
is large enough to limit pressure fluctuations)

Primary ventilated stack system

7 Sanitation

Ventilation of Sanitary Pipework

Relevant British Standard
BS EN 12056

It is essential that air is freely allowed to enter discharge and drainage pipes and thus help maintain an equilibrium of pressure within the pipe and the outside atmosphere; for example, should the pressure be greater outside the pipe than inside, the trap seals of sanitary appliances and gullies would be lost. By allowing a current of fresh air to flow through the whole system, any foul matter adhering to the insides of pipes would soon dry and be washed away; ventilating the pipes also prevents any build-up of foul (and possibly dangerous) gases. Air enters the drain at low level, via holes in inspection covers, etc., and rises up through the stack by convection currents.

The termination of a ventilating pipe into the atmosphere should be at a position that does not cause a nuisance or health hazard. It is recommended that if a ventilating pipe is within 3 m of a window opening it should be carried up above the window to a minimum height of 0.9 m. Ventilating pipes should be fitted with a domical cage or grating which does not unduly restrict the free air flow and prevents the nesting of birds, etc. The size of the ventilating pipe may be reduced in houses up to two storeys, but should be at least 75 mm in diameter.

Discharge stacks may terminate within the building when fitted with an air admittance valve. These valves are designed to allow fresh air to enter the ventilating pipe but prevent odours and gases escaping. Air admittance valves should not be used on discharge stacks connecting to drains which are subject to surcharging (filling with water) or on a drain which has an intercepting trap fitted, as they may cause the trap seal loss of the appliance.

The number of air admittance valves fitted to a drainage system should be limited to prevent excessive back pressures and, where five or more domestic dwellings are located in the same drainage run, a conventional open ventilating pipe will be required at the head of the system. Should the number of dwellings exceed ten, conventional venting will be required at mid-point and at the head of the system. Note that the valve must be located above the flood level of the discharge stack; where valves are fitted in roof voids, etc., they should be insulated, because any condensation forming within may freeze and prevent their use.

Stub stack

A discharge stack which is capped off with a rodding eye at its top end. Stub stacks are only permitted to be installed where they connect to a ventilated discharge stack or drain within 6 m from the base of the stack. However, this distance is increased to 12 m maximum where a group of appliances are fitted. If a stub stack is used, no branch waste connections may be made into the stack higher than 2 m above the invert level of the drain, and, in the case of a WC pan connection, this distance is reduced to 1.5 m maximum.

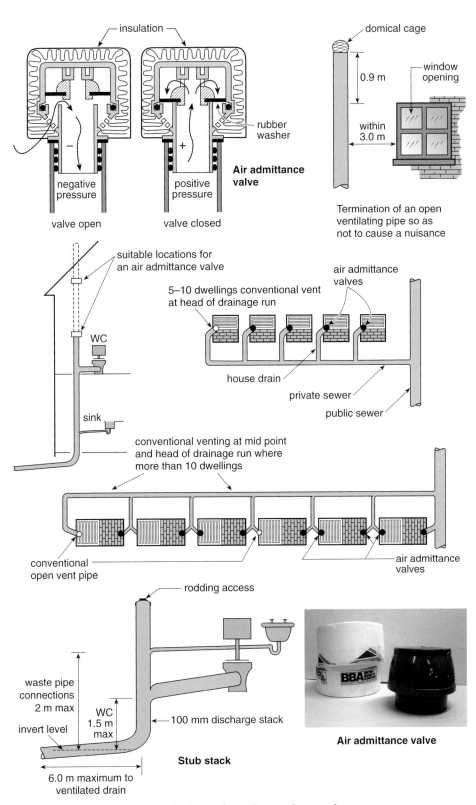

insulation

rubber washer

Air admittance valve

negative pressure

positive pressure

valve open

valve closed

domical cage

window opening

0.9 m

within 3.0 m

Termination of an open ventilating pipe so as not to cause a nuisance

suitable locations for an air admittance valve

WC

sink

air admittance valves

5–10 dwellings conventional vent at head of drainage run

house drain

private sewer

public sewer

conventional venting at mid point and head of drainage run where more than 10 dwellings

conventional open vent pipe

air admittance valves

rodding access

waste pipe connections 2 m max

invert level

WC 1.5 m max

100 mm discharge stack

Stub stack

6.0 m maximum to ventilated drain

BBA

Air admittance valve

Ventilation of sanitary pipework

Trap Seal Loss

Relevant British Standard
BS EN 12056

If the trap seal is lost, objectionable smells will enter the building; therefore, the water seal in the trap must be maintained under all circumstances. Trap seal loss can result from various unforeseen circumstances, such as leakages or evaporation. In designing any sanitary discharge system, special care will need to be taken to prevent pressure fluctuations occurring within the pipework itself. Typical design faults include:

- **Waving out**   Caused by the effects of the wind passing over the top of the ventilation pipe, bringing about pressure fluctuations; thus wave movements in the trap gradually wash over the outlet.
- **Compression**   This generally only occurs in high-rise buildings where the discharge of water down the main discharge stack compresses the air at the base of the stack, thus pushing the water out of the trap back into the appliance. This problem can usually be overcome by ensuring that a long radius bend is installed at the base of the stack and that no connections are made within 450 mm of the invert level of the drain; alternatively, a relief vent should be carried down to connect to the lowest part of the discharge stack.
- **Induced siphonage**   Caused by the discharge of water from another sanitary appliance connected to the same discharge pipe. As the water falls down the pipe and passes the branch pipe, it draws air from within, thus creating a partial vacuum; subsequently, siphonage of the trap takes place. To overcome this problem, trap ventilating pipes could be introduced into the system; these would permit air into the discharge pipe, preventing the development of a partial vacuum.
- **Self-siphonage**   Mostly caused in such appliances as washbasins; being funnel shaped, they tend to discharge their contents of water quickly. As the water discharges it sets up a plug of water, which, as it passes down the pipe creates a partial vacuum, thus causing siphonic action to take place. To overcome this problem of self-siphonage, a larger waste pipe is sometimes used, but in most cases a resealing trap cures the problem.

Resealing traps   A trap designed to maintain its water seal should a partial vacuum be created in the waste discharge pipe. There are various designs of resealing trap and the most common one used today incorporates an anti-vacuum valve; should the pressure drop inside the discharge pipe, this valve opens under atmospheric pressure, giving a state of equilibrium inside the pipe. Unfortunately these traps often tend to leak through this valve. The other types of resealing trap work on the principle of retaining the water in a reserve chamber should the conditions be right for siphonic action to take place.

Waste pipe valves   In addition to using a resealing trap to overcome trap seal loss, it may be possible to use a waste pipe valve, as identified on page 304.

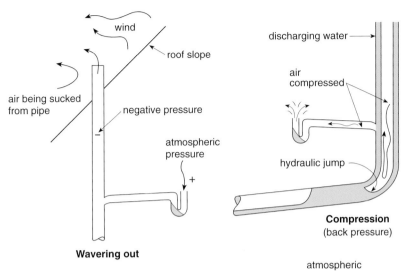

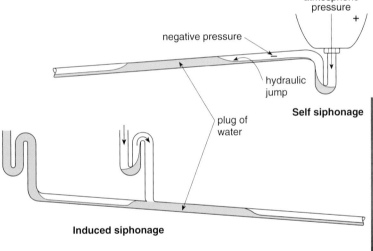

Wavering out

Compression
(back pressure)

Self siphonage

Induced siphonage

Resealing trap

valve lifts should there be a
negative pressure within the pipe

normal trap
seal

trap being
siphoned

remaining trap
seal

Resealing traps

Trap seal loss

Mechanical Disposal Units

Relevant British Standard
BS 3456

Food waste disposal unit

A mechanical device operated and fixed beneath sinks to macerate kitchen refuse into small fragments so that they can be flushed into the drainage system without causing blockages. A waste disposal unit cannot be fitted into a sink with the standard waste hole size as it would not be large enough to house the unit; therefore, to fit these devices a special sink is often required. When installing a waste disposal unit, a tubular trap must be fitted to its outlet because the units do not have their own integral trap fitted.

When using these units it must be remembered that water must be flowing down the waste pipe, otherwise the machine and pipe would soon become blocked up. Should the machine become jammed, a special tool is provided to turn the cutting head to loosen the obstruction; on many units there is a cut-off switch located on the machine which is provided to prevent the motor from becoming burnt out should it be jammed. Upon freeing the cutting head, the cut-off switch must then be pressed to reset the unit ready for use.

WC macerator pump

A special packaged unit consisting of a macerator and a pump which can be installed behind a WC pan to collect the discharge and macerate up any solid matter to allow it to be pumped vertically up to 4 m or horizontally up to 50 m and discharged into a small 19 mm nominal bore discharge pipe.

Macerator pumps are only permitted to be installed if there is also access to a WC discharging directly into a gravity system of drainage; the reason for this is that should there be an electrical failure the machine is put out of action. The holding tank to these machines needs to be ventilated to allow for its gravity filling and to facilitate emptying; therefore, to prevent unnecessary odours emptying into the room, it is recommended that the vent be extended and terminated externally to a safe position.

The electrical connections to any form of mechanical disposal unit must be via an unswitched fused spur outlet with the correct size fuse and a neon light indicator, and not simply connected to a conventional plug and socket.

Branch discharge pipes from waste disposal units and macerators require steeper gradients than is normal for waste appliances.

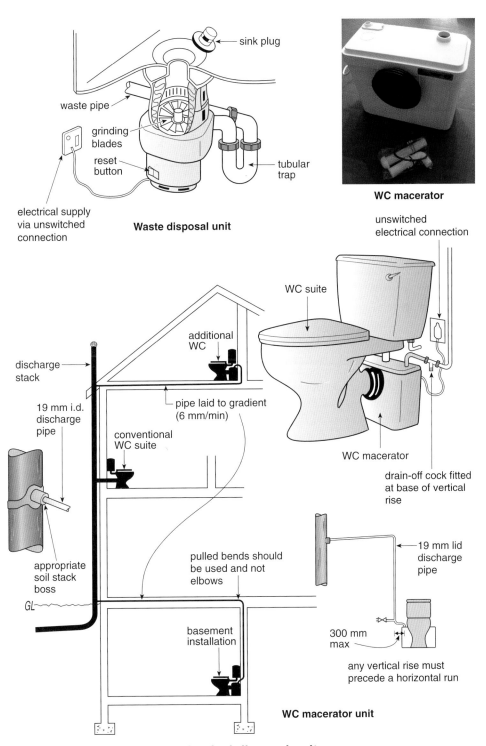

sink plug

waste pipe

grinding
blades

reset
button

tubular
trap

electrical supply
via unswitched
connection

Waste disposal unit

WC macerator

unswitched
electrical connection

WC suite

additional
WC

discharge
stack

19 mm i.d.
discharge
pipe

conventional
WC suite

pipe laid to gradient
(6 mm/min)

WC macerator

drain-off cock fitted
at base of vertical
rise

appropriate
soil stack
boss

GL

pulled bends should
be used and not
elbows

19 mm lid
discharge
pipe

basement
installation

300 mm
max

any vertical rise must
precede a horizontal run

WC macerator unit

Mechanical disposal units

7 Sanitation

Sizing of Sanitary Pipework

Relevant British Standard
BS EN 12056

The sizing of the branch discharge pipework has previously been discussed under the primary ventilated stack system (page 308); however, the main vertical discharge pipe size will be based upon the number of appliances connected and type of building. The sizing of discharge stacks for commonly used arrangements of pipework is often based on the following guidelines, allowing for one group of appliances on each level:

- 100 mm diameter in buildings up to 3 levels with no additional ventilation pipes
- 100 mm diameter in buildings up to 12 levels where the vertical discharge stack is ventilated with a 50 mm diameter ventilating stack (secondary ventilated stack system)
- 150 mm diameter in buildings up to 24 levels, again requiring the main discharge stack to be ventilated as above.

Note: The general guide above is based on the assumption that there are no offsets in the vertical discharge stack and that the drainage system is not prone to surcharging and no intercepting trap is fitted, in which case additional ventilation may be required.

In very tall buildings, or buildings which have groups of appliances connected to the main discharge stack, the pipe size is generally found by the discharge unit method: each appliance is given a discharge unit value to represent the average amount of water that would flow through a discharge pipe or drain. When all the required discharge units are added together, a calculation can be made to convert this figure to a total discharge volume, in litres per second (L/s), using the following formula:

$$\sqrt{\text{Discharge units}} \times \text{frequency factor} = \text{total discharge volume (L/s)}$$

With the discharge volume known, simply refer to Table 3, opposite, to identify the vertical discharge stack pipe size, or Table 4 to find the horizontal pipe size and gradient.

Example: Find the diameter of a vertical discharge stack serving a school with a total of 24 WCs (7.5 litre), 26 washbasins, 4 sinks and 1 washing machine.

WCs	24×2	$= 48$
Washbasins	26×0.5	$= 13$
Sinks	4×0.8	$= 3.2$
12 kg washing machine	1×1.5	$= 1.5$
	Total discharge units	$= 65.7$

$$\sqrt{\text{Discharge units}} \times \text{frequency factor} = \text{total discharge volume}$$
$$\text{Therefore}, \sqrt{65.7} \times 0.7 = 5.67 \, \text{L/s}$$

By referral to Table 3, a 125 mm diameter vertical discharge stack would be required.

Table 4 also suggests a 125 mm pipe would be required, laid at a gradient of 20 mm per metre run. Refer to the British Standard for further examples and tables.

Table 1 Discharge units

Sanitary appliances	Discharge unit
9 litre WC	2.5
6 or 7.5 litre WC	2.0
Sink	0.8
Washbasin or bidet	0.5
Bath	0.8
Washing machine (6 kg)	0.8
Washing machine (12 kg)	1.5
Dishwasher	0.8

Table 2 Frequency factor value

Intermittent use, e.g. dwelling	0.5
Frequent use, e.g. school or restaurant	0.7
Congested use, e.g. public toilets	1.0
Special use, e.g. laboratory	1.2

Table 3 Vertical stack size

Internal diameter (mm)	Square entry (L/s)	Swept entry (L/s)
60	0.5	0.7
70	1.5	2.0
80	2.0	2.6
90	2.7	3.5
100	4.0	5.2
125	5.8	7.6
150	9.5	12.4
200	16	21

Table 4 Discharge volume (L/s) of different horizontal pipe runs and drains

	Pipe or drain run						
Gradient (mm/m)	100 mm	125 mm	150 mm	200 mm	225 mm	250 mm	300 mm
5	1.8	2.8	5.4	10.0	15.9	18.9	34.1
10	2.5	4.1	7.7	14.2	22.5	26.9	48.3
15	3.1	5.0	9.4	17.4	27.6	32.9	59.2
20	3.5	5.7	10.9	20.1	31.9	38.1	68.4
25	4.0	6.4	12.2	22.5	35.7	42.6	76.6
30	4.4	7.1	13.3	24.7	38.9	46.7	83.9
35	4.7	7.6	14.4	26.6	42.3	50.4	90.7
40	5.0	8.2	15.4	28.5	45.2	53.9	96.9
45	5.3	8.7	16.3	30.2	48.0	57.2	102.8
50	5.6	9.1	17.2	31.9	50.6	60.3	108.4

7 Sanitation

Testing of Sanitary Pipework

Relevant British Standard
BS EN 12056

Generally, all inspection and testing should be carried out during installation, especially where the pipework will be inaccessible upon completion. Two tests are carried out to sanitary pipework:

Soundness test (air test)   A test to see if the system is airtight. The procedure is to insert testing bags or drain plugs into any open ends and fill all the appliance traps with water. It is also advisable to allow some water to cover the test plugs to provide a suitable seal. A rubber hose connecting a manometer and hand bellows is connected to one drain plug. By operation of the hand bellows, air is pumped into the system to give a water head pressure of 38 mm, as indicated on the manometer. A plug cock on the air inlet tube is now closed and the pressure should hold for 3 minutes with no pressure drop.

Should a leak exist, it will need to be found by maintaining a small air pressure within the system and applying leak detection fluid, such as washing-up liquid and water, to the joints; any leakage will be indicated by the formation of bubbles. The system may be charged with smoke, although this is not recommended for plastic pipework or where rubber components may be adversely affected.

Performance test   A test which is carried out to above-ground sanitary pipework to ensure that a minimum of 25 mm of water trap seal is retained in every trap when the pipework is subjected to its worst possible working conditions. Each of the following tests should be carried out a minimum of three times and the trap recharged before each test. The depth of the water seal should be measured with a dipstick (see figure).

Tests for siphonage in branch discharge pipes   The test for self-siphonage is carried out by filling the sanitary appliance to overflowing and removing the plug. To test for induced siphonage, several appliances should be discharged together and all the traps on the pipe run measured. The worst conditions occur when the sanitary appliances furthest away from the drain or discharge stack are discharged.

Tests for siphonage and compression in discharge stacks   A selection of sanitary appliances as indicated in the table should be discharged at the same time from the highest floor(s), thus giving the worst pressure conditions.

Example: A seven-storey block of flats has two sanitary appliances of each type on each floor, i.e. two baths, two sinks, two washbasins and two WCs. The amount and kind of appliances to be discharged to give a satisfactory test will therefore be:

$$2 \times 7 = 14 \text{ of each kind of appliance}$$

Thus from the table we see that one WC, two sinks and one washbasin should be simultaneously discharged from the top floor.

Performance testing in discharge stacks

Type of building	No. of each type of Sanitary appliance on stack	No. of appliances to be discharged together		
		WC	sink	basin
Domestic	1–9	1	1	1
	10–24	1	2	1
	25–35	1	3	2
	36–50	2	3	2
	51–65	2	4	2
Commercial and public	1–9	1–2	–	1–2
	10–18	1–2	–	2–3
	19–26	2–3	–	2–3
	27–50	2–3	–	3–4
	51–78	3–4	–	4
	79–100	3–4	–	5

Note: baths, showers, urinals and spray taps are ignored because their discharge does not adversely affect the normal peak flow load. In public buldings such as schools, where very congested periods can be expected, the larger figure should be chosen.

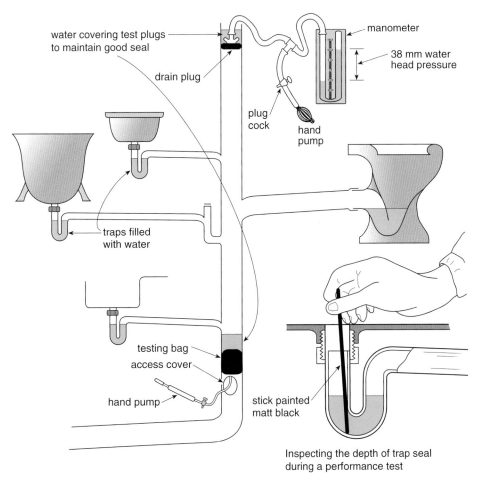

Testing of sanitary pipework

Inspecting the depth of trap seal during a performance test

7 Sanitation

Maintenance and Periodic Inspection

Relevant British Standard
BS EN 12056

Any drainage system should be kept in a clean, sound condition to maintain its maximum efficiency. During periodic inspections, access covers will need to be checked for operation; where cast iron or steel ventilating stacks have been used, a flush through with water or rodding may be required to remove rust accumulations at offsets and bends.

Occasionally, discharge pipes and drains become blocked due to an assortment of causes and it is for the plumber to overcome the problem of unblocking the pipe with the least inconvenience. The methods used can range, depending upon circumstances, from using drain rods or a suction plunger to chemicals.

Manual cleaning

Drain rods are passed down the pipe until the obstruction is met; then, by giving a few blows, the blockage is dislodged and often washed away by the pressure caused by the build up of water.

A variation of this is to pass a spring or drain auger down the pipe; this is rotated, thus dislodging the blockage. The spring can be either hand-held or machine operated.

Possibly one of the most frequently employed methods of removing blockages is to use a plunger. It works because it creates a pressure on the blockage of at least that of atmospheric pressure ($101.3\,kN/m^2$). As the plunger is withdrawn it leaves a void behind, creating a partial vacuum. When plunging sink wastes it is essential that the overflow pipe, if connected to the waste, is blocked up, otherwise air will travel down it and relieve the negative pressure within the pipe.

Chemical cleaning

When carrying out any work of this nature, protective clothing and eye shields should be worn and all work should be thoroughly flushed and washed down on completion.

A descaling fluid containing 15–30% inhibited hydrochloric acid and 20–40% phosphoric acid is poured into the pipes in small quantities at predetermined points. Alternatively, it can be applied via a drip feed into the pipe at a rate of about 4 litres every 20 minutes. When carrying out any descaling, all windows should be opened to ensure good ventilation. Most acid descaling fluid will attack linseed oil bound putty; therefore, prolonged contact with these jointing materials should be avoided.

Should the problem be less severe and the discharge pipework lined or blocked only with grease or soap residues, the pipework should be flushed with very hot water which has 1 kg of soda crystals dissolved in every 9 litres. **Note:** Soda crystals should not be confused with caustic soda.

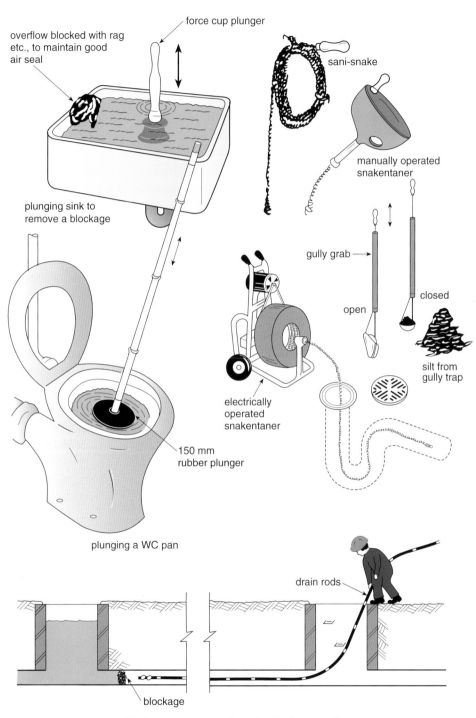

overflow blocked with rag etc., to maintain good air seal

force cup plunger

sani-snake

manually operated snakentaner

plunging sink to remove a blockage

gully grab

open

closed

electrically operated snakentaner

silt from gully trap

150 mm rubber plunger

plunging a WC pan

drain rods

blockage

Maintenance and periodic inspection

7 Sanitation

Part 8

Drainage

Plumbing, 4th Edition. R. D. Treloar.
© 2012 Blackwell Publishing Ltd. Published 2012 by Blackwell Publishing Ltd.

Below Ground Drainage

Relevant British Standard
BS EN 752

All drainage work must be carried out in accordance with Part H of the Building Regulations. Drainage is primarily divided into two groups: surface and foul water systems. *Surface water* is the water from roofs and the surrounding ground, whereas *foul water* is water which is contaminated by soil, waste and trade effluent. Clearly, surface water does not need to be treated and may discharge into a local water-course. Any drainage system must convey all surface water or liquid sewage away from the building in the most speedy and efficient way possible to the sewer or other discharge point without risk of nuisance or danger to health and safety. When designing any system, one must observe the following principles:

- Provide adequate access points
- Keep pipework as straight as possible between access points, and for all bends over 45° an access point should be provided
- Ensure all pipework is adequately supported
- Ensure the pipe is laid to a self-cleansing gradient
- The whole system must be watertight, including inspection covers
- Drains should not run under a building, unless this is unavoidable or in so doing they would considerably shorten the route of pipework.

There are three designs of below ground drainage:

Combined system   A system in which one pipe is used to convey foul and surface water. When this system is used, all points in the system open to the atmosphere must be trapped; the only exception to this would be ventilating pipes and fresh air inlets. The advantage of this system over separate and partially separate systems is that it is cheaper and easier to install; also, during periods of rain the whole system gets a good flush through. The disadvantages of this system are ones of cost, because all water (foul and surface) has to be treated at the sewage works. Also, at times of heavy rainfall, inadequately sized drains could be prone to surcharging, which makes water course pollution very probable.

Separate system   In this system the foul water is conveyed in one pipe to the sewage treatment works, and all the surface water is conveyed in another pipe completely independently of the foul water; thus there is no need for water treatment. Connections to surface water drains do not need to be trapped because there should be no unhealthy smells, but with the foul water drain, all connections must be trapped. One problem with this system is the danger of cross-connections, i.e. foul water being connected to a surface water drain.

Partially separate system   A compromise between the combined and separate systems. It consists of two pipes, one for surface water, and one for foul water plus a limited amount of surface water. This system allows the opportunity to overcome particular design problems, such as those which may occur where no local water-course is available and a soakaway may prove ineffective.

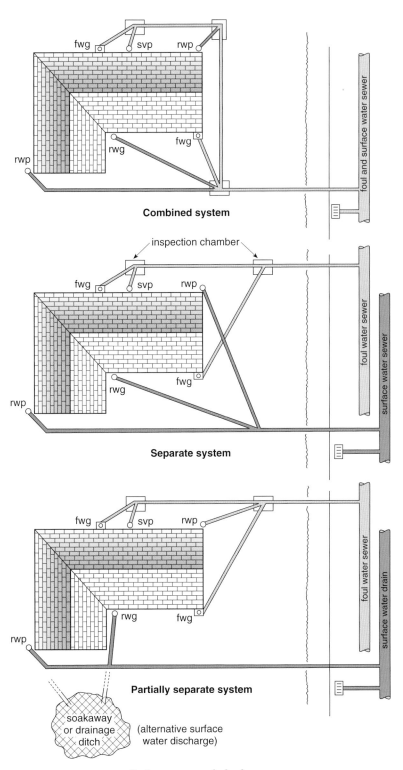

Combined system

inspection chamber

Separate system

Partially separate system

soakaway or drainage ditch

(alternative surface water discharge)

Below-ground drainage

Protection of Pipework

General provisions

When the drainage pipe is positioned in the ground it must be protected from damage due to ground movement, etc. This is usually achieved by laying the pipe on a granular bedding material and covering it with soil free from large stones or other such material. Different circumstances, such as the size of the pipe and the depth to which it is laid, call for different types of bedding. The bedding can be either for rigid or flexible pipes.

Rigid pipe materials include clayware, concrete and cast iron, whereas flexible pipe materials include the various plastics. If the ground is stable, some authorities will permit drainage pipes to be laid directly onto the trench bottom, although the main difficulty is ensuring a steady gradient. The purpose of the granular material is to distribute any excessive loads more evenly around the surface of the pipe, preventing its distortion and possible damage.

Where a pipe does not have the recommended cover, it may require additional protection from damage by several methods. These include encasement in concrete for rigid pipes, allowing for movement at joints, or covering the bedding material with some form of paving slab.

Allowance for settlement protection of buildings

Drains below buildings   Where a drain is run under a building it should be surrounded with at least 100 mm of granular material. If the crown (top) of the pipe is within 300 mm of the underside of the oversite concrete slab, it should be encased in concrete and be incorporated into the slab.

Drains penetrating walls   Should a drainage pipe have to run through a wall or foundation, special precautions will need to be taken to ensure the pipe does not fracture. This is best achieved by either of the following methods:

(1) Forming an opening through the wall giving a 50 mm space all around the pipe which is masked off with a rigid sheet material.
(2) Bedding into the wall a short length of pipe onto which are connected two 600 mm long pipes, either side of the wall (all joints being made good using flexible connectors). Should the pipe or wall move, the three pipes would act as rockers, allowing for movement.

Trenches close to building foundations   Where a drainage trench is excavated lower than the foundations of any building, and within 1 m, the trench should be filled with concrete up to the lowest level of the foundation. Distances greater than 1 m should be filled with concrete to a level equal to distance from the building, less 150 mm (see Building Regulations).

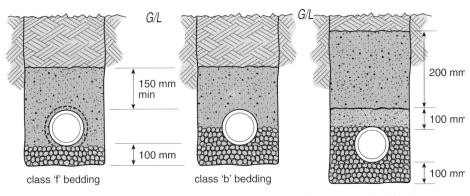

Bedding for rigid pipes

class 'f' bedding

150 mm min

100 mm

class 'b' bedding

Bedding for flexible pipes

200 mm

100 mm

100 mm

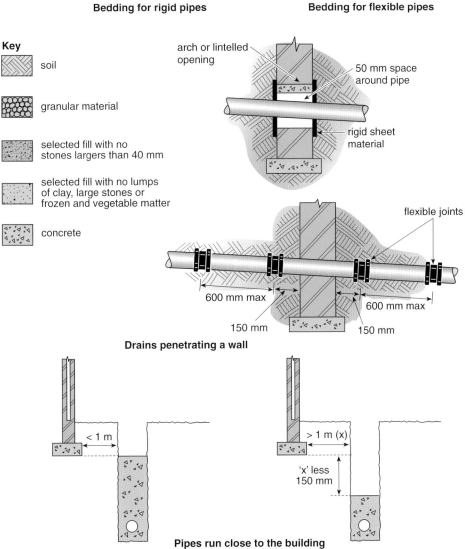

Key

soil

granular material

selected fill with no stones largers than 40 mm

selected fill with no lumps of clay, large stones or frozen and vegetable matter

concrete

arch or lintelled opening

50 mm space around pipe

rigid sheet material

flexible joints

600 mm max

600 mm max

150 mm

150 mm

Drains penetrating a wall

< 1 m

> 1 m (x)

'x' less 150 mm

Pipes run close to the building

Protection of pipework

8 Drainage

Gullies and Traps

The design of a drainage system may allow the incorporation of any one of a whole series of gullies or traps to help ensure that design performance is met. In general, the gully is a drainage fitting designed to receive surface and/or foul water from waste pipes. The purpose of a gully is to provide a trap preventing odours from the drain entering the atmosphere. All new works must discharge water from appliances below the cover or grating and, in many cases, a back or side inlet gully is used to assist in this design. The old method of discharging the water into a chamber above the grating is no longer permissible. Trapless gullies are also available but should only be used for surface water drains.

Yard and garage gullies A large gully which includes a sediment bucket to collect any grit or silt which might otherwise block the outlet pipe. The bucket should periodically be removed for cleansing purposes.

Anti-flood gullies and valves A fitting designed to prevent surcharging of surface water drains. With the ball type shown, when backflow or surcharging occurs, the ball rises onto the underside of the access cover until it finally sits in the rubber seating, preventing any further flow in either direction until the floodwater subsides. With the trunk valve, any backflow causes the float to rise, thus turning the flap valve to the closed position.

Grease trap and converters A grease trap is a device which houses a quantity of water and is designed so that, should grease from canteen kitchens be discharged into the drains, it would, when reaching the water, cool and solidify. Lighter solids would rise to the surface and collect in a solid cake; heavier solids would sink and collect in the galvanised tray. Periodically the tray should be lifted out, removing all the grease which has accumulated.

Some grease traps are designed to be fitted under sinks, but in most cases the trap is fitted externally to the building. The grease converter allows the effluent to pass over a series of baffles which cause the grease to form in globules on the surface. Applying a regular dosing of a mixture of micro-organisms, enzymes and food supplements, in powder or liquid form, to the globules of grease converts them into a water-soluble biodegradable product.

Intercepting trap A type of running trap (usually found in older properties) in the inspection chamber nearest the main sewer – sometimes called a disconnecting trap because it disconnects the drain from the sewer. The prime function of an intercepting trap is to prevent sewer gas from entering the house drains, but nowadays, because drainage systems are better designed and sewer gases seldom cause trouble, an intercepting trap is omitted. In the past, the intercepting trap proved a common cause of blockages owing to the collection of debris in the trap.

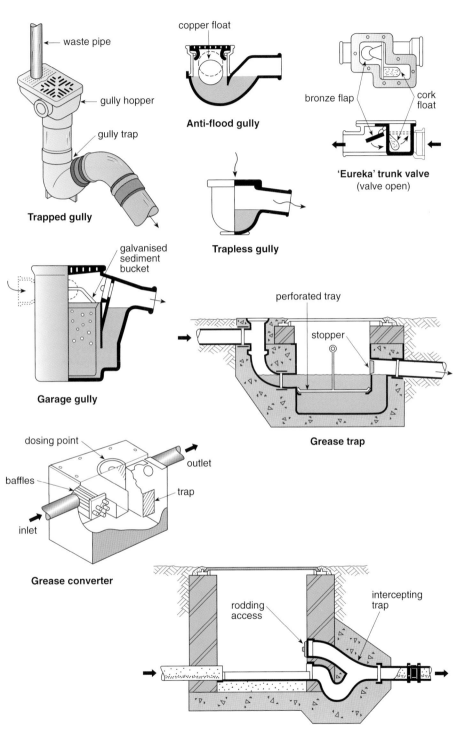

waste pipe

gully hopper

gully trap

Trapped gully

copper float

Anti-flood gully

Trapless gully

bronze flap

cork float

'Eureka' trunk valve
(valve open)

galvanised
sediment
bucket

Garage gully

perforated tray

stopper

Grease trap

dosing point

outlet

baffles

trap

inlet

Grease converter

rodding
access

intercepting
trap

Gullies and traps

Provision for Access

Relevant British Standard
BS EN 752

To enable internal inspection and testing of a drainage system, or to provide a route via which the clearance of blockages can be achieved, it is essential that sufficient provision is made for the internal access of the pipe. An access point should be provided at the following locations:

- At or near the head of a drainage run
- At changes of gradient or direction (i.e. bends)
- At junctions or branches, unless it is possible to clear blockages from another access point
- At changes in pipe diameter
- Between long drainage runs. This will be dependent on the type of access provided (see Building Regulations).

Three types of access will be found:

Rodding eyes   A capped extension on the pipe where access can be gained to a drain or any discharge pipe for inspection or for cleaning with rods. It is possible to design a 'closed' system of drainage to reduce the number of inspection chambers found in the more traditional systems. The rodding point system, as it is generally known, is often run in uPVC pipe and, with the omission of inspection chambers, considerably reduces the cost of installation.

Access fittings   A fitting, such as a bend, branch or gully, which has a cover fitted, usually bolted to the fitting. The cover may be located above ground or at ground level in the case of a gully. It may also be located below ground, in which case it will need to be incorporated into an inspection chamber or manhole; alternatively, a raising piece could be incorporated to allow termination at ground level. The access fitting, unlike the rodding eye, allows rodding in more than one direction.

Inspection chambers and manholes   A chamber constructed of brick, concrete or plastic and designed to expose a section of open pipe, in the form of a channel at its base. The definition of an inspection chamber or manhole is based on size. If the chamber is large enough, it is identified as a manhole, which would certainly be the case in all chambers over 1 m deep. All chambers are provided with some form of cover located at ground level and when positioned internally within a building, these should be bolted down with a greased double seal to prevent the passage of odours.

When a chamber is constructed in brickwork or concrete, its wall thickness should be adequate to resist any external pressures caused by the surrounding ground; in all cases they should be at least 200 mm thick. The base of the chamber should be benched up to allow any rising water flow back down into the channel. In chambers over 1 m deep, step irons or a ladder will need to be included.

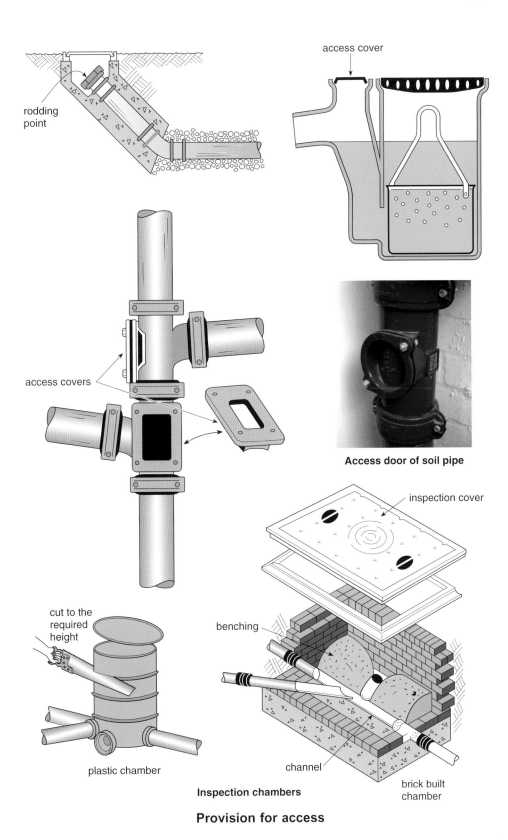

rodding point

access cover

Access door of soil pipe

access covers

inspection cover

cut to the required height

benching

plastic chamber

channel

brick built chamber

Inspection chambers

Provision for access

Connections to Existing Systems

Relevant British Standard
BS EN 752

When joining into an existing drain run several methods can be adopted. In most cases the most suitable method is to connect into an existing inspection chamber or construct a new one. Sometimes a direct connection has to be made, for example where a house drain is to join a public sewer.

The method of joining into the pipe depends on the material, but in many cases a junction block or saddle junction may be used which must be connected into the top half of the drain. If the pipe into which you are cutting is less than 225 mm in diameter, it is best to insert a new branch junction because a hole cut in such a small pipe would weaken it.

When a drainage pipe at high level is to join a drain at a lower level, where possible the drain should be run at a gradient adjoining the two. However, if the difference in heights between the two drains is excessive or space prevents the gradient being self-cleansing, access should be afforded to the pipe for future cleansing purposes. Either of two methods can be adopted for this: the ramp or the backdrop.

The ramp   This may be used where the difference in levels between the two drains does not exceed 680 mm. It consists of an open channel formed in the benching of an inspection chamber to join the two drainage pipes.

The backdrop or tumbling bay   This consists of a section of vertical drainage pipe joining the invert level in an inspection chamber and a drain pipe at higher level. A backdrop would be used if the vertical distance between invert levels of the drain exceeded 680 mm. The backdrop may be installed inside the inspection chamber if it is run in cast iron or plastic, but should clayware pipe be used it must be fitted externally to the chamber and encased in concrete.

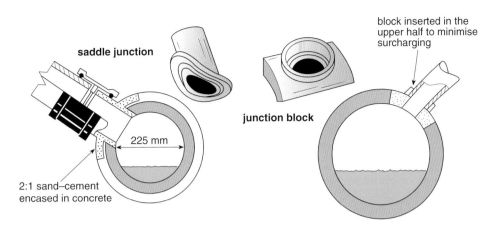

saddle junction

225 mm

2:1 sand–cement
encased in concrete

block inserted in the
upper half to minimise
surcharging

junction block

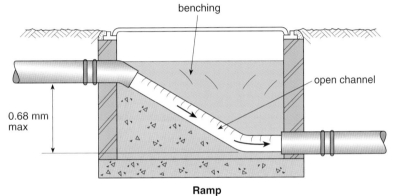

benching

open channel

0.68 mm
max

Ramp

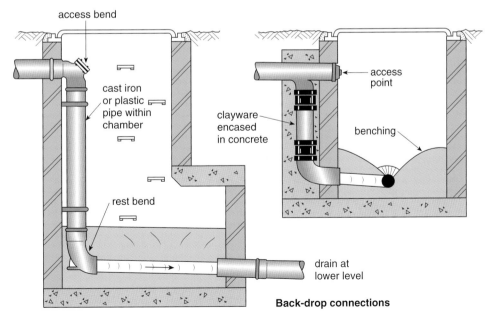

access bend

cast iron
or plastic
pipe within
chamber

rest bend

access
point

clayware
encased
in concrete

benching

drain at
lower level

Back-drop connections

Connections to existing systems

Determining Drainage Levels

Relevant British Standard
BS EN 752

Choice of gradient

The flow capacity of a pipe will depend upon the gradient, fall or incline on which the pipe is laid. Increasing the gradient increases the capacity. It is essential that the pipe is laid to a self-cleansing gradient with a flow velocity of between 0.75 m/s and 1.54 m/s. If the water flows too slowly, it will have insufficient velocity and impetus to carry with it any solid matter; however, where it flows too fast, the water will leave the solids behind.

The amount of flow passing through the pipe also has a bearing upon the gradient. Where intermittent flow is expected, giving shallow water, it is advisable to increase the gradient. *Maguire's Rule* is sometimes used to determine self-cleansing gradients with intermittent or low-flow rates and is based on a calculation in which the pipe diameter in millimetres is divided by 2.5: for example, 100 mm ÷ 2.5 = 40; therefore the gradient should be 1 in 40 (see table).

Where continuous flows can be expected, the gradient for 100 mm pipes can be reduced to 1 in 80, provided that one WC is connected; for 150 mm pipes, a gradient no flatter than 1 in 150 may be permitted, provided that at least five WCs are connected to the drainage run.

Maguire's Rule

Pipe diameter (mm)	Recommended fall
100	1 in 40
150	1 in 60
225	1 in 90

Setting out the fall

The method used to ensure the pipe is laid to the correct gradient will depend upon the length of drainage run, and for short sections a gradient board, or incidence board, is used. The gradient board is a plank of wood cut to the required gradient of a drain and used in conjunction with a spirit level. If a drain run is to have a fall of 1 in 40, the drop in depth over 1 m would be 25 mm (1÷40=0.025 m). So to make a simple gradient board, take a straight plank 1 m long and cut it at an angle (see figure).

Where a long drainage run is to be installed, the fall should be determined using sight rails and a traveller (sometimes called a boning rod). The site rails are positioned at either end of a drainage run at different heights, the difference being that of the required gradient. This is done by a site surveyor, using a site level, or by using a water level to give an accurate transference of levels between each end of the drainage run. The trench is then excavated to the required fall or backfilled as necessary. Two operatives are then used, one sighting his eye between the site rails and instructing his partner to raise or lower the traveller by lifting or lowering the pipe as necessary (see figure).

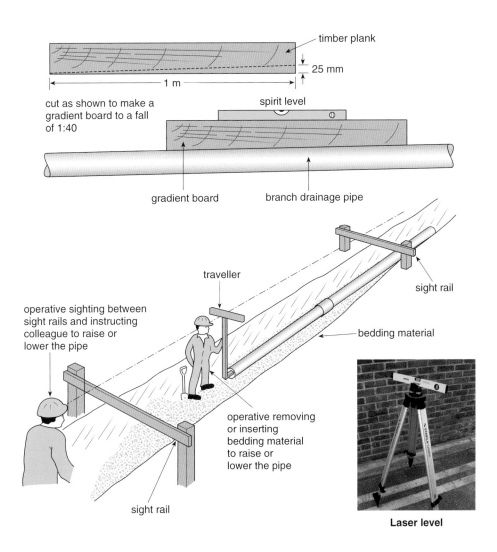

timber plank

25 mm

1 m

cut as shown to make a
gradient board to a fall
of 1:40

spirit level

gradient board

branch drainage pipe

traveller

sight rail

operative sighting between
sight rails and instructing
colleague to raise or
lower the pipe

bedding material

operative removing
or inserting
bedding material
to raise or
lower the pipe

sight rail

Laser level

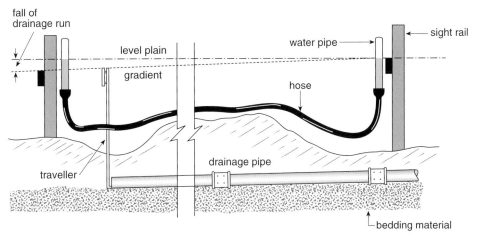

fall of
drainage run

level plain

gradient

water pipe

sight rail

hose

traveller

drainage pipe

bedding material

Determining drainage levels

Sizing of Drainage Pipework

Relevant British Standard
BS EN 752

The minimum diameter of any drainage pipe is 75 mm; however, if the pipe is to convey soil, water or trade effluent, the minimum size is 100 mm. In determining the pipe size for a drainage system, one must first calculate the probable expected flow rate of discharge water in litres per second.

In *soil and waste drainage systems* the flow rate is based upon the number of appliances connected to the system and its likely discharge, bearing in mind that not all appliances will be in use at the same time (see page 316, Sizing of Sanitary Pipework). For typical domestic dwellings, Table 1 can be used.

Table 1 Approximate flow rate from typical domestic households with minimal sanitary accommodation

No. of dwellings	Flow rate (litres/s)
1	2.5
5	3.5
10	4.1
15	4.6
20	5.1
25	5.4
30	5.8

In *surface water drainage systems* the drain should be large enough to carry the maximum flow from the paved and other surfaces, with an allowance for that which evaporates or soaks into the surrounding ground. This is known as an impermeable factor and is indicated in Table 2; in areas where water drains readily into the ground, the lower number is chosen.

Table 2

Surface	Impermeable factor
Roofs	0.8–0.95
Asphalt pavements	0.8–0.9
Jointed stone pavements	0.8–0.85

In the UK a rainfall intensity of 0.05 m per hour (50 mm/h) should be assumed and is used in the following calculation to find the flow in litres/s:

$$\text{Flow (litres/s)} = \frac{\text{Area (m}^2) \times \text{rainfall intensity} \times \text{impermeable factor}}{\text{seconds in one hour}} \times 1000$$

Example: The flow in litres/s from a roof measuring $72\,m^2$ would be:

$$\frac{72\,m^2 \times 0.05\,m/h \times 0.9}{3600} \times 1000 = 0.9\,litres/s$$

It is only in combined systems of drainage that both foul and surface water discharge are added together; otherwise they are simply two individual systems, each having its own requirements.

If we take as an example a group of five houses which are to connect to a combined system of drainage in which the total roof area measures $375\,m^2$, and the surrounding paths are to be $400\,m^2$, the probable flow to allow for (in litres/s) would be:

(1) Flow from sanitary accommodation (from Table 1 opposite) = 3.5
(2) Flow from roofs $375 \times 0.05 \times 0.95 \div 3600 \times 1000$ = 4.9
(3) Flow from paths $400 \times 0.05 \times 0.80 \div 3600 \times 1000$ = <u>4.4</u>
                                           Total = 12.8 litres/s

Having determined the flow rate, we can now refer to the graph below to find the minimum required size of pipe, bearing in mind the gradient at which the pipe is to be laid (see page 334). Thus, in this example, for a combined system serving five houses with a flow rate of approximately 13 litres/s, and with a gradient of between 1 in 60 and 1 in 150, a pipe size greater than 100 mm will be required: 150 mm pipe would therefore be chosen.

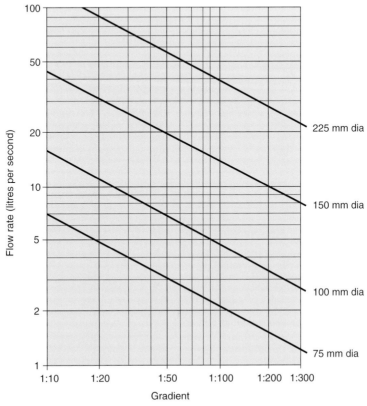

Discharge capacities of drains flowing 2/3 proportional depth

Pipe sizing of drainage pipework

Eaves Guttering

Relevant British Standards
BS 4576 and BS EN 607

The guttering fixed at the lower edge of a pitched roof. There are many designs of eaves gutter, the most common being half round, square line and 'O gee'. Probably the most common material used today for guttering is plastic, although it is also made in cast iron and galvanised pressed steel. The method of securing eaves guttering to the building depends upon the building design, but in general fascia brackets are used.

When installing any guttering, allowances for expansion must be taken into account, especially with plastic guttering materials as the finished gutter will be exposed to constant temperature changes. Occasionally putty is used to join the non-plastic materials, but this is not to be recommended because it dries hard and there is no allowance for thermal movement; instead, Plumbers' Mait or silicone rubber should be used, these being non-setting.

The fall or gradient of a gutter should be approximately 1 in 600. This fall would scarcely be noticeable to the human eye; for example, for a length of 10 m, the fall would only be 17 mm (10 ÷ 600 = 0.017 m). Once the fall has been determined, the brackets should be fixed at about 1 m intervals, as shown in the diagram. Once the gutter has been completed it is connected to the drain via a rainwater pipe.

Gutter size The size of the gutter is dependent upon the size of the roof area to be drained. The effective roof area for pitched roofs is different to that for flat roofs and needs to be calculated. Table 1 shows that for pitched roofs a greater 'run-off' or flow will be experienced and must be allowed for.

Table 1 Calculation of roof area

Roof pitch	Effective area (m²)
Flat	Plan area
30°	Plan area × 1.15
45°	Plan area × 1.4
60°	Plan area × 2.0
70° to vertical	Elevation area × 0.5

Example: If the plan area of a 45° pitched roof measures 6 m × 4 m, the effective area will be:

$$6 \times 4 \times 1.4 = 33.6 \text{ m}^2$$

Once the effective area has been determined, the gutter size can be found by referring to Table 2, or where different guttering materials are used, the manufacturer's data. Continuing with the above example, in which the effective area was 33.6 m², a half round gutter size of 100 mm minimum may be chosen.

Table 2 Half round gutter size (outlet at one end)

Max effective area (m²)	Gutter size (mm)	Outlet size (mm)	Flow capacity (litres/s)
18.0	75	50	0.38
37.0	100	63	0.78
53.0	115	63	1.11
65.0	125	75	1.37
103.0	150	89	2.16

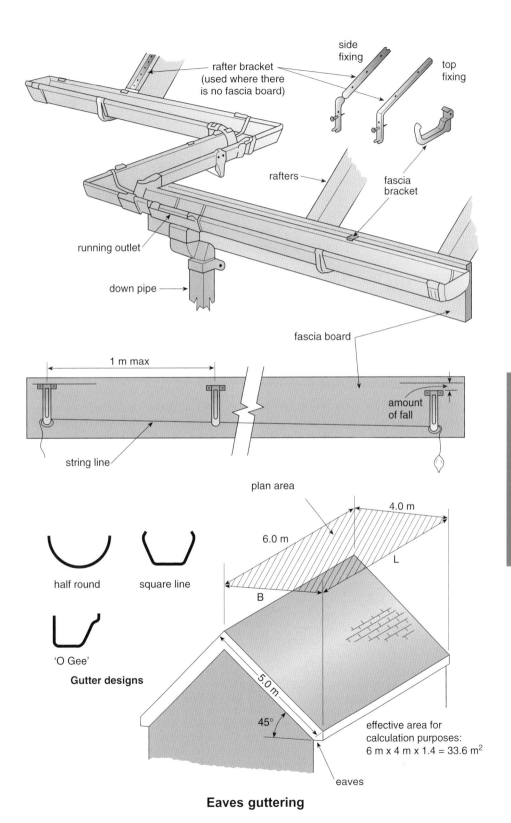

side
fixing

top
fixing

rafter bracket
(used where there
is no fascia board)

rafters

fascia
bracket

running outlet

down pipe

fascia board

1 m max

amount
of fall

string line

plan area

4.0 m

6.0 m

L

B

half round

square line

5.0 m

'O Gee'

Gutter designs

45°

effective area for
calculation purposes:
6 m x 4 m x 1.4 = 33.6 m²

eaves

Eaves guttering

Rainwater Pipes

Relevant British Standards
BS 4576 and BS EN 607

The pipe used to convey the surface water collected from the roof to the drain at ground level. In most cases this pipe is run down the outside of the building; normally, no jointing medium is used and the spigot simply enters the socket.

If the rainwater pipe is run internally, the joints must be made watertight. As with eaves guttering, it is essential to allow for the expansion of the material – otherwise buckling or cracking of the pipes will result. For the sake of appearance, it is essential to ensure that the rainwater pipes are fixed perfectly vertical; for this, it is best to use a plumb bob.

The rainwater pipe may terminate at the lower level in a rainwater shoe, discharging into a gully, or, preferably, may be run directly into a back inlet gully.

The size of the rainwater pipe should be at least that of the gutter outlet, and where a pipe serves more than one gutter, it should be as large as the combined areas of both outlets.

Rainwater connections to discharge stacks

In some areas with a combined system of drainage, the local authority will permit the foul and waste water discharge stacks to receive rainwater from roofs. Designing a system in this way can save on the cost of installing separate rainwater pipes.

To avoid excessive air pressure fluctuations within the discharge stack, this method of rainwater disposal is not to be recommended for buildings over 10 storeys in height, or for the removal of rainwater from roof areas exceeding $40\,m^2$ per stack. The main disadvantage of designing above ground drainage in this way is the problem of flooding if a blockage occurs in the discharge stack. The rainwater pipe should be connected to the stack via a branch connection (see figure); this ensures a free unrestricted flow of air through the ventilating pipe.

Selection of rain water and gutter fittings

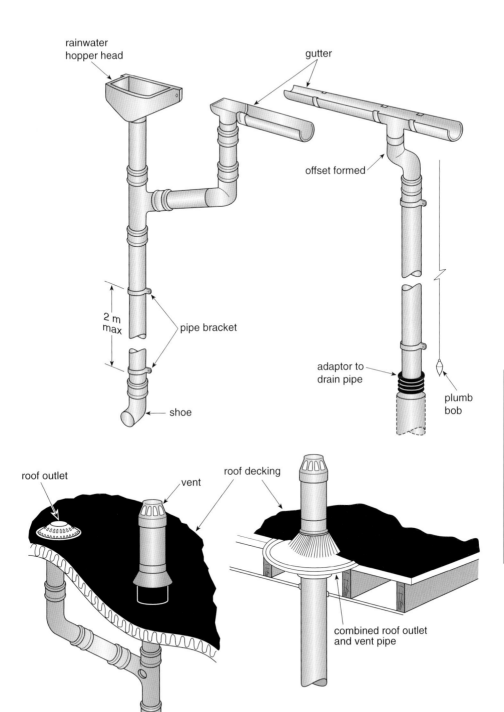

Rainwater connections to discharge stacks

Rainwater pipes

Soakaways, Cesspools and Septic Tanks

Relevant British Standard
BS EN 752

Soakaway   This is a hole, sited well away from the building, which is generally filled with bricks, rubble or stones to prevent the side walls caving in. The purpose of a soakaway is to receive surface water, and it is often used in country areas where there are insufficient surface water drainage pipes. The size and distance from the building are usually specified by the local authority, which must be notified of its use, but in general a soakaway should not be sited nearer the building than 5 m and its size should be enough to drain 12 mm of rainfall over the area. For example, the minimum size of a soakaway for an area of $125\,m^2$ should be $125 \times 0.012\,m = 1.5\,m^3$, measured below the drain connection invert level.

Cesspool (cesspit)   A watertight container used underground for the collection and storage of foul water and crude sewage. A cesspool might be used in areas where the main drainage has not been connected to the property and local sewage treatment such as a septic tank is impracticable. The contents from a cesspool will need to be removed regularly (and before completely full) for proper disposal. Where plastic cesspools are used, they will need to be anchored down with concrete to prevent a high water table causing them to be lifted out of the ground. The capacity of a cesspool should be at least 18000 litres $(18\,m^3)$ below the drain invert level.

Septic tank   A sewage disposal unit sometimes used in rural areas. The house drain is connected to a septic tank with a natural drainage irrigation trench or subsoil drain through which water can drain into the soil. Enough trenching must be constructed to ensure that flooding of the ground surface does not occur, because the effluent from septic tanks contains pathogenic bacteria (disease-causing bacteria); it also must not be allowed to flow directly into streams or underground rivers, etc.

The septic tank usually consists of a double compartment tank which can be constructed of concrete, but is nowadays more commonly made of reinforced glass fibre. As the liquid sewage flows into the tank, the solids settle out and are usually decomposed by anaerobic bacterial action, no chemicals being required. Periodically, the accumulated sludge and remaining solids must be removed by cleansing contractors.

Before the installation of a septic tank it will be necessary to obtain local authority approval and in most cases the siting must be a minimum distance from habitable buildings of at least 15 m, and preferably sloping away from the building. The capacity of a septic tank should be at least 2700 litres $(2.7\,m^3)$ below the drain invert level.

Both cesspools and septic tanks need to be sited within 30 m of vehicle access, to enable desludging as and when necessary.

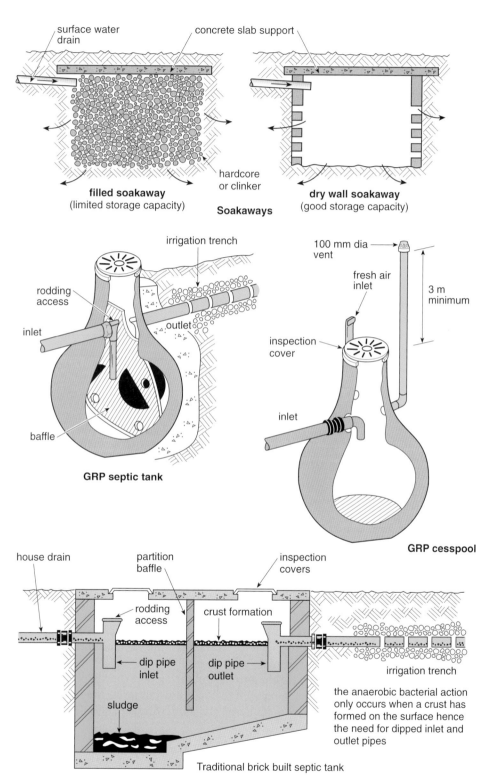

filled soakaway
(limited storage capacity)

Soakaways

dry wall soakaway
(good storage capacity)

surface water drain

concrete slab support

hardcore or clinker

irrigation trench

rodding access

inlet

outlet

baffle

GRP septic tank

100 mm dia vent

fresh air inlet

3 m minimum

inspection cover

inlet

GRP cesspool

house drain

partition baffle

inspection covers

rodding access

crust formation

dip pipe inlet

dip pipe outlet

irrigation trench

sludge

the anaerobic bacterial action only occurs when a crust has formed on the surface hence the need for dipped inlet and outlet pipes

Traditional brick built septic tank

Soakaways, cesspools and septic tanks

Soundness Testing of Drainage Systems

Relevant British Standard
BS EN 752

Generally, drainage works are tested in two stages: when the pipework is installed and the pipe exposed; and upon completion prior to handing over. The testing of pipes is carried out separately from manholes, etc.

Prior to soundness testing   A ball, 13 mm smaller in diameter than the pipe, should be passed down the pipe from the highest end, between access points and accessible branch drains, to check for correct directional gradient and obvious obstructions. It is possible to check for defects in a straight drainage run by an internal inspection, using a mirror and torch, as shown in the figure.

Water test (hydraulic test)   The testing equipment is set up (see figure) to give a minimum test head of 1.5 m. Note that the maximum head should not exceed 4 m at any point, which may mean carrying out the test in sections. When filling the pipe with water, it is essential to make sure there are no pockets of trapped air. When the 1.5 m head pressure has been achieved, the water should be allowed to stand for 2 h, topping up as necessary. After the 2 h have elapsed, a note should be kept of the amount of water needed to maintain the test head over the next 30 minutes.

   The pipe can be deemed sound provided that the water loss does not exceed that indicated in the table.

Acceptable water loss over 30 min test period

Pipe size (mm)	Loss (litres/m run)
100	0.05
150	0.08
225	0.12
300	0.15

Example: In testing a 20 m run of 100 mm pipe the water loss over the actual 30 min test period should not exceed $20 \times 0.05 = 1$ litre.

Air test (pneumatic test)   The air test should be carried out (see figure) by stopping all open ends with drain plugs or bags and applying an air pressure to the pipeline by blowing into the tube using a hand pump. The manometer and supply air, if possible, should be at separate locations; this ensures a positive test result over the whole system. The test pressure should be 100 mm water gauge; however, where trapped gullies and/or ground floor appliances are connected to the drain run, the pressure can only be 50 mm.

   Five minutes are allowed for temperature stabilisation and the air pressure readjusted. The head of water in the manometer should not fall more than 25 mm over a further 5 min test period. Where a 50 mm test is used, this head should not exceed 12 mm over the test period.

Testing inspection chambers and manholes   These tests are generally carried out by fitting a bag stopper in the outlet, from which the air can be removed at ground level, and stopping up the remaining connections. All plugs must be secured to ensure that they are not lost down the pipe. The chamber is filled with water and left to stand for 8 hours, topping up as necessary. The acceptable criterion is that no water should be observed issuing from the outside face of the chamber whilst it is exposed before backfilling with earth.

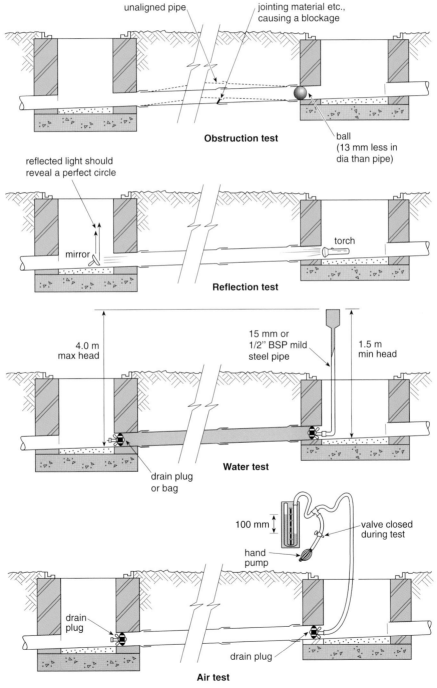

Soundness testing of drainage systems

Part 9

Sheet Weathering

Plumbing, 4th Edition. R. D. Treloar.
© 2012 Blackwell Publishing Ltd. Published 2012 by Blackwell Publishing Ltd.

Lead Sheet

Relevant British Standard
BS EN 12588

Sheet lead is available in almost any size, although the standard width of a roll is 2.4 m, lengths being up to 12 m. Sheets of this size are very heavy to handle; therefore, in most cases, it is worth the extra charge of buying it in smaller (strip) widths varying from 150 to 600 mm. Sheet lead is available in a range of thicknesses and is colour-coded accordingly.

Lead is highly malleable, and lead bossing can be carried out with ease. Due to its low tensile strength and lack of elasticity, lead will creep; therefore correct installation of the material is essential. It is important to adhere to the installation requirements given in the following pages.

Lead can very easily be welded together (see page 72). The thickness of the sheet used will depend upon its location and the type of building (e.g. historic or in a position of extreme exposure). Generally, for small flat roofs and for most flashing applications, codes 4 and 5 will prove adequate. Usually, where sheet is to be welded, code 4 is used, and if bossed, code 5 is chosen. In all cases, no flashing should exceed 1.5 m in length. Code 3 will be used for soakers and its maximum length must not exceed 1 m.

British standard thickness of sheet lead

BS 1178 Code No	Thickness (mm)	Mass per unit area (kg/m²)	Colour code
3	1.32	14.97	Green
4	1.80	20.41	Blue
5	2.24	25.40	Red
6	2.65	30.05	Black
7	3.15	35.72	White
8	3.55	40.26	Orange

When new lead is first laid an initial white carbonate is produced on its surface which will eventually be washed away by the rain but may result in the staining of brickwork, etc. To avoid this staining and provide a shiny appearance when completing the installation of any work, it is wise to apply a coating of patination oil with a cloth to the surface and lower edges, or between and below laps and clips.

Lead used on a large expanse will need to have an underlay to assist in allowing for thermal movement and to isolate the sheet material from the roof structure, thus preventing electrolytic corrosion between any steel roof fixing and the lead sheet. An underlay also acts as an insulator against heat transference and sound.

Traditionally impregnated inodorous felts have proved satisfactory on concrete and masonry surfaces; however, on timber decking where there is a risk of condensation or moisture forming below the sheet lead, a non-woven needle-punched polyester geotextile material should be used which will not rot and will allow the air to circulate. Building paper to BS 1521 class A can also be used on smooth and even surfaces as an alternative.

9 Sheet Weathering

Step flashings

Feature tops

Roof canopies and bay tops

Ornamental roof tops

Side cladding

Flat roofs

Parapet gutters

Lead sheet

9 Sheet Weathering

Lead Bossing

Bossing is the shaping of a malleable metal, and applies in particular to lead. Lead can be worked into many complicated shapes; although a little time, patience and practice is needed to master this skill, no book could possibly demonstrate it. A few guidelines upon the technique can, however, be given.

There are two golden rules. *Rule one*: remember you are trying to move lead from one place to another, either to gain lead or lose it; you are not trying to squeeze it into one area or stretch it out to fill another. So how do we move it? *Rule two*: set up nice curves, like waves, which can flow.

Losing lead (i.e. internal corner)

First set out your work and cut off the surplus lead (see figure). The creasing lines should be set in with a setting-in stick and the sides pulled up to the required angle. About 50mm from the edge of the corner, two temporary stiffening creases are put into the sheet; this helps the corner keep its shape while bossing.

After ensuring that your work has no high ridges (if so they should be rounded), a few blows are directed with a bossing mallet in a downward direction to set in the base of the corner. With the mallet being held on the inside, supporting the work, and by aiming blows from the bossing stick, the corner is then worked up. The blows should be directed at the metal, drifting it outwards progressively towards the top edge of the corner (watch those high ridges; keep the curves smooth). Finish by trimming off the surplus lead which is pulled out to form a tongue.

Gaining lead (i.e. external corner)

The lead should first be folded along its length to the angle required. This will cause a hump to occur which, if it forms a ridge, must be curved, as indicated above. Then the metal is bossed around, in and down, aiming all blows towards the new corner being formed; take care not to stretch the lead.

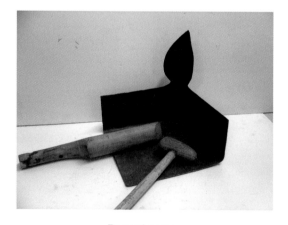

Bossed corner
Picture shows the excess lead that has been worked
up as the corner was moulded and bossed

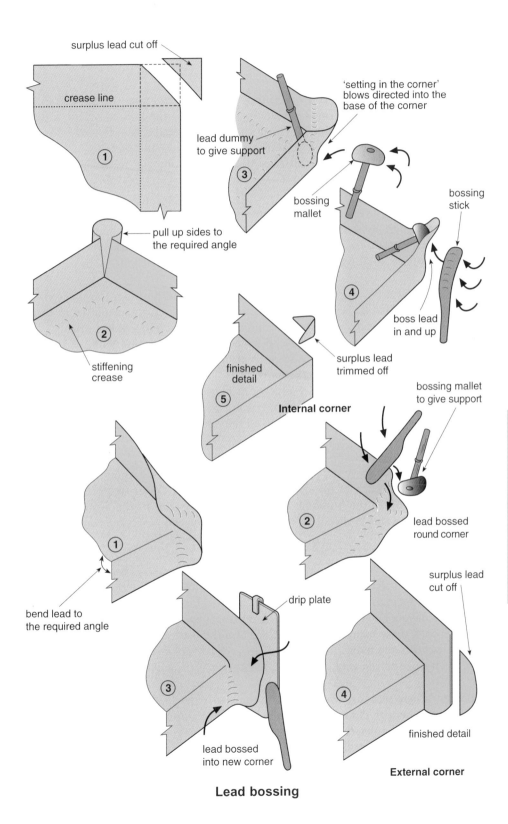

surplus lead cut off

crease line

①

pull up sides to the required angle

②

stiffening crease

'setting in the corner' blows directed into the base of the corner

lead dummy to give support

③

bossing mallet

bossing stick

④

boss lead in and up

finished detail

surplus lead trimmed off

⑤

Internal corner

bend lead to the required angle

①

②

lead bossed round corner

bossing mallet to give support

drip plate

surplus lead cut off

③

lead bossed into new corner

④

finished detail

External corner

Lead bossing

Sheet Fixing

When covering any roof or wall structure with metallic and non-metallic materials, it is essential to secure the sheet weathering material to the building. This is done by incorporating various fixing clips with the various details, as indicated below. In general, the spacing of any fixing should be between 300 and 500 mm, depending upon exposure. It is always worth remembering that one clip too many is better than one clip too few.

Wedges A fixing used to secure an abutment flashing turned into brick or block-work. The wedge consists of a strip of scrap lead, 20–25 mm wide, folded over several times to produce a thick section. The wedge is used in the way shown in the figure and the final front edge pointed up with cement mortar.

Clips A fixing used to secure the free edge of any lead flashing, thus preventing it lifting in high winds. Several designs of clip will be used to suit different situations, ranging from individual clips 50 mm wide to continuous fixing strips.

  To give adequate strength to any clip, especially in exposed locations, it is advisable to devise the clipping system to prevent undue lifting of the material. Copper and stainless steel make the most suitable clips, giving a stronger hold, but the difference in colour often gives rise to objection. This can easily be remedied by tinning its exposed surface, i.e. applying a thin coating of solder. There is no problem with electrolytic corrosion in using these 'mixed' metals as they are compatible.

Secret tack A special clip which allows the middle of a large section of lead to be held in position, preventing it sagging. First a strip of lead about 100 mm wide is welded onto the back of the sheet lead, leaving a long tail or free end. The sheet lead is offered up to the building surface and the free end of the 100 mm strip is passed through a slot in the structure. Finally the free end is secured to the internal surface.

Lead dot A sheet fixing used to secure sheet lead to masonry. To make a lead dot fixing, a dovetail-shaped hole is first cut in the brickwork; then, once the lead has been laid, a dot mould is placed over the hole and molten lead is poured into it, filling the dovetail cavity and leaving a dome-shaped head. There are two other types of dots, these being the soldered dot and the lead welded dot; they are often used to weather the fixing, securing lead to vertical timber cheeks or cladding, and they are formed in the way shown in the figure.

Nailing Where nails are used to secure the lead at its top edge or along one side, large-headed copper clout nails should be used, a minimum of 19 mm long and having serrated shanks; these or nails of a similar design are recommended because they cannot easily be pulled out. Note that where nailing is used, the fixing must be suitably covered and in all cases must not restrict the free movement of the lead.

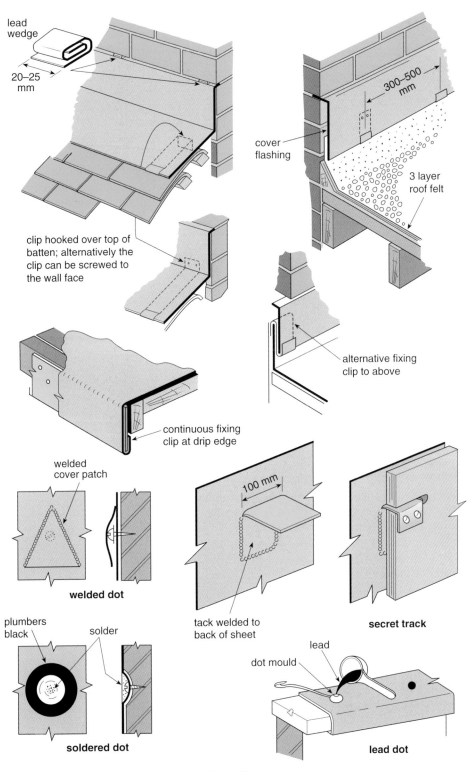

lead wedge

20–25 mm

clip hooked over top of batten; alternatively the clip can be screwed to the wall face

cover flashing

300–500 mm

3 layer roof felt

alternative fixing clip to above

continuous fixing clip at drip edge

welded cover patch

welded dot

100 mm

tack welded to back of sheet

secret track

plumbers black

solder

soldered dot

lead

dot mould

lead dot

9 Sheet Weathering

Sheet fixing

Lead Roof Coverings and Wall Cladding

Relevant British Standard
BS 6915

The maximum sizes of individual pieces of sheet lead must be restricted to allow for the continued expansion and contraction of the material, incorporating various joints (see figure). Large roofs are divided into a series of smaller areas called bays; the size of a bay will depend upon the thickness of the sheet lead and its location (see table). However, the width of bay may be increased providing the length is reduced, in each case ensuring the maximum size area remains about the same.

Maximum size of bay for all roof inclines

BS code no.	Flat and pitched roofs up to 80°		80° pitch to vertical cladding	
	Maximum spacing for joints with fall (e.g. rolls) (mm)	Maximum length between joints (e.g. drips) (mm)	Maximum spacing for joint with fall (e.g. welts) (mm)	Maximum length between joints (e.g. laps) (mm)
4	500	1500	500	1500
5	600	2000	600	2000
6	675	2250	600	2000
7	675	2500 < 10°*	650	2250
8	750	3000 < 10°†	700	2250

*This distance is reduced to 2400mm where the roof pitch is 10–60° and to 2250mm where the pitch is 60–80°
†This distance is reduced to 2500mm where the roof pitch is 10–60° and to 2250mm where the pitch is 60–80°

Method of fixing Apart from the sheet fixings previously mentioned, the bay is generally only nailed down at the highest end at 50 mm intervals, using copper clout nails, and for one-third of the length of the bay, nailing only to the undercloak (see figure). The rest of the lead is left to lie freely. Note that for pitched roofs and vertical cladding, two rows of nails will be required (three rows for codes 7 and 8).

In addition, vertical clad surfaces sometimes require intermediate fixings such as a lead dot or secret tack; caution must be exercised to ensure the thermal movement is not impeded, as experience has shown that with numerous intermediate fixings, fatigue cracking can result.

Preformed cladding Sometimes to assist on-site installation time, preformed lead-faced panels are produced in a workshop (see figure). The panels are simply hooked onto galvanised iron or stainless steel bars. The design of the fixing bracket is such that as one panel is engaged onto the bar, its angled design pulls the panel in close to the building. Note that the joint between the sides of each panel is weathered with a lead junction with a return welt and a lap incorporated at the top. The panels cannot be lifted off as this is prevented by the sheet above.

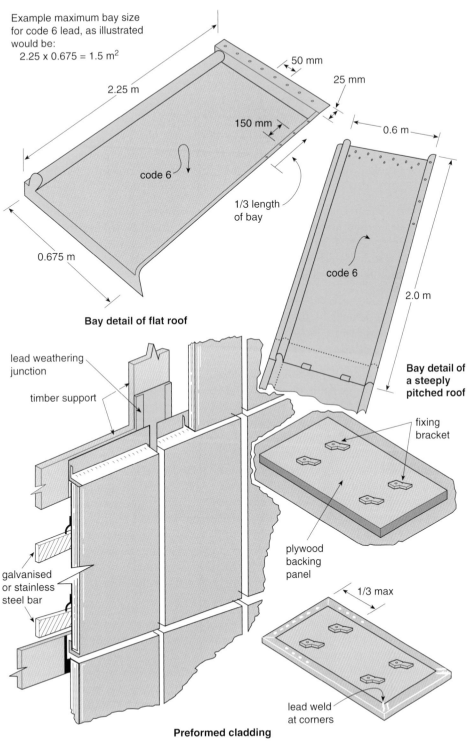

Example maximum bay size
for code 6 lead, as illustrated
would be:
  2.25 x 0.675 = 1.5 m²

2.25 m

50 mm

25 mm

150 mm

0.6 m

code 6

1/3 length
of bay

0.675 m

code 6

2.0 m

Bay detail of flat roof

**Bay detail of
a steeply
pitched roof**

lead weathering
junction

timber support

fixing
bracket

plywood
backing
panel

galvanised
or stainless
steel bar

1/3 max

lead weld
at corners

Preformed cladding

Lead roof coverings and wall cladding

Expansion Joints for Lead Roofs

Relevant British Standard
BS 6915

Joints running with the fall of the roof

Wood cored roll A joint used on flat and pitched roofs using a timber roll cut to the dimensions indicated in the figure. The undercut at the base is designed to prevent the lead being lifted by high winds. The splash lap is incorporated to stiffen the free edge and keep it in position; however, where for aesthetic reasons the splash lap is undesirable it may be omitted, as long as additional copper clips, at 450 mm centres, are provided to which the overcloak has been welted.

Hollow roll An alternative roll detail which can be used, omitting the timber core. The joint can be used for all roof inclines; however, it will be subject to damage on shallow pitched roofs due to foot traffic. The hollow roll will require the inclusion of copper or stainless steel clips and is formed by turning the roll over a wood core or spring which is removed upon completion.

Welts These joints are only suitable for steeply pitched roofs where they will not be vulnerable to damage by foot traffic. As with wood cored and hollow rolls, suitable clips will need to be incorporated. Welts are also occasionally used for joints running across the fall.

Joints running across the fall of the roof

Drip A joint used for roofs up to 10° and used in conjunction with a wood cored roll. The drip should be 50–60 mm in height and be designed to include a splash lap which prevents the lead lifting in strong winds. The larger drip will be employed where a roll abuts the drip.

Two designs of drip are recommended: those for flat roofs up to and including 3°, in which the undercloak is rebated into the roof deck above and nailed at 50 mm spacing; and those for roofs which exceed 3°, for which the undercloak should terminate with two rows of copper clout nails at 75 mm intervals (see figure).

This second method is designed to prevent the lead slipping because of creep. Where this second method is adopted it will be necessary to seal over the nail fixings with a welded dot. In the past, a 40 mm drip incorporating an anti-capillary groove was sometimes used; however, because the groove gradually becomes blocked up with dust and dirt, etc., it proves ineffective and is also vulnerable to water penetration in storm conditions.

Lap A joint used where the roof pitch exceeds 10°, up to and including vertical surfaces. Any lap joint will require an effective 75 mm vertical cover. The top end of the undercloak is usually fixed down with two rows of copper clout nails and the overcloak lapped to incorporate a continuous fixing clip to prevent wind lift. For roofs below 30° pitch the continuous clip usually takes the form of a piece of lead welded to the underlap. For roofs over 30°, a copper or stainless fixing clip will suffice. When lap joints are employed the joints are usually staggered to avoid excessive thickness and over-complicated intersections with other joints.

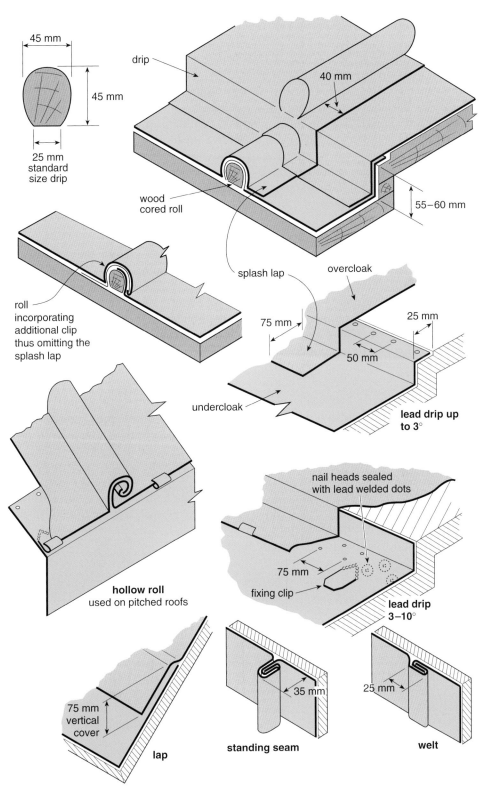

45 mm

45 mm

25 mm
standard
size drip

drip

40 mm

wood
cored roll

55–60 mm

roll
incorporating
additional clip
thus omitting the
splash lap

splash lap

overcloak

75 mm

25 mm

50 mm

undercloak

**lead drip up
to 3°**

nail heads sealed
with lead welded dots

hollow roll
used on pitched roofs

75 mm

fixing clip

**lead drip
3–10°**

75 mm
vertical
cover

lap

35 mm

standing seam

25 mm

welt

Expansion joints for lead roofs

Abutment Flashings in Lead

Cover flashing A strip of lead which is designed to weather the upstand of a roofing material such as that from a flat lead roof or bituminous felt type covering. The cover flashing is turned into the brickwork at the brick course above the upstand and allowed to hang down to give a minimum of 75 mm vertical cover. It is essential that the cover flashing is suitably clipped at 450 mm centres via lead wedges into the brick joints and hanging clips along its lower edge. The maximum length of any flashing is not to exceed 1.5 m; therefore, where lengths longer than this are needed, a lap joint will be required at adjoining sections, giving a minimum of 100 mm lap (a minimum of 150 mm in exposed locations).

Apron flashing Where a pitched roof abuts a wall, a lead flashing will be required to give a minimum of 75 mm vertical upstand (see figure). The amount of cover given to lie down over the roof slope is to be a minimum of 150 mm; however, where the roof pitch is below 25° or in an exposed location, this distance should be increased to 200 mm. Sufficient hanging clips are essential to prevent the lead sagging and eventually pulling away from the brickwork. It is usually possible to boss the lead *in situ* over the contours of most designs of roof tile; however, where sharp raised edges are encountered, the lead can be cut and welded, thus avoiding excessive thinning.

Stepped flashing A special cover flashing which weathers a pitched roof to brickwork. The flashing could be either single steps or a continuous running strip, not exceeding 1.5 m in length; longer lengths are achieved by overlapping further strips. There are two basic ways in which step flashings are designed: by using the step flashing in conjunction with soakers; or by using what is called step and cover flashing.

The setting out of step flashing is illustrated in the figure opposite using a piece of material 150 mm wide, which allows for a water line of at least 65 mm. In the case of step and cover flashing, add to this width 150–200 mm, depending upon the tile profile and its exposure, to lay out across the tiles.

Soakers The soaker is designed to give an upstand at the abutment of a pitched roof to a wall, being incorporated with the laying of the tiles or slates, and is made on site from sheet lead.

The width of a soaker used to join up to an abutment must be a minimum of 175 mm; this would allow for an upstand of 75 mm against the wall and the remaining 100 mm can lay out across the roof and under the tiles or slates. The length of the soaker varies and depends upon the length of the roof slates. The calculation used to find the length is: gauge + lap + 25 mm. The gauge is the distance between the roof slate battens; the lap is the distance one slate overlaps the slate next but one below it; and the 25 mm is optional and is purely for fixing purposes, being bent over the battens to prevent them slipping.

9 Sheet Weathering

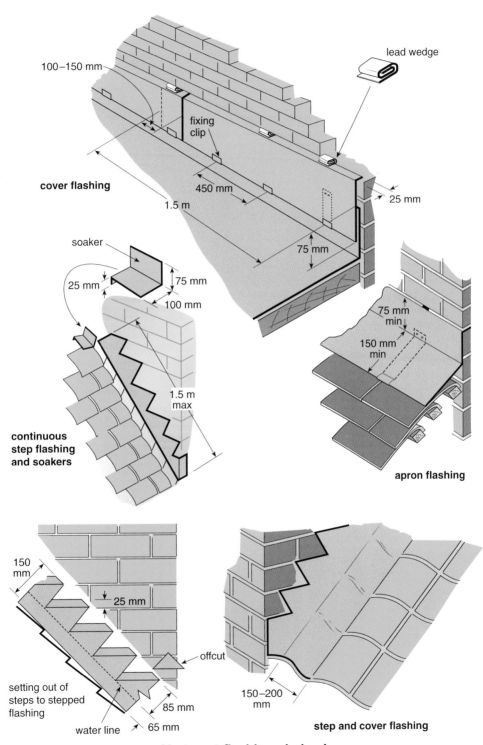

100–150 mm

lead wedge

fixing
clip

cover flashing

450 mm

1.5 m

25 mm

75 mm

soaker

25 mm

75 mm

100 mm

1.5 m
max

continuous
step flashing
and soakers

75 mm
min

150 mm
min

apron flashing

150
mm

25 mm

offcut

setting out of
steps to stepped
flashing

85 mm

water line

65 mm

150–200
mm

step and cover flashing

Abutment flashings in lead

Chimney Flashings in Lead

The roof weathering which prevents rainwater penetrating the building where a chimney stack passes through the roof structure. Chimney flashings consist of: a front apron, soakers, step flashing, back gutter, and a cover flashing. With all sheet roof work, the lowest pieces are positioned first, working up the roof and allowing the higher pieces to overlap those below.

Front apron   The lowest piece, being made from a piece of lead 300–350 mm wide; its length is the width of the stack plus 300 mm, allowing 150 mm each side of the roof. In some situations, 200 mm may be required on each side to allow for deep-profiled tiles or exposure. The front apron can be bossed or welded (see figure) to produce a detail which maintains the required fixing and weathering cover, as previously described.

Stepped flashing and soakers   In addition to what has been said already (see page 358) the front edge is turned round the chimney-stack to give a 75 mm cover along the corner detail. It is possible to finish the side flashing at the corner, i.e. not turning the front edge, provided that the previous front apron maintains a 150 mm minimum turn to the stack.

Back gutter   As with the front apron, it is possible to boss this detail; however, it does involve considerably more work than that for the apron. The approximate width of material is 500 mm, allowing for 100 mm minimum upstand to the brickwork, 150 mm for the gutter sole and the remainder to lay up under the tiles or slates. The length is the same as that for the front apron. As with the front apron, the details are best explained in the figure.

The final, and highest, piece of chimney flashing is the cover flashing which is designed to cover the vertical upstand of the back gutter. Cover flashings are described on page 358.

Sometimes the chimney is located at the apex of the roof (ridge), in which case no back gutter will be required; instead, the stepped flashing is terminated at the ridge and a saddle piece used to provide suitable cover at the highest point.

Chimney flashings

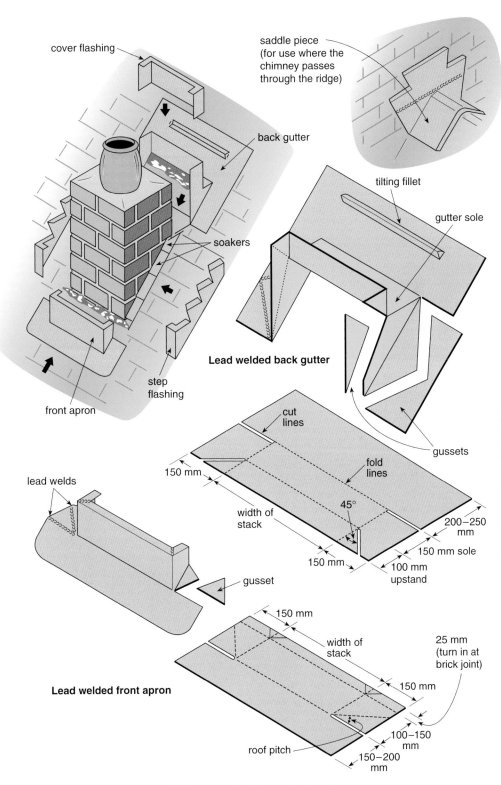

cover flashing

saddle piece
(for use where the
chimney passes
through the ridge)

back gutter

soakers

step
flashing

front apron

tilting fillet

gutter sole

Lead welded back gutter

gussets

cut
lines

fold
lines

150 mm

45°

width of
stack

200–250
mm

150 mm

150 mm sole

100 mm
upstand

lead welds

gusset

Lead welded front apron

150 mm

width of
stack

25 mm
(turn in at
brick joint)

150 mm

100–150
mm

roof pitch

150–200
mm

Chimney flashings in lead

9 Sheet Weathering

Lead Slates and Pitched Valley Gutters

Lead slates   A special flashing used to weather any obstruction passing through a roof – usually a ventilating or flue pipe. The size of the lead required to produce the slate will be dependent upon the roof pitch, the size of the pipe and the type of roof tiles. Generally the slate should be made to give 150 mm minimum distance cover at all points, i.e. sides, back, and up the pipe itself (see figure).

When positioning the slate, the gap between the pipe and the lead also needs to be weathered; this is achieved by using a solvent welded collar for plastic pipe, or by dressing the lead into the top of the pipe opening. Sometimes a mastic or silicone sealant is used to make this seal, although only as a last resort.

The lead slate can be formed by bossing, although it would be much more practicable to lead weld as necessary. Firstly, a piece of lead wide enough is turned around a piece of rigid pipe and its meeting edges are butt welded together. The pipe is then cut to the required pitch, positioning the weld to the back; the raking edge is dressed to form a flange which is placed on the base. The hole is marked and cut and the sleeve is lap-welded as necessary.

Sometimes, overflow pipes passing through a steeply pitched roof need to be weathered in a similar fashion to the slate. This is known as a lead sleeve.

Pitched valley gutters   Where a pitched roof requires to turn an internal angle, the intersection will need to be weathered. Several methods can be adopted for this purpose, including using lead linings laid onto valley boards. For most gutters, the valley boards should extend 225 mm up each side from the centre of the gutter, with a tilting fillet positioned 150 mm up from the base.

The tilting fillet is designed to tilt the free edge of the tile or slate less steeply than the rest of the roof to ensure that the tiles or slates bed tightly on one another; it also helps prevent capillary attraction between the sheet metal and the general roof covering material. A gap of 125 mm should be maintained between the two roof pitches to prevent the build-up of leaves, etc., which may cause a blockage.

Where bituminous sarking felt is used beneath the tiles, it should not be allowed into the sole of the gutter, below the lead because, should the felt soften in hot weather, the lead may stick to the substrate, restricting its free movement. The fixing of the lead itself should be limited to two rows of copper nails across the top of each piece and on gutters over 60°, additional fixing used only at the sides of the top third. No piece of lead should exceed 1.5 m in length and a 150 mm minimum lap is required at each piece, this distance being increased to 220 mm where the roof pitch is lower than 20°. Should two valleys meet, a saddle will be required at the highest point (see figure).

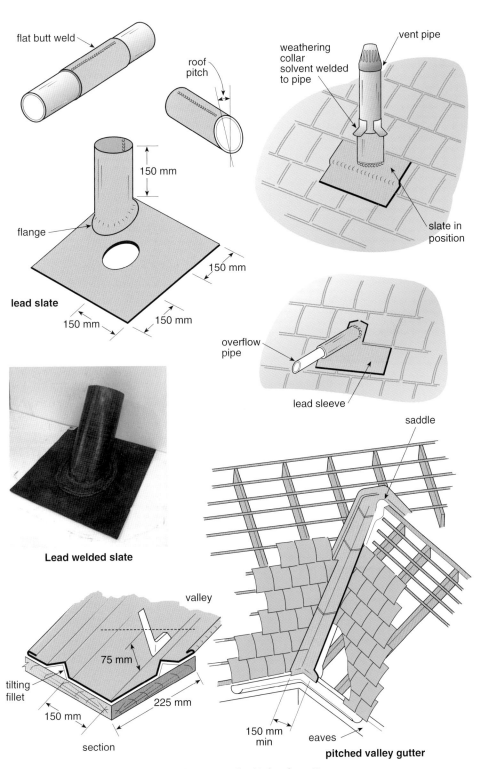

flat butt weld

roof pitch

weathering collar solvent welded to pipe

vent pipe

slate in position

150 mm

flange

lead slate

150 mm

150 mm

150 mm

overflow pipe

lead sleeve

Lead welded slate

saddle

valley

75 mm

tilting fillet

150 mm

225 mm

section

150 mm min

eaves

pitched valley gutter

Lead slates and pitched gutters

Gutter Linings

Where a lead covered roof terminates, a sheet lead gutter will usually be formed at its lowest end. Lead-lined gutters will also be found where a tiled or slated roof is designed with a parapet wall or where two pitched roofs meet at a horizontal valley. The fabrication of the gutter is for the most part identical to the installation of a small flat roof, the size of bay being that previously advised for lead roof coverings.

Drip design The splash lap of a drip may be omitted in gutter linings because the detail will be less vulnerable to lifting in strong winds. It is often desirable to omit this splash lap, not only to simplify installation, but also because grit and dirt tend to collect under the lap and slow the flow of water through the gutter. Where the splash lap is omitted, it is essential that the side lap is maintained.

Gutter design The position of drips in a gutter will be dependent on the sheet thickness, which is often governed by the outlets and amount of fall. For new work, the designer will allow sufficient height for the required fall to suit the drip spacing for a given thickness of lead. However, with existing gutters the outlets are usually fixed and the amount of height for the fall governed by the existing wall height. Thus it may be necessary to choose thicker lead, allowing for longer bay lengths (see figure).

Box gutters A gutter lining, usually with parallel sides, with a minimum width of 225 mm. All upstands from the gutter must be a minimum height of 100 mm. Where it meets a tiled or slated roof, the lead is carried under the tiles up the roof slope to maintain the minimum 100 mm depth. It is permissible to increase the upstand to 200 mm for codes 5 and 6 or 300 mm for codes 7 and 8 where necessary, but with excessively deep-sided box gutters a separate piece of lead will be required. It is important not to nail the taller upstands as it will restrict their free thermal movement.

Tapered gutters A tapered gutter will occur where two pitched roofs meet or where a pitched roof abuts a vertical wall. A tapered gutter is distinguished by the sole (base) of the gutter becoming progressively wider at each drip, becoming its widest at the highest point. The minimum width at the lowest end should be 150 mm and for long gutters a roll, or several rolls, may be required to divide excessively wide sections into two or more bays.

Outlets

Chute A chute outlet is where the sole of the gutter is taken through the parapet wall to discharge into a hopper head. This form of outlet is not prone to blockages by leaves, etc., and allows for a good flow capacity.

Catchpit A design of outlet which allows for an internal downpipe. The minimum depth should be 150 mm and its length equal to that of the width of the gutter. Smaller depths can be used provided the length is increased to prevent surcharging in storm conditions. When a catchpit is chosen it should incorporate an overflow to warn of blockages.

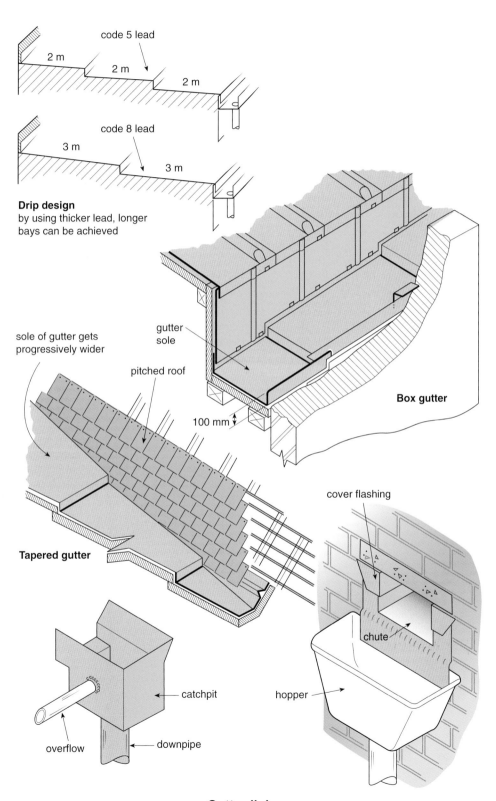

code 5 lead

2 m
2 m
2 m

code 8 lead

3 m
3 m

Drip design
by using thicker lead, longer
bays can be achieved

sole of gutter gets
progressively wider

gutter
sole

pitched roof

100 mm

Box gutter

Tapered gutter

cover flashing

chute

catchpit

hopper

overflow

downpipe

Gutter linings

Dormer Windows

The dormer window is a typical leadwork detail which will be found as a common architectural feature in a pitched roof. Making a covering involves developing the leadwork skills previously described and adapting those processes to form a covering consisting of the following three parts.

Tray and apron flashing   A detail in which the lead is bossed or welded to fit at the base of the window opening. The window sill sits on the tray. The lead extends out to lie down the roof slope and around the front face and side of the vertical timber (see figure). The apron detail is similar to that of the chimney front apron.

Side cheek   This is the side cladding on which the lead hangs vertically. The most important consideration is the size of the panel and its fixing, and to make this decision, the page on roof cladding should be reviewed (see page 354). The lead is turned round the front edge and a return welt formed upon which a timber covering bead is placed.

The base of the side cheek is weathered using soakers where the dormer protrudes from a tiled or slated roof, or by using a continuous side flashing. Where a side flashing is used, it is either welted to a lead covering, should this be the main roof covering material, or it is simply laid out across the roof a minimum distance of 150 mm, and fixed as necessary. **Note:** The maximum flashing lengths are not to exceed 1.5 m.

Dormer top   The top of the dormer is covered in the same way as any small flat roof, divided with wood-cored rolls if necessary. Where the lead terminates at the sides and front of the top, it should be turned down to give a minimum 75 mm vertical cover to the side cheek, fixing clips being incorporated at 450 mm centres.

The back lays up under the roof covering above (usually 200 mm, minimum) depending upon the roof pitch. Generally the direction of fall from the dormer top is made to incline to the sides, although a small dormer may have no fall at all. However, where the dormer is large, comprising three or more bays, the fall will need to flow to the front where an eaves gutter may be positioned to collect the water run-off – or directed backwards, in which case a gutter lining will be needed.

As an alternative to a flat-topped dormer, a curved top may be designed for small or medium sized roofs. This may overcome the need for a front or rear gutter. Where a curved top is included, the sheets of lead are generally joined using welts running with the curve over the dormer top. Should the length of the curve exceed 1.5 m, a cross joint will be needed.

Sometimes an internal or inset dormer will be found. These may be dealt with in a simple variation of the process described above.

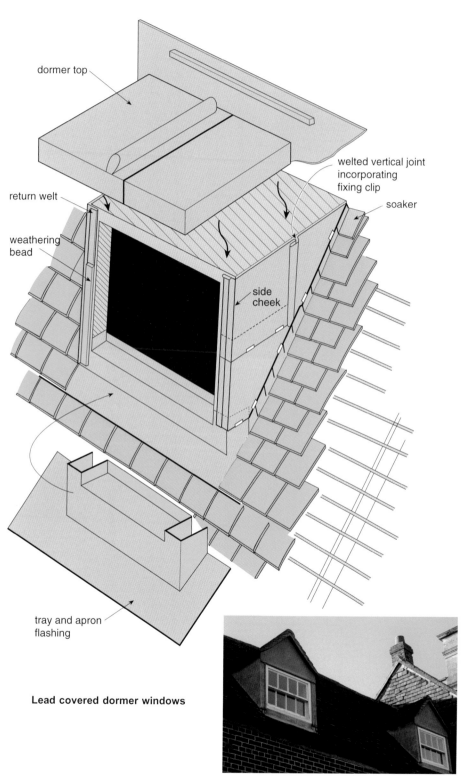

dormer top

welted vertical joint
incorporating
fixing clip

return welt

soaker

weathering
bead

side
cheek

tray and apron
flashing

Lead covered dormer windows

Dormer windows

Part 10

Energy Conservation & Sustainability

Plumbing, 4th Edition. R. D. Treloar.
© 2012 Blackwell Publishing Ltd. Published 2012 by Blackwell Publishing Ltd.

Environmental Awareness

Environmental awareness relates to the conservation of energy and the impact that gas emissions into the atmosphere have on climate conditions. The emission of the various 'greenhouse gases' is believed to cause climate change. Greenhouse gases include: sulfur dioxide (SO_2), oxides of nitrogen (NO_X) and carbon dioxide (CO_2). Each of these is produced by burning fossil fuels (coal, gas and oil, etc.). By far the largest emission is of carbon dioxide; on average, 5 tonnes is produced from each home per year, which equates to around $300\,m^3$ of CO_2 gas being discharged into the atmosphere. 16% of CO_2 emitted in the UK comes from domestic heating. There are other gases that destroy the ozone layer, such as halon (used in fire extinguishers) and chlorofluorocarbons (CFC) (used as a refrigerant in older types of refrigerator and on earlier types of foam pipe insulation).

The British Government is tackling this problem by introducing all sorts of legislation restricting what material can and cannot be used. For example, halon and CFC gases are now banned from general use. The Energy White Paper, published in 2003, indicates that the UK aims to reduce its overall CO_2 emissions by 60% by the year 2050. In order to achieve this goal many changes are being implemented in the way we use and produce our fuel. For example, many wind farms are being established around the country to produce electrical power, thereby limiting the combustion of fossil fuel; combined heat and power (CHP) units are being developed which use the heat produced as a by-product of the electrical generating process. Fuel prices are constantly rising, forcing individuals to turn off appliances or cut down on the transport they use, e.g. by sharing car journeys. When undertaking new building works, high levels of insulation need to be provided in order to ensure that heat is not lost through the fabric of the building. Plumbing systems are also affected by legislation, dictating the design the system must follow and what appliances can be installed, etc.

For example:

- **Part L1 of the Building Regulations:** This controls the type of domestic central heating system or hot water system to install.
- **Water Supply Regulations:** Identify the need for water conservation and loss of heat through pipework.

In order to ensure compliance, systems today need to be certified by the installer to confirm that they comply with the requirements laid down by law. Only approved installers can certify that their work complies. An approved installer would have undertaken an assessment course to confirm that they are competent in the area of plumbing works which they are certifying.

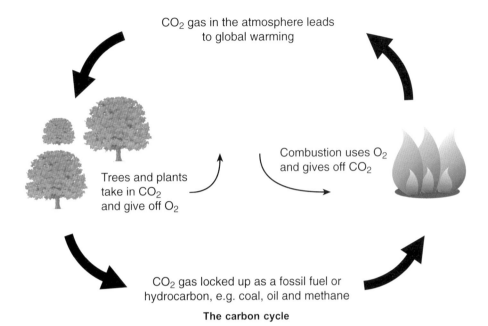

CO$_2$ gas in the atmosphere leads to global warming

Combustion uses O$_2$ and gives off CO$_2$

Trees and plants take in CO$_2$ and give off O$_2$

CO$_2$ gas locked up as a fossil fuel or hydrocarbon, e.g. coal, oil and methane

The carbon cycle

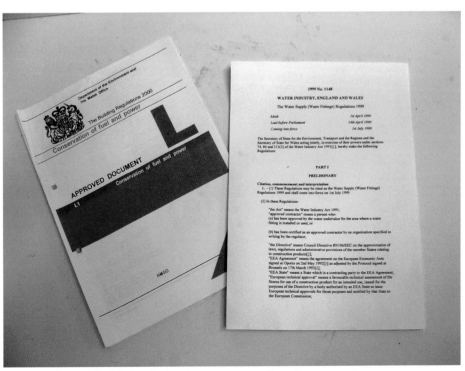

Typical legislation that enforces conservation of energy and resources

10 Energy Conservation

Energy Efficiency in Domestic Dwellings

Standard Assessment Procedure (SAP)

The energy rating of a home, as recommended by the Government, is determined by a simple method referred to as the Standard Assessment Procedure (SAP). It estimates the energy efficiency of a dwelling on a scale of 1 to 120, although it is possible to achieve a higher number. The SAP rating depends upon the building design, including the insulation, solar heat gains, ventilation, heating and hot water efficiency and controls. The higher the SAP rating, the better is the energy efficiency of the dwelling. Basically, SAP points are awarded for the level of performance offered by the building, e.g. so many points for the insulation, so many points for the type of boiler, etc. The method of calculation for the SAP rating is set out on a worksheet or computer program and is based upon standard occupancy and heating patterns, price of fuel, etc., over the whole year. The SAP rating is then converted to produce a carbon index, which is used for Building Regulation approval. The carbon index ranges from 0 to 10: the higher the number, the greater the efficiency. Currently a minimum of 8 is needed for Building Regulation approval.

In order to achieve a high SAP rating, specific measures need to be adopted. For the plumber this means the installation of a well-designed heating system, with all the appropriate controls, as identified over the following pages.

Product Characteristics Data File

Manufacturers test and quote full-load efficiencies for boilers. In reality, boilers operate mostly on part load, typically designed with outside temperatures of $-1°C$; they also have significant heat losses from:

- The boiler casing (standing loss)
- Cool air passing through the boiler by natural means (draught loss)
- Forced draught through the appliance during a pre-purge of air.

To give a more realistic operating efficiency averaged over a whole year, the Government introduced a system of identifying the average annual efficiency for individual boilers. Domestic boilers are now very efficient, operating at over 90% efficiency. In 1993, the Government introduced the Boiler (Efficiency) Regulations, and in 2002 introduced legislation in the form of Building Regulations, specifying a minimum seasonal efficiency requirement. By 2005 the minimum domestic boiler rating was to be 86% and, as boiler design improves, so will the need to install better, more efficient systems. The best independent source of boiler efficiencies, for both new and old boilers, is the Boiler Efficiency Database, accessed at the following web address: www.boilers.org.uk.

Typical annual running costs for different types of domestic boiler

Fuel and boiler type	Flat	Bungalow or terraced	Semi-detached	Detached
Natural gas				
Old cast iron (band G)	£230	£300	£350	£500
New boiler (band D)	£170	£225	£260	£360
New boiler (band A)	£140	£190	£220	£310
LPG				
Old cast iron (band G)	£500	£660	£770	£1100
New boiler (band D)	£370	£495	£560	£800
New boiler (band A)	£315	£420	£480	£680
Oil: 28 second/kerosene				
Old cast iron (band G)	£230	£310	£360	£510
New boiler (band D)	£185	£245	£285	£400
Coal				
Old cast iron	£340	£450	£520	£735
New boiler	£310	£415	£480	£680
Electric				
'Off peak' storage heaters and dhw	£280	£370	£430	£610
Direct acting on demand, heating and dhw	£690	£920	£1070	£1520

The above table clearly shows the cost savings that can be made by selecting the right boiler type. LPG and direct-acting (on peak) electric boilers are clearly currently very expensive appliances to run in comparison to other fuels. Likewise the cost benefits of installing a high-efficiency appliance can also be seen.

Renewable Energy

Renewable energy relates to energy that is replenished at a rate equal to that at which it is being consumed. It has gained in popularity over the last few years owing to its low contributions to the carbon cycle and its sustainability. Many renewable forms of energy stem from solar energy. Solar energy is said to produce both direct and indirect forms of energy.

Direct forms of energy include:

- solar energy, to include domestic hot water heating and photovoltaic cells
- ground source heat pumps.

Indirect forms of energy include:

- biomass *(plants grow as a result of photosynthesis)*
- wind and hydro power *(winds are produced as a result of the differential heating and cooling of the atmosphere at 2 points [winds create waves])*

In addition to solar energy, other forms of usable energy are being developed, including: Tidal power; Geothermal; Biological Hydrogen Production; and Ocean Thermal Energy Conversion (OTEC). These methods of power production are beyond the scope of this book.

Why are renewable forms of energy so important?

In a nutshell, all forms of fossil fuel are running out and figures such as 40–50 years of fossil fuel stocks are all that remains if the current rate of consumption continues. Fossil fuels include coal, oil and natural gas, each with a high percentage of carbon, which resulted from the decomposition of buried dead organisms millions of years ago. In addition it is feared that releasing the CO_2 and other pollutants into the atmosphere, from burning these fuels, could result in **'climate change'**, raising the temperature of the Earth due to an increase in the 'greenhouse effect'.

The greenhouse effect

This relates to the process in which the energy leaving the Earth's surface is absorbed by some of the gases in the atmosphere, called greenhouse gases, which include: water vapour; carbon dioxide, methane and ozone. They re-radiate this energy in all directions, including back down towards the Earth's surface, causing additional heating.

Impacts of climate change

- Sea level would rise *(during the last century they rose by 150–200 mm)*
- Extreme weather could be experienced *(e.g. drought or flooding)*
- Extinctions and loss of biodiversity will occur.
- Increased levels of disease would affect human health.

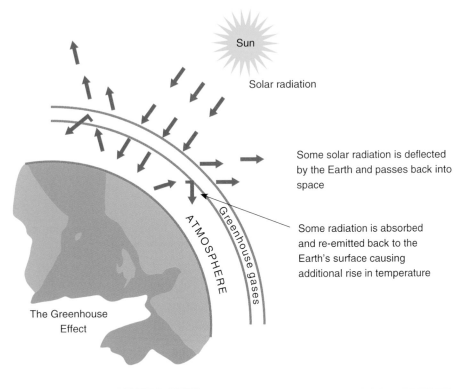

Sun

Solar radiation

Some solar radiation is deflected by the Earth and passes back into space

Some radiation is absorbed and re-emitted back to the Earth's surface causing additional rise in temperature

ATMOSPHERE

Greenhouse gases

The Greenhouse Effect

Light energy

Wind energy

Wave energy

Renewable energy

10 Energy Conservation

Energy Costs & Payback Period

Calculating energy costs and the kilowatt hour (kWh)

A kilowatt hour is the unit of measurement used by gas/electricity companies to measure the amount of energy we have used in our homes. A one-bar electric fire typically uses 1 kW of energy for each hour it is switched on, and that would be 1 kWh. If the fire was left on for 6 hours, it would use 6 kWh of energy. A 100 W light bulb if left on for 10 hours would use 1 kWh of energy. If you replaced this light by a 50 W bulb, it could remain on for 20 hours before it used the same amount of energy.

Running costs and energy savings

Assuming the energy cost for 1 kWh of electricity is 15 pence; the resultant cost of running a 100-watt light bulb, assuming it is left on for 4 hours a day, every day for a year, would be:

$$0.1 \text{ kW} \times 4 \text{ hr} \times 365 \text{ days} \times 15 \text{ pence} = 2190 \text{ pence or } £21.90$$

Should you replace this bulb with an energy-efficient light bulb that operates at, say, 20 W (i.e. $\frac{1}{5}$ of the running cost!), how much will you save each year?

$$£21.90 \div 5 = £4.38 \text{ running cost}$$

$$\text{Therefore } 21.90 - 4.38 = £17.52 \text{ saving!}$$

Over 8 years, assuming you get that long from the bulb and energy costs do not change, you will have a total long-term saving of $£17.52 \times 8 = £140.16$ over that of using the old design of 100 W light bulb. Multiply this by a number of bulbs and the saving in costs can be staggering.

Payback period

You may see no point in replacing a perfectly good, standard 100 W light bulb with a new energy-efficient bulb. However, if the cost of a replacement energy-efficient bulb is £5, the payback period would be:

$$\text{Cost of new bulb} \div \text{Savings each month}$$

$$\text{So: } £17.52 \text{ saving a year} \div 12 \text{ months} = £1.46 \text{ saving each month.}$$

$$\text{Therefore: } £5 \div £1.46 = 3.42$$

Thus the new bulb would take a $3\frac{1}{2}$-month period to pay for the extra cost of purchasing the new light bulb. This exercise demonstrates the term 'payback period', which is often used when discussing renewable energy technologies.

    For example: If you install a solar system at a cost of £3500 and you save £310 on your energy bill per year, the payback period will be (3500 ÷ 310), or around 11 years.

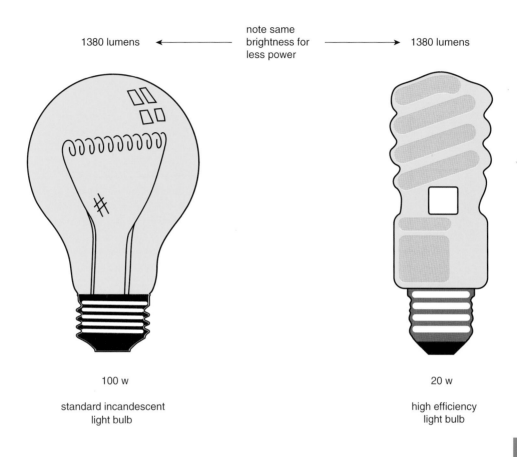

1380 lumens ← note same brightness for less power → 1380 lumens

100 w

standard incandescent light bulb

20 w

high efficiency light bulb

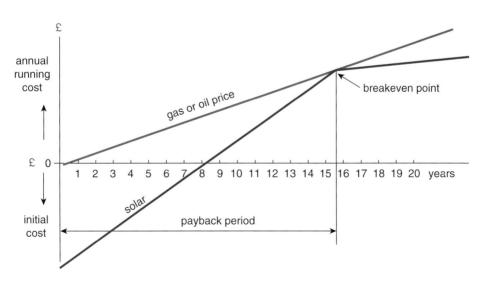

£

annual running cost

£ 0

initial cost

gas or oil price

breakeven point

1 2 3 4 5 6 7 8 9 10 11 12 13 14 15 16 17 18 19 20 years

solar

payback period

Solar Energy

Solar energy is the radiant heat and light from the sun. Two distinct approaches have been adopted to capture solar energy; these are:

- Solar heating
- Solar power.

Solar heating

With this system water is allowed to get hot within a thermal collector as it is warmed up by the radiant heat from the sun. Direct and reflected radiant heat from the sun falls upon the surface of the collector and in so doing warms the water within. This water is transferred to a hot storage vessel to be used as and when required. Transferring the heated water may take place within either an open (direct) or closed (indirect) circuit, by either natural convection currents being set up or by the use of a circulating pump.

There are two basic designs of solar collector:

- Flat panel
- Evacuated tube.

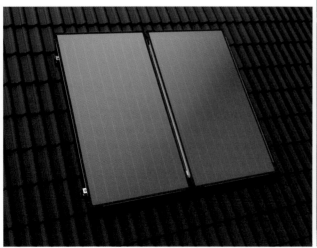

Flat plate collector

Evacuated tube collector

With the flat plate collector the water is directly heated as it flows through the simple, glass-topped insulated box with a flat solar absorber made of sheet metal attached to copper pipes and painted matt black. The evacuated tube, on the other hand, contains a fluid that heats up and vaporises; the water within the dhw system is heated indirectly as it flows past the heat-exchanger part of the solar collector, located at one end, turning the vapour back into a liquid for reheating. These solar collectors are illustrated on page 381.

Solar power

Solar power is the conversion of sunlight into electricity. Two basic methods can be employed, to include:

- Photovoltaic (PV), which directly transforms solar energy into electrical energy
- Concentrating solar power (CSP). This utilises lenses or mirrors and tracking systems to focus a large area of sunlight into a small beam, which in turn is used to heat up water to boiling temperature to provide power to turn a Stirling engine (similar to a steam engine), which turns a generator to produce electrical power.

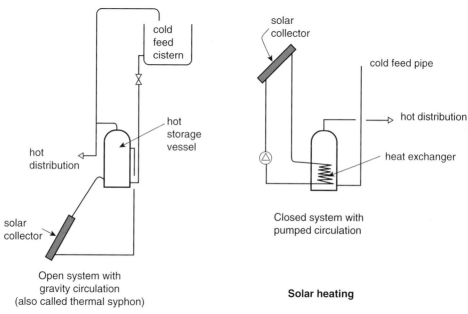

Open system with gravity circulation (also called thermal syphon)

Closed system with pumped circulation

Solar heating

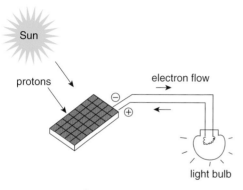

Solar power

Solar thermal heating and photovoltaic panels

Solar Hot Water Heating Systems 1

Relevant British Standard
BS 5918 & BS EN 12976-1

A solar water heating system for domestic hot water comprises three main components:

1. Solar collector
2. Circulatory pipework to and from the solar collector & hot store
3. Hot water storage vessel.

Solar collector

Two distinct types of solar collector will be found, as shown on the previous page. The evacuated tube collector is the most efficient; however, it is not as robust as the flat plate collector and is often subject to being easily damaged. The amount of solar heat one can expect to achieve from a solar collector would be dependent on your location, for which the map opposite gives some indication as to the annual expectation. For example, in the south-east of Britain you would get approximately 1100 kWh/yr for every m^2 of solar panel surface, which should deliver around 40–45 litres of hot water at 60°C, although this would be dependent upon the collector selected. It is generally recommended that you have at least 3 m^2 where an evacuated tube collector is used and at least 4 m^2 should a flat plate collector be selected.

Circulatory pipework to and from the solar collector

The solar heat from the sun is collected by the solar collector and warms up the water within. This heated water is then passed to the hot storage vessel, where it is stored for use when required. The method of transferring the water to the hot store is usually by means of a circulating pump, although systems relying on gravity circulation will be found (i.e. allowing the hot water to rise by convection currents). Where a pump is to be used for this purpose sensors located on the hot store and solar collector allow the pump to switch on and off as necessary, allowing for water circulation only at times that the water in the solar collector is warmer than that located within the hot store.

Hot storage vessel

For a new system this would typically be a well-insulated cylinder, of no less than 300 litres. The cylinder itself would have two heat exchangers located inside, one to be used to pass the water from the solar collector through and the other to be used to pass water from a boiler that has been installed as backup.

On average, each individual person uses 40 litres of hot water per day and such a large vessel would be more than sufficient to meet the needs of the customer, but where the cylinder is of smaller size the potential for the freely heated water would be missed, as the circulation would stop once the temperature reached 60°C, being the maximum recommended temperature of the stored water. Therefore the storage tank should ideally exceed by 1.5–2 times the daily water use to maximise the benefits of the system. Where the water temperature of 60°C is not achieved, then some form of backup, such as a boiler, should be used to raise the temperature, thereby limiting the risk of Legionella forming.

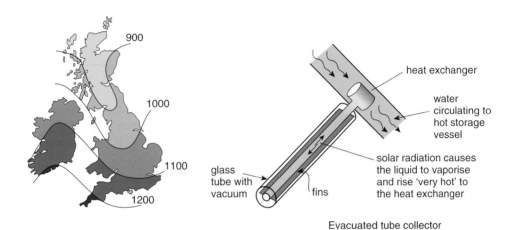

Solar Hot Water Heating Systems 2

Design considerations

Owing to the changes in temperature, which may range from −15°C to as high as 200°C, the liquid used within the circulatory pipework needs to be protected from causing damage. In the case of the lower temperatures an antifreeze solution could be added to the liquid within the closed circuit. As far as the higher temperatures are concerned, pressurising the system prevents the water from turning to steam, but it should be noted an expansion vessel would be required to take up the additional expansion – plus a pressure relief valve would be incorporated where expansion of the water exceeds the size allowed for in the system. Alternatively a drain back system could be installed, whereby the water in the collector is allowed to drain out of the panels to prevent over-temperature situations or frost damage.

Mounting the solar panels

Generally planning permission is not required, but you would need to confirm that to be the case with the local planning officer, as within certain areas, such as a conservation area, permission may be required. Building Regulation approval will be required, however, because the roof structure will need to be inspected to ensure it will take the additional loads imposed by the solar collectors.

The panels may be either mounted onto an existing roof, or may be built in with the roof and in effect act as the roof covering, made good where necessary by flashing into the tiles used for the remainder of the roof.

Roof orientation

The roof should ideally be south-facing to maximise the solar energy it receives; however, providing it faces between SE and SW it should be acceptable. The angle at which the panels are fixed should ideally be between 30 and 40°, which is fortunate because in the UK this is the average tilt of a house roof. Ideally you need to ensure surfaces are not affected by shading from trees etc. as that can cause a loss in performance.

Maintenance

Bird excrement and general air pollution etc. can cause losses of between 2 and 10% if they are never cleaned off; therefore the panel will require cleaning. In addition the expansion vessel and pressure relief need to be checked, as does the level of antifreeze solution. In the case of drain back systems, they will require draining in winter.

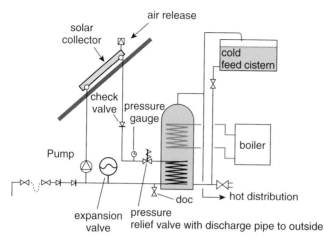

Pressurised system of
solar dhw

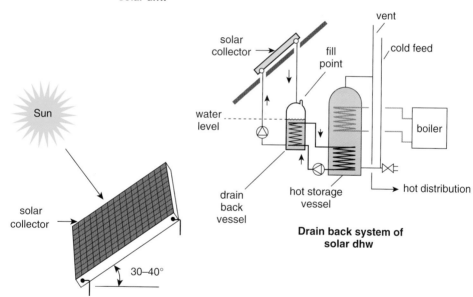

Drain back system of
solar dhw

Solar Power

The most common method of converting solar power into electricity is to use a photovoltaic (PV) cell. These are essentially silicon-based semiconductors that, when exposed to the photons in light rays from the sun, cause free electrons within the material to travel along a conductor around an electrical circuit.

Three distinct types of PV cell will be found, including:

- Monocrystalline silicon crystals, with an efficiency of between 15 and 18%. These are the most expensive PV cells to produce.
- Poly-crystalline silicon, with an efficiency of between 8 and 12%
- Amorphous silicon, which is the cheapest to manufacture and is found both in rigid and flexible form. These have the lowest efficiency, of around only 4 to 6%

Photovoltaic panels in use

Ideally the PV cells should be located so that they are facing due south, and ideally at an angle of about 30° so as to ensure the maximum amount of light falls onto them. The system could be a standalone set-up, where all the electricity generated is used directly, or electricity can be stored within a series of batteries for use as and when required. Alternatively the system could be connected to the national grid, whereby the excess electricity generated is fed back into the supplier's network, for which you will be paid.

Where a standalone system is employed, it would be necessary to have some form of backup generator to use for power when all the battery charge has been used.

An array of PV cells would be rated in watts or kilowatts peak (kWp), which relates to the power it would generate at midday. A 2 kWp unit covering an area of about 20 m^2, if poly-crystalline silicon was used, would typically supply about 1600 kWh per year, which would meet about ⅓ of a typical domestic household's demands.

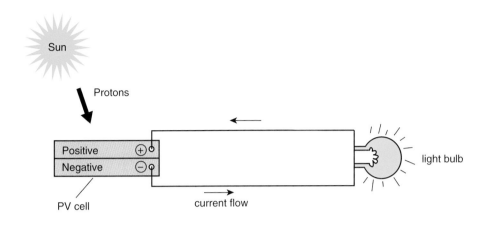

Sun

Protons

Positive ⊕○
Negative ⊖○

PV cell

current flow

light bulb

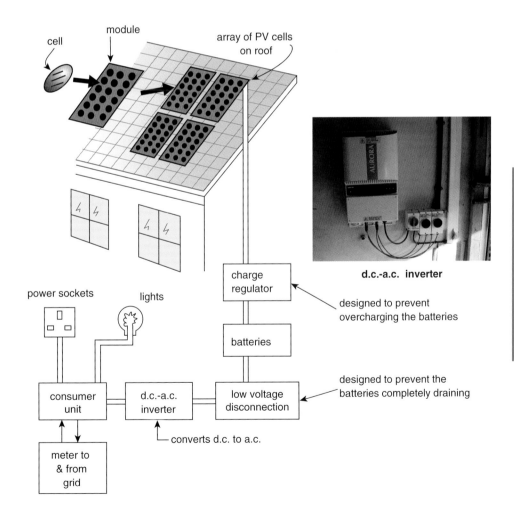

cell

module

array of PV cells
on roof

d.c.-a.c. inverter

charge
regulator

designed to prevent
overcharging the batteries

power sockets

lights

batteries

designed to prevent the
batteries completely draining

consumer
unit

d.c.-a.c.
inverter

low voltage
disconnection

converts d.c. to a.c.

meter to
& from
grid

The Heat Pump

A heat pump is a device that utilises the heat extracted from one area to heat another. The heat pump is like a refrigerator; note how the area at the back of a fridge is hot, as it has taken the heat from within the cooled compartment. There are three key elements that go to make up a working heating system:

1. Heat collector (e.g. a ground loop or air source)
2. Heat pump
3. Heat distribution system (e.g. an underfloor heating system).

Heat collector

Solar energy heats up the ground to a reasonably constant temperature of around 8–10°C, with seasonal variations close to the surface of 1–2°C. If sufficient pipe of, say, 200 to 400 m in length was run within the ground and cold water was pumped through the pipe slowly but surely, the water within the pipe would rise in temperature to that of the area surrounding the pipe. And this temperature is all that is required for the heat pump to work.

Heat pump

The heat pump itself consists of four components:

- Evaporator
- Condenser
- Compressor
- Expansion valve

The **evaporator** is basically a heat exchanger where the pre-warmed water from the heat collector passes and in so doing heats up the refrigerant contained within the closed circuit of the heat pump. This liquid has a very low boiling point and rapidly heats up as it boils, changing to a vapour.

The vapour formed is then pressurised by the **compressor**. Increasing the pressure of the vapour causes it to rise in temperature, and it is pressurised sufficiently to increase its temperature to a point at which it is hot enough to heat the water within a water-filled heating system.

The hot vapour next passes into the **condenser**, which is another heat exchanger, where the heat distribution pipe passes through a coil within. The hot vapour passes over the coil and in so doing cools as it heats up the water within the heating system. As the vapour cools, it condenses back to a liquid.

Finally the cooled vapour passes through into the **expansion valve**, which in effect is a small orifice causing a big drop in pressure and temperature, making the refrigerant ready for reheating within the evaporator by the water taken from the outside heat collector. The process is simply repeated.

Heat distribution

The heating system water that was heated within the condenser, identified above, is now simply pumped through the circulatory pipework to warm the building as necessary.

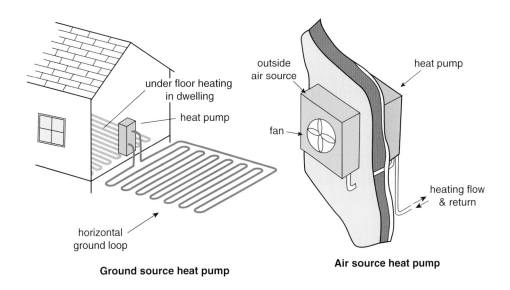

Ground source heat pump

Air source heat pump

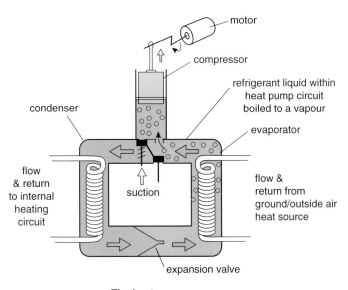

The heat pump

Installation of a Heat Pump

The choice of whether to install a heat pump or not would depend upon several circumstances, including two key factors: namely, the:

■ Location of the heat source (i.e. from the ground, water or air)
■ Type of heat distribution system (underfloor heating or radiators?)

Heat source
The method of heating the water that is to be fed to the heat pump ideally would be taken from the ground, as this would be the most stable source available; however, where space is limited it is possible to take the heat from the outside air. During very cold conditions this would be unacceptable, though; alternatively a lake or pond nearby would be acceptable, but again due to seasonal variations problems may arise in achieving the required temperature rise of the water.

Where a ground source heat pump is employed the pipe loop is typically installed at 1 m depth, with trenches about 3–5 m apart and ideally pipes in each trench 0.3 m apart to prevent cooling the ground too much. All pipe runs should be at least 1–1.5 m away from the building. Note that the water used to be circulated through the heat collector needs to be treated with an antifreeze solution to prevent it freezing when exposed to 0°C or below.

Ground conditions
Firstly, is there room to install a pipe loop within the ground? If this method is selected, it must be understood that the area chosen could not be built upon, as this would prevent the heat from the sun re-warming the ground as required. Also the type of soil would have an influence on the length of the pipe coil used: for example, moist, packed soil would have a greater conductivity than dry, loose soil, which traps air. If the soil is dry and loose you may need up to 50% more ground loop. If space is restricted, it is possible to drill a borehole down 100 m or so. The advantages of boreholes – apart from utilising less space – is that ground temperature conditions are much more stable, but a borehole costs much more to install. A good alternative where space is limited is to install what is referred to as a 'slinky', which is loops of pipework, but care needs to be maintained to ensure the ground temperature can recover.

Coefficient of Performance (CoP)
The CoP is a ratio of useful heat energy output to the amount of electrical energy input, and the greater it is, the more efficient your system would be. For example, a CoP of 3 means that for every 1 kWh of electricity input you will get 3 kWh of heat output. A well-designed system would achieve a CoP of 3–4. Electricity is required for running pumps and time clocks etc., and where photovoltaics are used to supply this energy greater savings can be made. The CoP attainable is limited ultimately by the temperature rise needed to make the heat useful, and it is for this reason that underfloor heating is ideally suited to the design of technology where heating system temperatures are kept to a minimum, usually around 30 to 45°C.

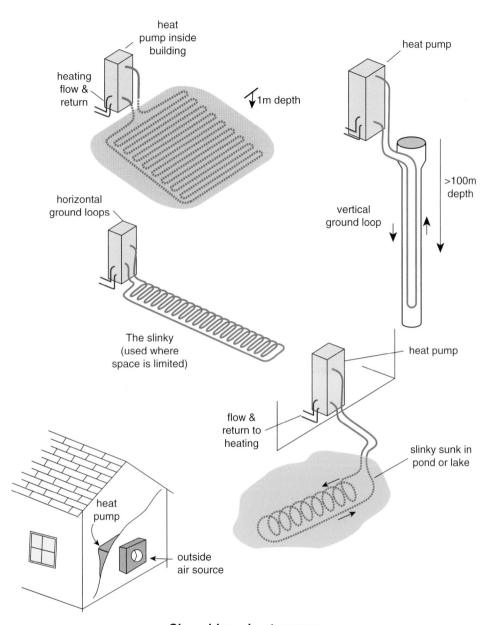

heat
pump inside
building

heating
flow &
return

↓ 1m depth

horizontal
ground loops

The slinky
(used where
space is limited)

heat pump

> 100m
depth

vertical
ground loop

heat pump

flow &
return to
heating

slinky sunk in
pond or lake

heat
pump

outside
air source

Closed loop heat pumps

10 Energy Conservation

Biomass

The term Biomass relates to a renewable source of energy that is, in effect, 'carbon neutral' and therefore will not increase the carbon dioxide (CO_2) levels in the atmosphere. It is derived from recently living organisms, unlike fossil fuels, which were derived from ancient living organisms. Fossil fuels took their CO_2 from the atmosphere a long time ago and the carbon they contain has been locked up for a very long time. Burning fossil fuels increases the CO_2 content in the atmosphere.

Carbon neutral

From an environmental point of view, burning wood from a sustainable source neither adds to nor reduces the overall level of CO_2 in the environment. Burning wood will produce only as much CO_2 as was absorbed when the tree was growing, and no more than that produced by the tree if it was left rotting on the forest floor. So, as long as the wood used comes from a managed forest, i.e. one tree planted for every one felled, then CO_2 emissions are in balance. It would be a different situation if the trees were simply cut down to be burned and not replaced. Alternatively, fast-growing trees like willow; poplar; hazel and chestnut are being grown on a short-rotation coppice cycle every 3–4 years and processed into wood chips. (Coppicing refers to cutting the trees down to ground level, but leaving the tree stump to regrow.)

Solid biomass

In solid form biomass can be found produced from wood; straw and grasses, etc. and is available as either logs, pellets, bricks or wood chip.

Liquid bio-fuel

Crops such as oilseed rape and sugar cane can be fermented to produce *biodiesel* and transportation fuel. Biodiesel can also be made from used vegetable oils collected from restaurants and fish-and-chip shops, etc.

Biogas

This is often produced by digesting human/animal waste anaerobically (in the absence of oxygen). Rotting garbage also releases methane gas naturally, such as is found in landfill sites, and is collected as a landfill gas.

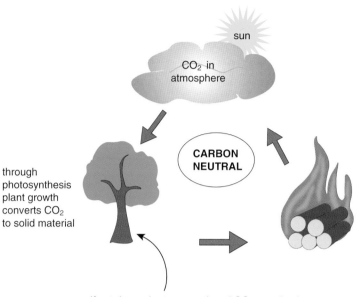

sun

CO_2 in atmosphere

CARBON NEUTRAL

through photosynthesis plant growth converts CO_2 to solid material

burning solid material releases the CO_2 back into the atmosphere

If cut down plants are replaced CO_2 remains in balance and therefore will not contribute to global warming

Biomass

Biofuel

Biogas plant

Biomass/Wood Burners 1

A modern wood burner is typically 80–90% efficient, which is a far cry from the old, traditional open fire where no more than 20% of the energy is utilised. The wood pellet and wood chip boilers are the most versatile and function much like any fossil fuel boiler, feeding the burner automatically with the fuel with ignition being achieved by the use of a small electric hotplate. The pellets or wood chips are supplied in bags of up to 20 kg in weight for hand feeding into the hopper head, or the fuel can be bulk-purchased and stored adjacent to the boiler, being automatically fed to the hopper head via a vacuum transfer or auger-fed connection tube.

Wood logs

These may be from slow-growing hardwoods, from broad-leaf or deciduous soft-wood trees, or from conifers and fast-growing evergreen trees. Because hardwoods are denser than softwoods they generally produce about twice as much heat as the same volume of softwood.

Logs need to be fully seasoned, i.e. must have been cut at least 1–2 years prior to use, so as not to contain too much water; the moisture content should ideally be less than 20%. Heat energy is wasted in burning timber where the water must first be driven off. The best way to dry timber is to split the logs rather than simply cutting them to length.

Wood chips

These are simply small pieces of timber 20–50 mm in length that have been generated by passing waste wood and tree saplings etc. through a shredder. Wood chippers cut across the grain by the action of the blades rotating round the surface of a disc. The wood is either fed into the shredder manually or by hydraulic-powered feeding rollers. In order to prevent decomposition of the wood chips they must not be stored with a moisture content of greater than 25%. One of the biggest problems with wood chips is the unknown nature of the timber used and the possible contaminants, such as wood preservative, which may affect the combustion process of the fuel. Wood chip boilers are generally only available with heat inputs greater than 30 kW.

Wood pellets

Pellets are made from wood shavings and sawdust that has been compressed under high pressure and pressed into small cylindrical shapes, approximately 6 mm in diameter and 20 mm long. The pellet is held together by lignin, which is naturally contained in the wood. A pellet of good appearance will have a surface that is smooth and shiny, with no dust. Poor-quality pellets will be seen to have longitudinal cracks and there will be a high dust content.

Other materials, such as hemp and straw, are now being produced to form pellets. Therefore if you change the type of pellet the boiler was set up to burn, it may require the air intake to be adjusted in order to ensure good combustion.

Increasing costs

Logs   Chips   Pellets

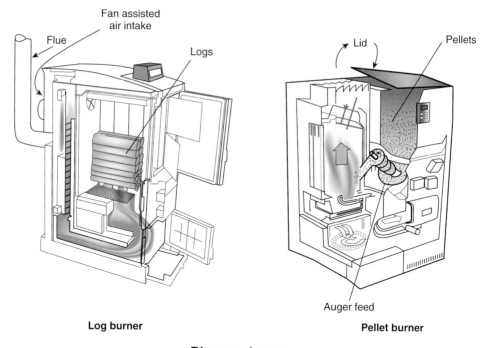

Fan assisted air intake

Flue

Logs

Lid

Pellets

Log burner

Auger feed

Pellet burner

Biomass burners

10 Energy Conservation

Biomass/Wood Burners 2

Fuel storage

The provision of a dry storage area is essential, and where automatic feed is required the store should be designed to prevent pockets of pellets/chips from accumulating where they cannot fall into the collecting point of the feeder. The store would also need to be located where delivery can be easily made.

Corresponding approximate volume of wood to heating oil		
1 m³ stacked softwood logs	120 litres or 1200 kWh	1 kg = 4 kWh
1 m³ stacked hardwood logs	220 litres or 2200 kWh	1 kg = 4 kWh
1 m³ wood pellets	325 litres or 3250 kWh	1 kg = 4.9 kWh
1 m³ wood chips	104 litres or 1040 kWh	1 kg = 3.4 kWh
1 m³ heating oil	1000 litres or 10,000 kWh	1 kg = 11 kWh

It should be noted that for the same amount of heat energy, the storage space required for wood pellets would need to be more than 3 times greater than that required for heating oil, and this storage would be much greater again where logs or chips are to be used.

The moisture content of the wood supplied will have a big impact if it is purchased by weight, for timber with a high water content will weigh more and energy is wasted in converting the water to steam to drive it out.

The buffer or thermal storage unit

Traditional wood-burning installations used open-circuit designs of plumbing for the domestic hot water (dhw) and central heating, only allowing the energy to be transferred directly into the heating or dhw circuit as applicable. Today's modern systems allow the energy to be stored in accumulators, buffer tanks or thermal storage units for use later when the stove or boiler is not in operation. Biomass boilers work most efficiently and cleanly when they are working at their maximum output, and because of this it is essential not to oversize the appliance – another reason for the inclusion of a buffer tank to allow the boiler to run for longer periods without short cycling. The installation of a buffer/thermal storage unit also allows for the integration of a fossil fuel appliance into the installation to act as a backup or for the inclusion of a solar panel, although the solar panel is generally connected directly to the dhw cylinder to ensure that the solar energy is not being used for the space heating and to overcome the greater losses from passing through several heat exchangers.

The formation of ash will occur in the boiler, but owing to the high level of efficiency found within a modern appliance cleaning this out is generally limited to about once every 3 weeks. Some appliances do not require cleaning out for 2–3 months at a time, but a lot will depend upon the fuel used.

The flue system used generally needs to be double wall insulated stainless steel, with the flue height at least 1 m above the roof line. Wood burning temperatures are far greater than those of gas or oil.

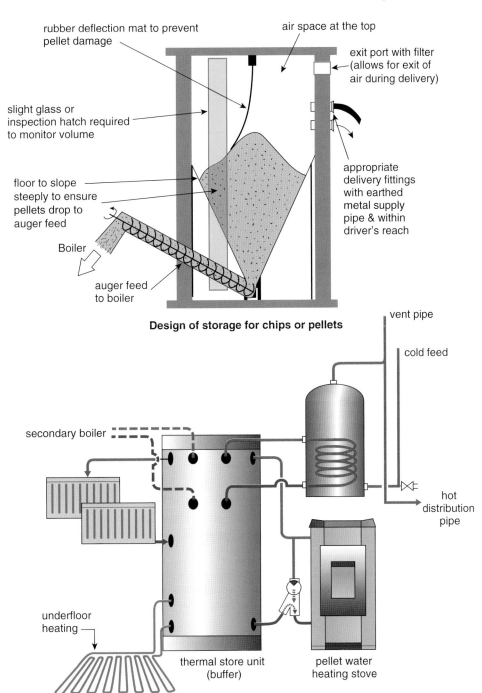

Note: chamber is to be free of electrical works to avoid dust explosions

rubber deflection mat to prevent pellet damage

air space at the top

exit port with filter (allows for exit of air during delivery)

slight glass or inspection hatch required to monitor volume

appropriate delivery fittings with earthed metal supply pipe & within driver's reach

floor to slope steeply to ensure pellets drop to auger feed

Boiler

auger feed to boiler

Design of storage for chips or pellets

vent pipe

cold feed

secondary boiler

hot distribution pipe

underfloor heating

thermal store unit (buffer)

pellet water heating stove

10 Energy Conservation

Wind & Micro Hydro (Water) Power

When the wind blows or water flows downhill there is an ability to produce electrical energy. This is simply generated by allowing the wind or water to turn a turbine. The turbine consists of a rotor assembly with a shaft and blades attached. As fluids pass over the blades, they impart rotational energy to the rotor. The rotor is the part of an electric generator that turns, allowing a magnet to rapidly pass a coil of wire which in effect forces the free electrons in the windings to flow out through the external electrical circuit.

Wind power

Wind is created by the sun warming up the Earth's surface, which causes the air to become heated and in so doing it expands as it warms up. This hot column of air now rises and is replaced by denser, cold air that travels across the land from a column of descending air that is cooling down many miles away. It is this movement of air from one column to the next that results in a wind, and its speed depends upon the rate of the air rising and falling.

As the wind passes the blades of the wind turbine they turn as a result of aerodynamics. Air passes more rapidly over the longer side of an aerofoil blade, creating an area of lower pressure to that side. Because the blades of a turbine are constrained, they are forced into a rotation. The siting of a wind turbine needs to be such that it can collect as much of the wind potential as possible, and therefore a site with buildings and trees within close proximity would be inadequate. For an effective system, typical annual average wind speeds at the height of the generator should be 5–6 m/s, with speeds up to 7 m/s for commercial wind farms. Ideally the blades should be located as high as possible, and the larger in size they are, the better; for example, doubling the size of the blade increases the power that can be harvested fourfold.

Water power

The potential for converting flowing water into electrical energy depends on the amount of water flowing per second (the flow rate) and the height from which the water falls (the head or pressure generated). If you had a very good flow rate, you would still require at least 1 m of head, as fast-flowing water in itself does not contain sufficient energy for useful power production. The potential power output in kW from a water turbine is found using the following formula:

$$5 \times \text{flow (m}^3\text{/s)} \times \text{head (m)} = \text{P (kW)}$$

So assuming the flow rate was to be 0.4 m/s and the head was 1 m, the potential output would be $5 \times 0.4 \times 1 = 2$ kW. Increase the head to, say, 10 m and the output would be 20 kW. Two distinct designs of water turbine will be found, impulse and reaction turbines, as shown in the illustrations opposite.

Generating your own electricity by using wind or water power can be undertaken using a standalone system with some form of backup generator, for the power supply when all the battery charge has been used. As with PV systems, any additional power generated could be sold back to the electrical supply company.

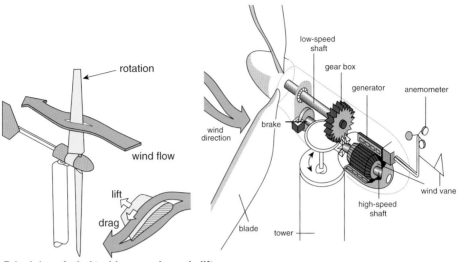

rotation

wind flow

lift

drag

Principles of wind turbine aerodynamic lift

low-speed shaft

gear box

generator

anemometer

wind direction

brake

blade

tower

high-speed shaft

wind vane

Flow over hills and obstacles

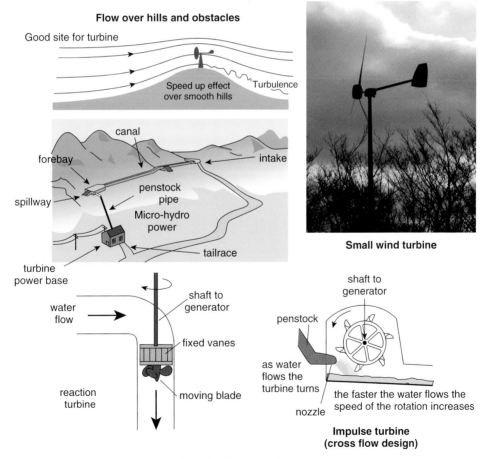

Good site for turbine

Speed up effect over smooth hills

Turbulence

canal

forebay

intake

spillway

penstock pipe

Micro-hydro power

turbine power base

tailrace

Small wind turbine

shaft to generator

water flow

fixed vanes

reaction turbine

moving blade

shaft to generator

penstock

as water flows the turbine turns

nozzle

the faster the water flows the speed of the rotation increases

Impulse turbine (cross flow design)

Wind and micro hydro power

Combined Heat & Power (CHP)

Combined heat and power (CHP) integrates the production of usable heat and electrical energy into one process, rather than allowing the excess heat to be wasted.

Packaged and mini-CHP

Packaged and mini-CHP units are designed to meet the heat requirements of a medium-sized to large standalone building, such as a block of flats or a hospital, although it could be used for a whole range of numerous other examples. These units can also be designed to feed back excess power generated into the electricity network, or to contribute excess heat to a district heating network. Once installed, they can cut energy bills by thousands of pounds per annum and will reduce CO_2 emissions considerably. There is a specific form of micro-CHP that has been designed for individual households. It acts as a replacement for a standard domestic gas boiler and will generate around 1 kW/h of electrical energy, mainly for consumption in the home. CHP domestic boilers can potentially reduce yearly fuel bills by up to £600, and reduce household carbon emissions by up to 40%. Electricity generation within the CHP unit generally uses either the principle of the Stirling engine or the organic Rankine cycle.

The Stirling engine

This system operates by cyclic compression and expansion of nitrogen gas contained within a sealed cylinder. Heated gas expands and cooled gas contracts. When a burning mixture of gas is played onto the top part of the cylinder the gas inside heats up, expands and in so doing forces a piston inside to move downwards; the hot, expanded gas is then exposed to the cooled water surrounding the lower part if the cylinder and subsequently it contracts. This continued expansion and contraction of the gas and subsequently the lowering and raising of the piston is used to turn the crankshaft and rotor of an alternator.

The water that passes the lower part of the cylinder to cool the nitrogen is used for the heating circuit.

The organic Rankine cycle CHP unit

The organic Rankine cycle (ORC) is broadly similar to a heat pump, with the exception that when the refrigerant is turned to a vapour it is used to turn the rotor of an alternator. The ORC consists of an organic fluid of high molecular mass, which is pumped around a closed circuit that is indirectly heated by the boiler to produce a vapour. The organic vapour, once it has passed through the turbine, passes onto a condenser/heat exchanger where it cools and condenses back to a liquid and in so doing heats water from the central heating circuit. The condensed fluid is pumped back to the evaporator and so the process continues.

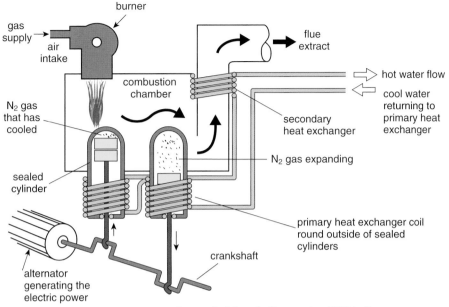

Concept of the sterling engine CHP boiler

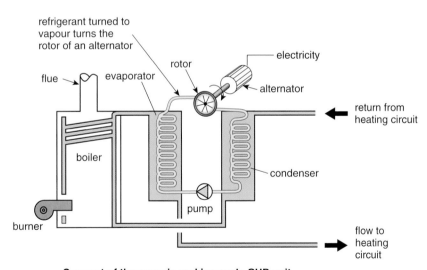

Concept of the organic rankine cycle CHP unit

Combined heat and power

Rainwater Harvesting 1

Relevant British Standard
BS8515

Rainwater harvesting simply refers to collecting the rainfall that drops onto the roof of your property. Collecting the water in this way can alleviate the effects of localised flooding, where the existing drainage system is struggling to cope, and in addition the water collected can be reused for the garden or, increasingly, for WC flushing and washing machines without the need for treatment. Around 50% of water in a domestic dwelling and over 60% of water in offices can be used for this purpose. Cutting down on water usage ultimately saves energy on water treatments. Harvested water is also free from chlorine and limescale.

Sustainable drainage systems (SUDS)

SUDS aim to use the rainwater around a property and ultimately dispose of it within the site boundaries, thus preventing downstream pressure that could lead to drains overcharging and flooding. Such systems often use the water collected to support rooftop gardens or 'green roofs' and provide greater biodiversity at a site, attracting many different species of insect, bird and animal life.

Potential rainwater to be harvested and cost saving

To find the volume of rainwater that can potentially be collected, you carry out the following calculation:

Collection Area (m^2)	×	Yearly Rainfall (mm)	×	Drainage Factor	×	Filter Efficiency	×	Total Yield (litres)

The yearly rainfall and drainage factors to be used can be taken from the tables opposite, and the filter efficiency is given by the manufacturer, but will usually be between 0.8 and 0.9.

Example: A tiled, pitched roof of $125\,m^2$ in the Midlands can expect to harvest:

$$125 \times 755 \times 0.75 \times 0.85 = \underline{60,164\ \text{litres}}$$

In round figures this is just over $60\,m^3$ and, if the cost for the water in this area was to be £2.20/m^3, there would be an annual saving on the water bill of around $60 \times 2.2 = \underline{£132}$.

The size of a storage tank for a small dwelling would typically be 3000 litres, while for larger properties 15,000 litres and over would be normal.

Preventing bacterial contamination

Salmonella is found in the faeces of birds and small mammals and therefore is very likely to be found on a roof. Survival in water and bio-films is low and these pathogens are bound to particles, so good filtration to 1250 microns (1.25 mm) maximum before storage ensures that bacteria proliferation is not maintained. Pathogens collecting on roofed areas are also exposed to UV light, heat and drying out, which destroys many bacteria. Note that the filter is installed prior to the storage tank to allow the filter to occasionally dry out. A continual wet, slimy location is a breeding ground for bacteria.

E. coli is another pathogen and is found within the faeces of domesticated animals. Because these faeces are not found on roofed areas they are less of a problem, which is one reason why rainwater harvesting for reuse is not recommended from paved areas. Water collection from paved areas may also contain petrol and oil spills.

Tanks may be positioned above or below ground; above-ground tanks, however, must maintain a dark environment so that algae growth cannot proliferate. It will also be necessary to keep the temperature down to an average of <20°C, again to prevent any bacteria from multiplying. Additional UV treatment may be required where above-ground tanks are used, as UV light destroys certain bacteria.

Because the harvested water is not potable it must be noted that where water is to be supplied to an outside tap, for watering or car washing purposes, it is essential that a "Not Drinking Water" sign is affixed above the tap. In addition, to ensure no cross-contamination, under the Water Regulations pipes that are to be used for a rainwater supply must be identified as a non-potable supply by marking the pipe with a band of green-black-green along its length and at no greater than 0.5 m intervals.

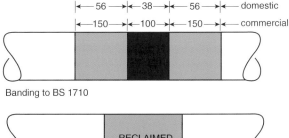

Method of marking pipework

Yearly rainfall	(mm)
Scotland (N)	1671
Scotland (E)	1135
Scotland (W)	1732
N & NE England	755
NW England & N Wales	1291
Midlands	755
East Anglia	606
SW England & S Wales	1247
SE England	777
Northern Ireland	1111

Drainage factors for roofs	
flat roof	0.5
pitched roof with tiles	0.75
pitched roof with slates	0.8
large metal roofs	0.85–0.9

Green roof

Rainwater harvesting 1

Rainwater Harvesting 2 (System Design)

Relevant British Standard
BS8515

Two distinct systems will be found: the direct and indirect systems. The basic difference between the two systems is that with the direct system the rainwater is distributed directly to the appliances from the storage tank, whereas with the indirect system the water is supplied to the appliances via a header cistern. Note: to ensure that water is always available, even when there has been little or no rainfall, mains water is supplied as a backup, as shown. Rainwater is classified under the current Water Regulations as Fluid Category 5; therefore it must have a type AA or type AB air gap or conform to EN1717 to ensure no back siphonage into the mains supply.

Excess rainwater filling the storage tank is run off to a soak-away or drain. A special anti-rodent trap should be located on this exit pipe to prevent the entry of vermin and also smells from the drain. Also, if this discharge is to be connected to an existing drainage system, an anti-surcharge valve needs to be installed downstream of the trap to prevent the tank being filled via this route should the drain become overloaded.

Stored water
Water supplied to the header cistern may be supplied with a float-operated valve or via a float switch, as shown in the illustration. Using a float switch in this way reduces wear and tear on the pump by restricting the potential for continued on–off switching of the pump. As with all storage cisterns, these should be insulated to protect against heat loss and heat gain. The growth of the *Legionella* bacteria can occur if temperatures rise above 25°C.

Aesthetic qualities
Harvested rainwater needs to be clear and odourless if it is to be accepted by the end user. Water collected from 'green roofs' should be avoided, as the water would undoubtedly be stained due to passing through the soil, as would water collected from a dirty roof: it would look unsightly within the toilet bowl and be of no use for washing purposes, causing staining to the clothes. Black specks floating within the water suggest that there is a build-up of sediment at the base of the tank or cistern and could be the result of not installing the inlet with its outlet facing upwards, as shown, so as not to stir up this sludge. Smells would not normally be present, and if they are experienced it suggests a possible build-up of organic material within the storage tank.

Maintenance
Clearly all roof areas and gutters need to be kept clean. It is recommended that all filters are cleaned every 3 months, although it is possible to have automatic backwashing for commercial system filters. If the filter is not clean, less water will be harvested. The storage tanks etc. should be inspected annually, and cleaned if necessary. On a final point of safety, never enter the confines of an enclosed tank, as unknown gases might be present, which could lead to rendering you unconscious.

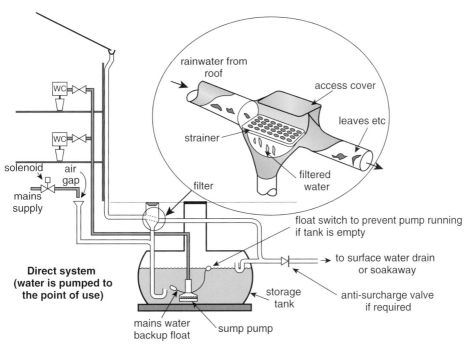

rainwater from roof

access cover

leaves etc

strainer

filtered water

filter

float switch to prevent pump running
if tank is empty

**Direct system
(water is pumped to
the point of use)**

solenoid

air gap

mains
supply

WC

WC

to surface water drain
or soakaway

storage
tank

anti-surcharge valve
if required

mains water
backup float

sump pump

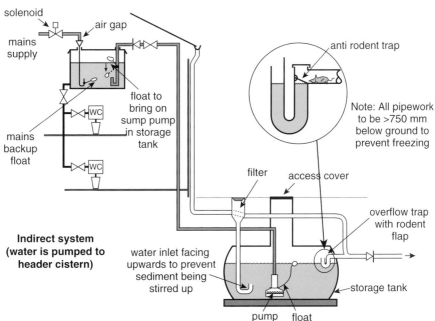

solenoid

air gap

mains
supply

float to
bring on
sump pump
in storage
tank

mains
backup
float

WC

WC

anti rodent trap

Note: All pipework
to be >750 mm
below ground to
prevent freezing

filter

access cover

overflow trap
with rodent
flap

**Indirect system
(water is pumped to
header cistern)**

water inlet facing
upwards to prevent
sediment being
stirred up

storage tank

pump   float

Rainwater harvesting 2

Grey Water Recovery

Relevant British Standard
BS8525

Grey water refers to the waste water from baths, sinks, basins, dish washers and washing machines that has not been contaminated by faecal matter or urine. It gets its name from the fact that it is generally cloudy in appearance, unlike that of freshly collected rainwater or the potable water from the supply mains. Water that does contain faecal matter or urine is referred to as black, or sometimes brown, water.

Grey water can be collected, cleaned and filtered and the water can be reused for WC flushing and constructed wetlands, such as green roofs and irrigation systems. It is particularly useful when used for irrigation, owing to the high levels of nitrogen it contains, which is beneficial to plant growth. Up to a third of domestic mains water usage can be saved in the reuse of grey water, but energy use is increased in cleaning the water to prevent bacterial growth. It is important that grey water is not kept for longer than 48 hours otherwise anaerobic decomposition can occur, resulting in foul-smelling water being produced. To prevent this from happening a sensor is located within the storage tank to purge the water to waste should there be no water flow through the unit during this time. Where grey water is used in an industrial situation, or in car washes, it is used almost immediately and therefore grey water recycling is a good environmental solution. Most grey water systems use bromine or chlorine to disinfect the water and possibly UV light treatment to kill off any remaining bacteria. Sometimes the water is purposely coloured during its treatment with a blue or green vegetable dye, thereby ensuring there is no possibility that it would be drunk, as no one would care to drink grey water from a tap. Any hair or soap that flows into the holding tank floats to the surface and is skimmed off as the tank fills; whereas any solids, such as sand, sink to the bottom and are drained when the unit is purged as identified above.

If you were to combine grey water recovery with a system of rainwater harvesting, you could possibly meet up to 95% of your total water needs. For drinking water purposes you would only require around 2 litres per person per day, and this would amount to the bulk of the cost of purchasing water from the local water company. By looking carefully at the illustration you will see that the harvested rainwater is used to supply the storage cistern feeding the washing machines. This water is in turn collected and used yet again for flushing the WCs. The careful design of the float switches allows water to fill the storage cistern that supplies the WCs. First it takes any water from the grey water storage tank; if that is empty it then uses any water from the harvested rainwater tank; and finally, if that also has no storage, the third and lowest set of float switches come into operation and allows the cistern to fill from the mains supply.

Note that as with the harvested rainwater, all pipework installed needs to be readily identifiable as reclaimed water. Larger grey water systems that supply more than one property tend to use more sophisticated Biological treatment, similar to a sewage treatment works.

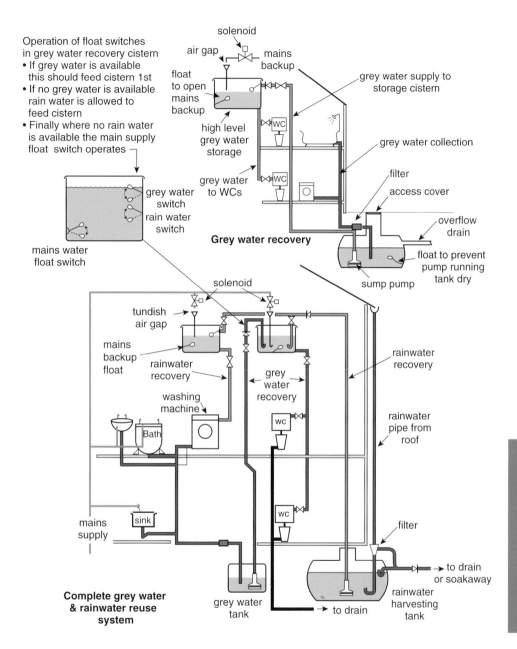

Operation of float switches
in grey water recovery cistern
• If grey water is available
  this should feed cistern 1st
• If no grey water is available
  rain water is allowed to
  feed cistern
• Finally where no rain water
  is available the main supply
  float switch operates

solenoid

air gap

mains
backup

float
to open
mains
backup

solenoid

high level
grey water
storage

grey water supply to
storage cistern

grey water
to WCs

grey water collection

WC

WC

filter

access cover

overflow
drain

grey water
switch

rain water
switch

mains water
float switch

Grey water recovery

float to prevent
pump running
tank dry

sump pump

solenoid

tundish
air gap

mains
backup
float

rainwater
recovery

washing
machine

Bath

grey
water
recovery

rainwater
recovery

rainwater
pipe from
roof

WC

WC

mains
supply

sink

filter

to drain
or soakaway

**Complete grey water
& rainwater reuse
system**

grey water
tank

to drain

rainwater
harvesting
tank

10 Energy Conservation

Part 11

Assessing Your Knowledge

Plumbing, 4th Edition. R. D. Treloar.
© 2012 Blackwell Publishing Ltd. Published 2012 by Blackwell Publishing Ltd.

Assessing Your Knowledge

This section of the book is designed to allow you to check your knowledge of the skills of plumbing. The assessments have been divided into three areas as follows:

(1) **Self assessment:** 100 multiple choice questions on pages 409–423. Answers are to be found at the end of the book (page 476).

(2) **Supplementary assessment:** 100 short-answer questions (pages 424–463) typical of the sort you are likely to be given in college-based training programmes leading to a qualification in plumbing. No direct answer is given here to any question, although the answer to each lies somewhere in the book. To a large extent, the questions follow the order in which information is given in this book; hence it is possible to work through the book to find the answers. However, it is probably best to tackle the questions as your knowledge grows.

Unfortunately, the length of this book dictates that there is no space left to answer all of the supplementary questions. However, you should be able to confirm the answers with your college tutor and I have given the page numbers where the information can be found to help you solve each question; alternatively, the answers can be found at the following web address:

www.blackwellpublishing.com/treloar

For those with no internet access I would also be pleased to pop an answer sheet to the supplementary questions in the post, for the cost of £5.00 to cover the cost of copying, post and packaging. Should you need this option, please send a cheque made payable to R. Treloar at the following address: Roy Treloar, Colchester Institute, Sheepen Road, Colchester, Essex, CO3 3LL.

(3) **Problem solving:** ten short problems (pages 464–473) typical of those encountered in plumbing systems; a trained plumber should be able to find the solutions without too much difficulty. The answers to these questions may need some thought and transference of knowledge, or indeed consultation with others, such as your college tutor. The answers are given to the problems on pages 474–475.

Good luck.

Self Assessment

(100 multiple choice questions)

Check your responses to these questions by referring to the answer sheet on page 476. The main type of multiple choice question will be the 'four option multiple choice' kind. These comprise either a direct question or statement, known as the stem, followed by a choice of four different answers, called the responses. Only one of these responses is the correct answer (the key); the others are incorrect (but plausible) distractors to the key. Candidates are required to select their response by choosing either (a), (b), (c) or (d).

Example:

When running an overflow pipe from a storage cistern it is necessary to:

(a) Ensure a trap is fitted in the pipe-run
(b) Ensure the connection to the drain is watertight
(c) Discharge the pipe into a gutter, if possible
(d) Terminate the pipe in a conspicuous position

d

The correct answer is to terminate the pipe in a conspicuous position; therefore the response should be (d).

(1) Which of the following is a non-ferrous metal?

(a) Aluminium
(b) Cast iron
(c) Low carbon steel
(d) Wrought iron

(2) What is meant by the term plumbo-solvent?

(a) The use of lead-free solder fittings
(b) The addition of chlorine to the water supply
(c) The ability of water to dissolve lead
(d) The ability of water to dissolve copper

(3) What is meant when water is said to be at its maximum density?

(a) It will expand no more
(b) It exerts its greatest amount of pressure
(c) It occupies the smallest volume
(d) It is at its highest temperature

(4) At what temperature is the maximum density of water?

(a) −273°C
(b) 0°C
(c) 4°C
(d) 100°C

11 Assessing Your Knowledge

(5) Which of the following metals is an alloy?

(a) Brass
(b) Iron
(c) Aluminium
(d) Copper

(6) Heavy grade low carbon steel tube to BS 1387 is identified by which colour coding?

(a) Red
(b) Blue
(c) Brown
(d) Green

(7) Electrolytic corrosion (galvanic action) occurs when:

(a) Two dissimilar metals are in contact with an inert gas
(b) Two dissimilar metals are in contact via an electrolyte
(c) A metal is exposed to stormy weather conditions
(d) Two dissimilar metals are placed in a liquid incapable of passing an electric current

(8) One method of joining to polyethylene pipe is carried out by:

(a) Solvent weld cement
(b) Fusion weld cement
(c) Compression joints with inserts
(d) Capillary joints

(9) 28 mm diameter copper tube, when fixed horizontally, should be supported at intervals of:

(a) 0.5 m
(b) 1 m
(c) 1.8 m
(d) 2.7 m

(10) Bends are often used on pipes rather than elbows because:

(a) The frictional resistance is reduced
(b) The frictional resistance is increased
(c) The pipework is easier to run
(d) Insulation material is easier to apply

(11) The type of thread cut onto low carbon steel tube is:

(a) British Standard Whitworth
(b) British Standard Pipe thread
(c) British Standard Fine
(d) Metric

(12) Annealing a metal will:

(a) Make it softer and more workable
(b) Give it a protective coating of aluminium
(c) Increase its tensile strength
(d) Remove surface blemishes

(13) Rippling to bends on copper tube when using a bending machine is caused by:

(a) The backguide being adjusted too tightly
(b) The backguide being adjusted too loosely
(c) The bend being pulled too quickly
(d) Loss of hydraulic fluid

(14) Copper and galvanised steel should not be used in the same water system because:

(a) Limescale will form in the pipework
(b) A fungus will grow within the system
(c) Electrolytic corrosion will occur
(d) Erosion of the galvanised steel will occur

(15) The term calorific value of a fuel relates to the:

(a) Energy produced from a given quantity of fuel
(b) Mass of the fuel compared to that of water
(c) Temperature at which the fuel burns
(d) Gas/air ratio

(16) Which of the following is the most important commissioning procedure to carry out upon completion of a hot or cold water installation:

(a) Tidy up and ask the client to inspect the work
(b) Work out the final account to be paid for the work
(c) Fit the insulation material to ensure compliance with the water supply regulations
(d) Check for leaks, adjust water levels and check flow rates

(17) Continuity bonding is carried out to pipework to:

(a) Ensure that any earth leakage is conveyed to the ground via the water pipes
(b) Prevent overloading of the household electrical system
(c) Prevent an electric shock should a live wire touch any pipe
(d) Reduce the accumulation of scale in water pipes

(18) The correct procedure to adopt when lifting heavy objects is:

(a) Feet together and lift using the back to take all the strain
(b) Feet slightly apart, bend at the knees and keep a straight back
(c) Feet wide apart and lift using the back to take all the strain
(d) Feet slightly apart, bend at the knees and keep a curved back

11 Assessing Your Knowledge

(19) The person who is employed by the local authority to ensure that the Building Regulations are observed is called the:

(a) Ombudsman
(b) Clerk of works
(c) Building control officer
(d) Local byelaws officer

(20) The colour code of wires to a 13 amp 3-pin plug is:

	Live	Neutral	Earth
(a)	Blue	Brown	Green
(b)	Red	Blue	Green & Yellow
(c)	Brown	Blue	Green & Yellow
(d)	Black	Red	Green

(21) Which of the following should not be used on fires involving flammable liquids?

(a) Water
(b) Dry powder
(c) Carbon dioxide (CO_2)
(d) Blanket

(22) The mass of 1 litre of water is:

(a) One kilogram (1 kg)
(b) One gram (1 g)
(c) One pound (1 lb)
(d) More when heated

(23) To obtain a suitable soft soldered capillary joint on copper tube the joint should:

(a) Have a small space between the adjoining surfaces
(b) Have a large space between the adjoining surfaces
(c) Be painted with an oxide solution
(d) Be suitably belled out to allow a molten pool to be formed

(24) A flux is used to:

(a) Burn into the base metal to ensure the complete removal of all oxides and dirt, etc.
(b) Assist the insertion of the pipe into the fitting
(c) Prevent the solder vaporising and escaping from the joint
(d) Prevent the oxidation of the metal to be soldered

(25) When bronze welding it is necessary to use a:

(a) Neutral flame
(b) Carburising flame
(c) Oxidising flame
(d) Large spreading flame

(26) The type of thread found on acetylene welding equipment is:

  (a) BSP
  (b) Recessed
  (c) Right-handed
  (d) Left-handed

(27) Acetylene gas should never be conveyed in pipelines made of or incorporating:

  (a) Copper
  (b) Mild steel
  (c) Cast iron
  (d) Rubber

(28) A leak from a propane gas blow torch, or similar equipment using fuel gas, should be found using a:

  (a) Match or lighted taper
  (b) Manometer
  (c) Soap solution
  (d) Coloured die, added to the gas

(29) The process of bronze welding can be said to be a type of:

  (a) Hard soldering
  (b) Soft soldering
  (c) Brazing
  (d) Autogenous welding

(30) When welding low carbon steel or lead, using a neutral flame:

  (a) Flux must be applied to the cleaned joint
  (b) The filler rod must contain the necessary flux
  (c) No flux is required
  (d) A porous weld would result

(31) Hard waters contain:

  (a) Calcium salts
  (b) Epsom salts
  (c) Acidic salts
  (d) Saline salts

(32) An air gap must always be maintained when supplying a sink with water from a supply pipe in order to prevent:

  (a) Back-siphonage
  (b) Self-siphonage
  (c) Aeration of the water supply
  (d) Legionnaires disease

11 Assessing Your Knowledge

(33) Where different pressures are at the hot and cold connections to a mixer tap serving a sink or bath, the tap should:

(a) Be fitted with a pressure-reducing valve
(b) Have a divided outlet
(c) Have a combined outlet
(d) Be of a mono-block design

(34) The service valve fitted on the cold distribution pipe from a storage cistern should be a fullway gatevalve or fullway quarter-turn ballvalve in order to:

(a) Make installation easy
(b) Prevent water hammer
(c) Prevent blockages due to debris from the cistern
(d) Keep frictional resistance to a minimum

(35) When water freezes it:

(a) Expands about 1600 times
(b) Decreases in size by about one-tenth
(c) Increases in size by about one-tenth
(d) Goes through a chemical change

(36) An underground water supply pipe should be a minimum depth below ground of:

(a) 350 mm
(b) 750 mm
(c) 1000 mm
(d) 7500 mm

(37) Should the ballfloat of the float operated valve in a flushing cistern become full of water:

(a) The cistern would continuously flush
(b) The cistern would fail to flush
(c) Water would discharge from the overflow
(d) Water hammer would be apparent

(38) The main difference between a high pressure float operated valve and a low pressure float operated valve (both to BS 1212) is:

(a) The size of the ballfloat
(b) The size of the valve body
(c) The size of the seating orifice
(d) The length of the lever arm

(39) When running a hot water pipe through a wall it should be sleeved to:

(a) Allow for thermal movement of the pipe
(b) Allow for thermal movement of the wall
(c) Allow settlement of the building
(d) Make maintenance easier

(40) Secondary circulation is carried out to convey hot water to:

(a) The cylinder from the boiler
(b) A bathroom towel rail
(c) A point close to the draw-off point, thus preventing a wastage of water
(d) Radiators

(41) The feed cistern supplies water to:

(a) The cold distribution pipework
(b) The hot storage vessel
(c) A WC pan or urinal
(d) A set of booster pumps

(42) Primary circulation is the water which circulates:

(a) Between the cold feed cistern and the dhw cylinder
(b) Between the hot storage vessel and the boiler
(c) Between the hot storage vessel and the furthest draw-off point
(d) Around inside the hot water cylinder

(43) A dhw system in which the water drawn off from the taps has passed through the boiler is known as a:

(a) Primary system
(b) Single feed system
(c) Direct system
(d) Indirect system

(44) Unvented dhw supply refers to a system with:

(a) A hot stored capacity less than 15 litres
(b) A hot stored capacity greater than 15 litres
(c) A single feed cylinder
(d) A storage cistern vented to the atmosphere

(45) The water in the primary flow and return is prevented from mixing with the water in a single feed indirect cylinder by:

(a) The trapping of an air pocket
(b) Ensuring the system has a good head pressure
(c) Fitting a pump on the primary flow
(d) Having separate cold feed connections

(46) A secondary return should be returned to the:

(a) Boiler
(b) Top third of the hot storage cylinder
(c) Bottom third of the hot storage cylinder
(d) Last radiator

(47) In a vented low temperature hot water heating system the position of the feed and expansion cistern determines the:

(a) Static head
(b) Circulating head
(c) Circulating pressure
(d) Amount of water which will expand upon heating

(48) An air separator is a device which:

(a) Allows cold feed and vent connections to be closely grouped in a fully pumped system and permits air bubbles to form which simply rise up out of the system
(b) Allows the cold feed and vent connections to be made anywhere in the system as any air drawn into the system is expelled
(c) Prevents boiler noises
(d) Is fitted to condensing boilers to allow the condensate to be expelled from the flue gases

(49) A combination boiler:

(a) Allows the dhw and c.h. waters to mix
(b) Heats both dhw and c.h. waters independently
(c) Is a boiler which uses two fuels (e.g. gas and oil)
(d) Is a design of boiler used for sealed heating systems

(50) When calculating the heat loss through the building fabric the external temperature is usually taken to be:

(a) 4°C
(b) 0°C
(c) −1°C
(d) Irrelevant

(51) The transference of heat by convection currents is particularly important in:

(a) Space heating
(b) Soldering copper tubes
(c) Refrigerators
(d) 'U' value calculations

(52) When a new heating installation has been completed it should be commissioned by the:

(a) Building control officer
(b) Householder
(c) Site agent
(d) Installer

(53) The size and therefore capacity of a feed and expansion cistern is based upon the:

(a) Amount of water in the heating system
(b) Amount of water in the hot storage vessel
(c) The size of the loft hatch
(d) Number of occupants using the building

(54) A room sealed (balanced flued) heater draws the air required to support combustion of the fuel from:

(a) The room only in which the heater is fitted
(b) Outside the building at a point adjacent to the outlet
(c) Outside the building at a point some distance from the outlet
(d) An air line

(55) A steel panel radiator connected to a secondary return would:

(a) Act as a heat leak
(b) Prevent the circulation of water
(c) Be suitable providing it was chromium plated
(d) Cause the discoloration of the hot water

(56) Condensation in a conventional flue is the result of:

(a) Excessive amounts of fuel being used
(b) Excessive flue gas temperatures
(c) Excessive cooling of the flue gases
(d) Insufficient air being supplied to support combustion

(57) A thermocouple is:

(a) A device which generates a small electric current
(b) A flame failure device
(c) A fusible link, designed to break and close a gas valve in the event of a fire
(d) The point where the pilot flame burns

(58) Which of the following is the most important task to carry out, when leaving a partially completed gas installation:

(a) Ensure the gas supply is turned off at the mains
(b) Ensure the safety of all the gas pipe runs, irrespective of whether the gas is in the pipeline or not
(c) Purge the pipeline of air before leaving
(d) Inform the client when you should return

(59) The test point located close to a gas appliance is to enable you to:

(a) Check to see the pressure at the burner
(b) Check the gas rate
(c) Take a sample of gas
(d) Check for earth leakages

(60) The purpose of a flue terminal is to:

(a) Minimise down draughts
(b) Filter out any remaining undesirable gases from the products of combustion
(c) Do away with the need for a down draught diverter
(d) Provide a neater finish to the flue

(61) Increasing the loading to a gas regulator would:

(a) Decrease the outlet pressure
(b) Increase the outlet pressure
(c) Decrease the inlet pressure
(d) Increase the inlet pressure

(62) The term gas-to-air ratio refers to the:

(a) Calorific value of the gas
(b) Relative density of the gas compared to air
(c) Amount of gas in a volume of air to allow ignition
(d) Volume of air required to give complete combustion of the fuel

(63) The products of complete combustion of natural gas are:

(a) Carbon dioxide and carbon monoxide
(b) Oxygen and carbon monoxide
(c) Hydrogen and methane
(d) Water vapour and carbon dioxide

(64) The recommended fuel supplied to an oil-fired pressure jet burner should be:

(a) Class D 35 second fuel
(b) Class C2 28 second fuel
(c) Gas oil
(d) Red diesel fuel oil

(65) The term viscosity refers to:

(a) The amount of hydrocarbons in the fuel
(b) The oil's ability to ignite
(c) The oil's ability to flow easily
(d) The fact that the amount of smoke produced is high

(66) An oil level indicator will be found:

(a) Inside the boiler casing
(b) On or adjacent to the oil storage tank
(c) Connected to a constant pressure control
(d) Fitted on the oil supply line close to the boiler

(67) A constant oil level control is used:

(a) On the supply line to a vaporising burner
(b) On the supply line to an atomising burner
(c) On the supply line to a pressure jet burner
(d) On the supply line to an oil storage tank

(68) The flue/appliance efficiency is found by:

(a) Reducing the smoke reading to its minimum and taking the temperature of the flue gas
(b) Calculating the oil consumed over a three-month period
(c) Referral to the manufacturer's instructions
(d) Comparing the CO_2% against the flue temperature

(69) A test for earth continuity is carried out in order to:

(a) Ensure the phase conductor is connected to its correct location
(b) Ensure all exposed metalwork is connected to a suitable earth
(c) Check all three phases are located within the outlet socket
(d) Ensure the temporary bonding wire is working effectively

(70) A typical example of an electrical circuit wired in series would be:

(a) Christmas tree lights
(b) The ring main
(c) The lighting circuit
(d) A battery

(71) The rating to which a fuse will blow can be found using the following simple formula:

(a) Volts ÷ amps = ohms
(b) Volts ÷ watts = amps
(c) Watts ÷ volts = amps
(d) Watts × volts = amps

(72) In running the electrical supply to an appliance a spur outlet refers to:

(a) A socket run from the lighting circuit
(b) A supply run from the ring main to an isolated socket
(c) A term used to indicate that a boiler has been fitted with a 13 A 3-pin plug
(d) A radial circuit run from the consumer unit

(73) A solenoid is:

(a) A device which uses the current from one circuit to switch on the current to another circuit
(b) A device which converts a.c. to d.c.
(c) A device used to hold a valve open
(d) A device which stores a charge of electrical energy

(74) The trap to a bath which discharges into a primary vertilated stack system must have a minimum depth of seal of:

(a) 38 mm
(b) 50 mm
(c) 65 mm
(d) 75 mm

(75) The test to ensure that the trap seals are maintained in a discharge system is known as:

(a) Performance test
(b) Bourbon test
(c) Soundness test
(d) Hydraulic test

(76) Compression is most likely to occur in a discharge stack:

(a) When a connection is made within 200 mm of a WC branch
(b) At the foot of the stack
(c) When two branch connections are opposite
(d) When no allowance has been made for expansion

(77) Self-siphonage is most likely to occur in a trap which has water discharged from a:

(a) Bath
(b) Basin
(c) Sink
(d) WC

(78) Water which is removed from a trap due to the discharge of water from another appliance, further down the discharge pipe, is called:

(a) Self-siphonage
(b) Induced siphonage
(c) Back-siphonage
(d) Momentum

(79) The contents of a washdown WC pan are removed by:

(a) Induced siphonage
(b) Siphonic action
(c) Compression
(d) Momentum

(80) An above-ground sanitary discharge system is required to withstand a soundness test of:

(a) 25 mm water gauge
(b) 38 mm water gauge
(c) 100 mm water gauge
(d) 1.5 m head of water

(81) A sparge pipe will be found fitted to a:

  (a) High or low level WC suite
  (b) Slop sink or hopper
  (c) Bowl urinal
  (d) Slab urinal

(82) The minimum internal diameter of a drain or private sewer used to convey foul water should be:

  (a) 50 mm
  (b) 75 mm
  (c) 100 mm
  (d) 150 mm

(83) The instrument used to measure the air pressure within a drainage pipe during a test is called a:

  (a) Manometer
  (b) 'U' tube
  (c) Viscometer
  (d) Absolute pressure gauge

(84) The base of a soil discharge stack should be fitted with a:

  (a) Duck bend
  (b) Back or side inlet gully
  (c) Long radius bend
  (d) Knuckle bend

(85) When the main drainage is not connected to a dwelling and a septic tank is being used to receive all the foul water, the type of drainage system to be chosen is known as:

  (a) Combined
  (b) Separate
  (c) Partially separate
  (d) Surface disposal

(86) To prevent foul water drains surcharging during heavy rainfall, which of the following systems of drainage should be chosen?

  (a) Combined
  (b) Separate
  (c) Relief
  (d) Pumped

(87) What is meant by the term 'cladding'?

  (a) A board used to assist levelling a drain pipe
  (b) The weathering of a soil stack as it passes through a pitched roof
  (c) The weathering of a vertical surface
  (d) The sheet lead weathering a chimney to a pitched roof

11 Assessing Your Knowledge

(88) A drip is provided to lead-lined gutters in order to:

(a) Permit thermal movement
(b) Make fixing easier
(c) Slow down the flow of water
(d) Make it possible to achieve the required fall

(89) What is meant by the term 'carbon neutral'?

(a) The overall level of CO_2 in the environment is not increased by burning the fuel.
(b) The overall level of CO_2 in the environment is lowered by burning the fuel.
(c) The appliance burning the fuel produces only Carbon Monoxide (CO)
(d) The appliance burning the fuel produces only Carbon Dioxide (CO_2)

(90) When flat lap welding to sheet lead the weld should consist of:

(a) One loading of filler metal
(b) Two loadings of filler metal
(c) Three loadings of filler metal
(d) Four loadings of filler metal

(91) The colour code to identify code 5 sheet lead is:

(a) Green
(b) Red
(c) Blue
(d) Black

(92) A weathering slate piece is used for:

(a) Weathering a chimney
(b) Weathering a pipe as it passes through the roof
(c) Terminating the top edge of a sheet roof covering
(d) Replacing a broken tile on a roof

(93) The recommended maximum length of a sheet lead flashing is:

(a) 1 m
(b) 1.5 m
(c) 2 m
(d) 2.5 m

(94) The SI unit for calorific value is measured in:

(a) Megajoules per kilogram
(b) Kilograms per cubic metre
(c) Kilowatts per second
(d) Kilowatts per hour

(95) The formula used to find the cross-sectional area of the end of a cylinder is:

(a) $\pi\,d$
(b) $\pi\,r^2$
(c) $2\,\pi r$
(d) $\pi\,d^2$

(96) Given that the coefficient of linear expansion of copper is 0.000016, what would the increase in length of a 150 m long copper pipeline be when subjected to an increase in temperature of 10°C?

(a) 0.2 m
(b) 0.24 m
(c) 2 mm
(d) 24 mm

(97) How many litres would be contained in a tank with a volume of $2.76\,\text{m}^3$?

(a) 27.6
(b) 276
(c) 2760
(d) 27 600

(98) The intensity of pressure acting at a boiler base 3 m below the water level in a feed cistern is:

(a) $29.43\,\text{kN/m}^2$
(b) $29.43\,\text{kN/m}^3$
(c) $29.43\,\text{N/m}^2$
(d) $29.43\,\text{N}$

(99) A tank measures 1 m high with a base of 2 m × 0.5 m; the mass of the tank itself is 10 kg. What is the total mass of the tank and its contents when completely filled with water?

(a) 110 kg
(b) 1010 kg
(c) 1100 kg
(d) 1110 kg

(100) The quantity of heat required to raise the temperature of 1 litre of water by 1°C is:

(a) 1 BTU
(b) 1 therm
(c) 4.186 kJ
(d) 4.2 kW

The answers to these multiple choice questions can be found on page 476.

11 Assessing Your Knowledge

Supplementary Assessment

(100 short answer questions)

Note: The page number given at the end of each question gives an indication where the answer can be found.

(1) The safety signs illustrated in Figure 1 indicate things the operative must or must not do. Identify the meaning of each sign. (Page 19.)

A	
B	
C	

Ⓐ   Ⓑ   Ⓒ

Figure 1

(2) State how hazard signs differ from prohibition and mandatory signs. (Page 18.)

(3) Identify the statutory document which must be observed on site at all times to ensure safety. (Page 12.)

(4) Identify the contents of each of the fire extinguishers shown in Figure 2 and give examples of their fire preventive uses. (Page 27.)

Label	Contents	Uses (i.e. type of fire)
Blue		
Red		
Cream		
Black		

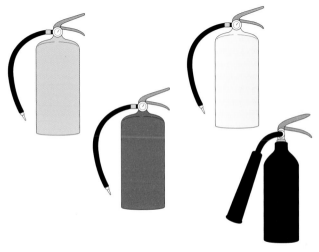

Figure 2

(5) Identify the statutory document which should be observed when carrying out work involving the following (Page 28):

Water Supply	
Gas Supply	
Electricity	
Building	

(6) State the contents of the series of pipes which have been colour-banded, as shown in Figure 3, found in a boiler house running along a wall. (Page 30.)

Pipe No	Contents
1	
2	
3	
4	
5	

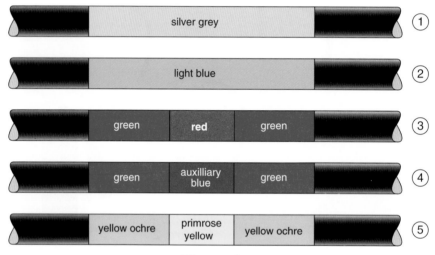

Figure 3

(7) Convert the following SI units into their British Imperial equivalents. (Page 32.)

14 kg =	_____ lb
50 litres =	_____ gal

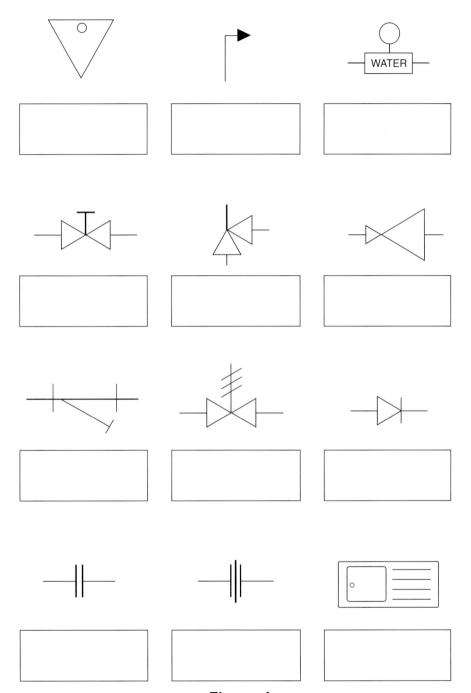

Figure 4

(8) In the space provided below each symbol, indicate the name given to the graphic symbols shown in Figure 4. (Page 31.)

(9) Convert the following British Imperial units into their SI equivalents. (Page 32.)

5 yds =	_____	m
60 000 Btu/h =	_____	kW

(10) What special precautions need to be observed when using LPG in or around trenches and drains, and why? (Page 14.)

(11) With reference to the following materials, complete the table, listing the materials in order of their correct position within the electrochemical series, placing the cathodes above the anodes. (Page 41.)

Materials: aluminium, copper, lead, tin, zinc.

Physical properties of typical metals

Material in order of the electrochemical series	Chemical symbol	Melting point (°C)	Relative density (kg/m³)	Specific heat capacity (kJ/kg K)

(12) Calculate (in mm) the amount of expansion that would occur in a length of copper pipe 30 m long when subjected to a temperature rise of 62°C. (Page 43.)

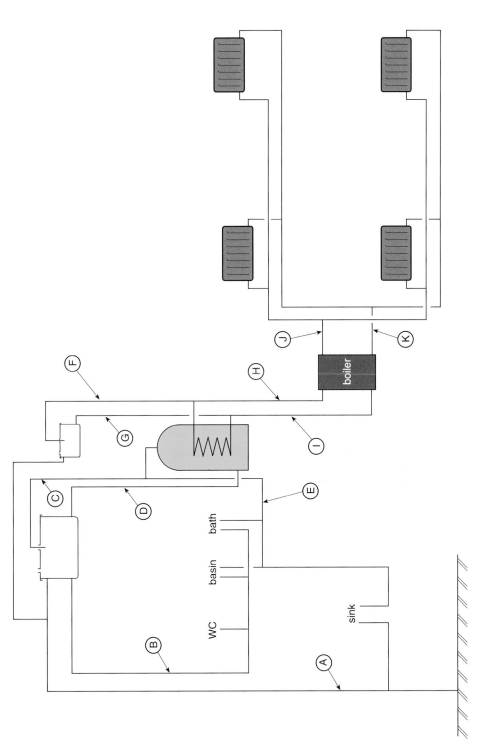

Figure 5

(13) Identify the name given to the design of systems shown in Figure 5. (Pages 89, 105 and 141.)

Cold water system:	
Hot water system:	
Central heating system:	

(14) Complete Figure 5 (using correct graphic symbols) by indicating all the necessary valves, etc., which would be included in the design of the various systems. Also complete the following schedule to identify the name given to each pipe indicated and give a suggested pipe size. (Pages 89, 105 and 141.)

Section	Name	Pipe size
A		
B		
C		
D		
E		
F		
G		
H		
I		
J		
K		

(15) Define what is meant by relative density and specific heat capacity. (Pages 40-42.)

Relative density:

Specific heat capacity:

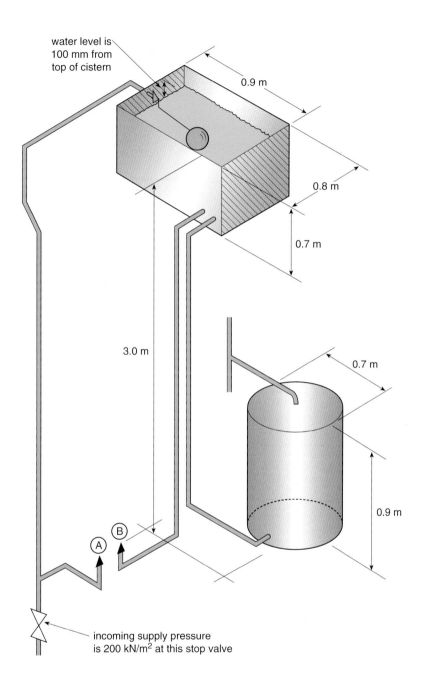

water level is 100 mm from top of cistern

0.9 m

0.8 m

0.7 m

3.0 m

0.7 m

0.9 m

Ⓐ Ⓑ

incoming supply pressure is 200 kN/m² at this stop valve

(16) With reference to the dimensions given in Figure 6, calculate the actual capacity of the storage cistern and the capacity of the dhw storage cylinder. (Page 34.)

Actual storage capacity of cistern	

Storage capacity of dhw cylinder	

(17) With reference to the dimensions given in Figure 6, calculate the intensity of pressure at tap B and compare this to the mains pressure at tap A. Also find the height to which the mains pressure would rise above the stop valve in a vertical pipe, ignoring the frictional resistance. (Page 38.)

Pressure at tap B =

Height to which mains pressure would rise above the stop valve	

(18) Find the total pressure at the base of the dhw cylinder in Figure 6. (Page 38.)

(19) In the space provided illustrate a flat dresser. (Page 51.)

(20) Give the name of each of the two pipe cutters, illustrated in Figure 7, used to cut low carbon steel tube. Each has an advantage over the other. State what this advantage is. (Page 49.)

Cutter A

Cutter B

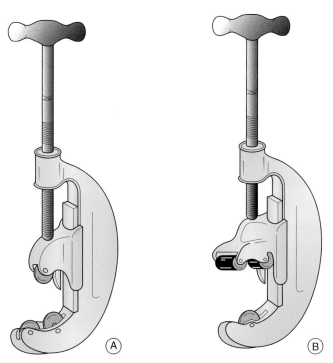

Figure 7

(21) When soft soldering, why is the application of a flux required? (Page 64.)

(22) Figure 8 illustrates oxyacetylene welding equipment. In the spaces provided round the figure, name each of the components indicated. (Page 67.)

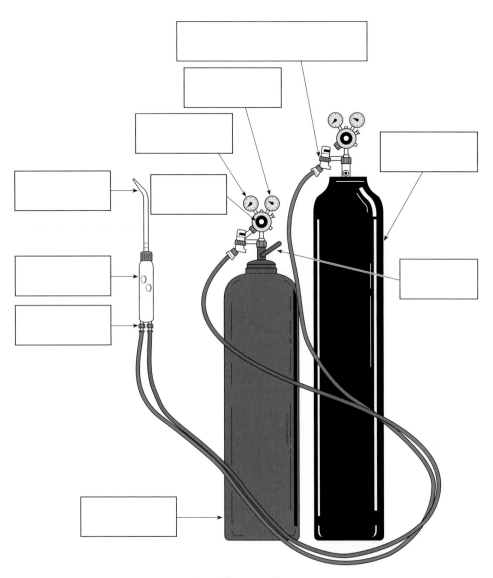

Figure 8

(23) When autogenous welding on materials such as lead and steel, no flux is required. State why this is so. (Page 68.)

(24) Figure 9 shows two completed bronze welded joints. In the spaces provided near the figure give the name used to describe each joint. (Page 71.)

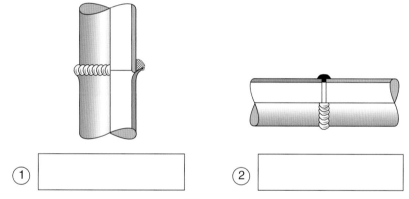

①

②

Figure 9

(25) Although the joints indicated in question 24 are referred to as welded joints they are not truly welded. State why this is so and identify what type of joints they are. (Page 70.)

(26) Shown in Figure 10 is a section through a screwdown valve. State the name given to the tap illustrated and in the spaces provided name the arrowed components. (Pages 75-77.)

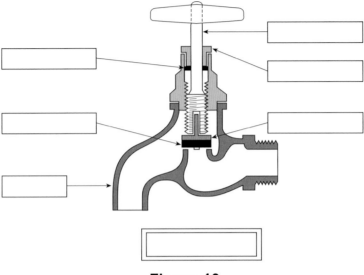

Figure 10

(27) State why a gatevalve or fullway ballvalve is fitted in the cold feed pipe to a vented system of dhw in preference to a stopcock. (Page 74.)

(28) Under current Water Regulations, only float-operated valves conforming to BS 1212 Parts 2 or 3, may be fitted to cisterns in domestic premises; give the reasons for this. (Page 78.)

11 Assessing Your Knowledge

(29) Complete Figure 11 using correct graphic symbols to show an indirect system of cold water supply. (Page 89.)

(30) Illustrate in the space provided how two small cisterns may be coupled together in a domestic situation to give a larger capacity, ensuring stagnation of the water will not occur. (Page 91.)

(31) With reference to Figure 12 name the components, numbered 1–6, fitted to the cold supply of the unvented system of dhw. (Page 107.)

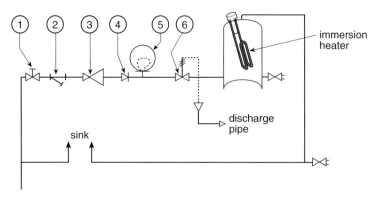

immersion heater

discharge pipe

sink

Figure 12

1.	4.
2.	5.
3.	6.

(32) Name the valve missing from Figure 12 which ensures complete safety regardless of excessively rising water temperatures caused by thermostat failure. (Without this device the system does not comply with the Building Regulations.) (Page 106.)

(33) Each of the dhw systems illustrated in Figure 13 (over) shows major errors; state what these faults or contraventions are. (Pages 102-107.)

System A

System B

System C

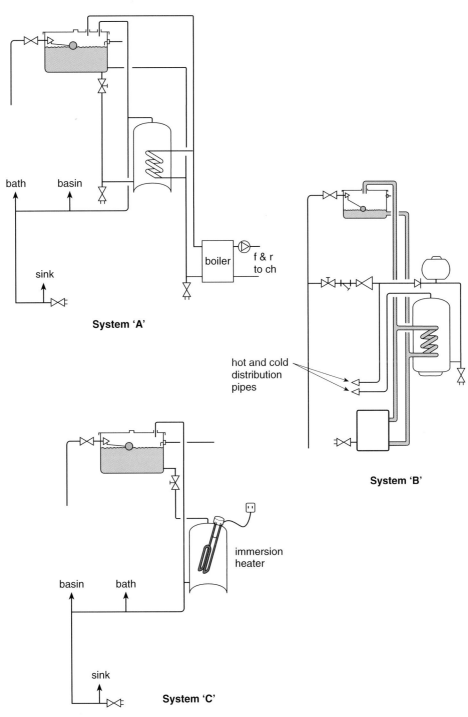

Figure 13

(34) Name the instantaneous water heater shown in Figure 14 and state the purpose of the valve indicated at 'X'. (Page 113.)

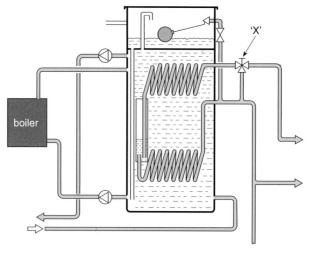

Figure 14

System referred to as:
Purpose of valve 'X':

(35) With reference to Figure 15, indicate in the space provided the maximum permissible depth which can be cut into the floor joist. Also indicate the maximum permissible distance at which the notch can be cut from the supporting wall. (Page 121.)

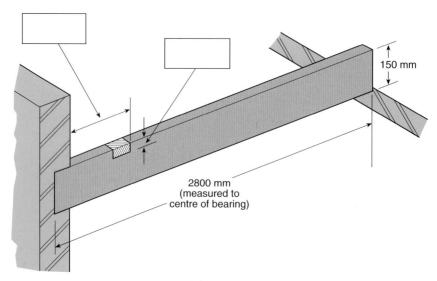

Figure 15

(36) Complete the illustration in Figure 16 to show (Page 141):

(a) A one pipe wet c.h. system
(b) A two pipe wet c.h. system
(c) A two pipe reversed return wet c.h. system

(37) State why the water level is adjusted low down in the f & e cistern feeding a wet c.h. system. (Page 144.)

(38) With reference to the bungalow illustrated on page 167 under the entry of Radiator and Boiler Sizing, complete the table below to find the heat emitter requirements for the bathroom. (Page 168.)

Heat requirements. Location: bathroom

Fabric loss Element	Area L × B = (m²)	Temp. diff. (°C)	U value (W/m²°C)	Rate of heat loss (W)
Window	×	×		
External walls	×	×		
Internal walls	×	×		
Floor	×	×		
Roof	×	×		
	Fabric loss			
Ventilation loss				
volume × air change × temp. diff. × factor =				
fabric loss + ventilation loss =				
plus 15% for intermittent heating =				
Total rate of heat loss =				

(39) What dangerous gas is produced as a result of incomplete combustion of natural gas? (Page 186.)

(40) What do the initials LPG stand for? (Page 224.)

11 Assessing Your Knowledge

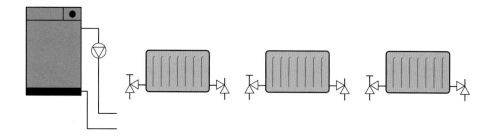

'A' one pipe wet c.h. circuit

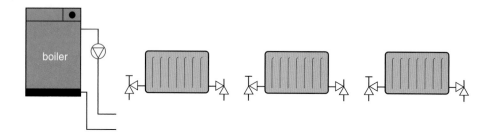

'B' two pipe wet c.h. circuit

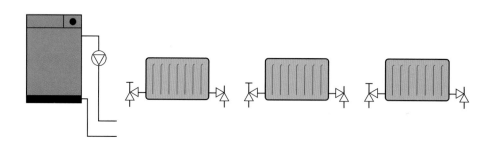

'C' two pipe reversed return wet c.h. circuit

Figure 16

(41) When carrying out a tightness test to an existing natural gas installation what is the pressure (in mbar) to which the system should be tested? (Page 196.)

(42) What is meant by the term let-by when testing gas installations? Describe how this test is carried out. (Page 196.)

(43) What is the name given to the device fitted in the pipeline to a gas appliance which maintains the gas at a constant pressure as recommended by the manufacturer? (Page 200.)

(44) Complete Figure 17 to show the flue pipe, terminal and minimum distances to be observed when terminating the natural draught open flue from a natural gas appliance. (Page 209.)

(45) What is meant by the term 'room sealed appliance' and why should these appliances generally be fitted in preference to open-flued appliances? (Page 216.)

(46) Calculate the effective free air ventilation requirements for an open-flued appliance of 18 kW natural gas net heat input. (Page 220.)

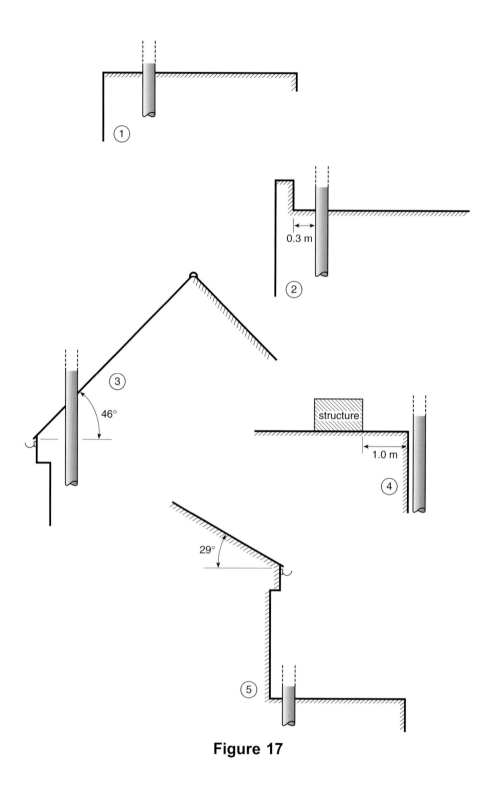

Figure 17

(47) With reference to the air grille illustrated in Figure 18, calculate its effective free air size. (Page 221.)

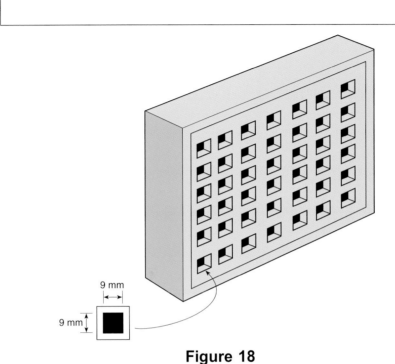

Figure 18

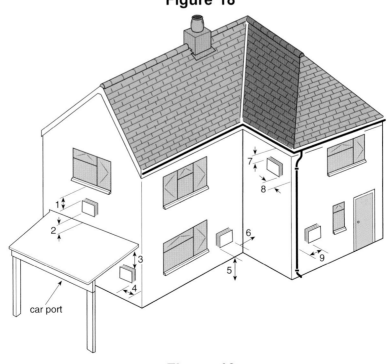

Figure 19

(48) Identify the minimum terminal location dimensions indicated for the 22 kW natural draught balanced flue appliances shown in Figure 19. (Page 217.)

1.	4.	7.
2.	5.	8.
3.	6.	9.

(49) State the minimum ventilation requirements (in mm²) for a decorative fuel effect gas fire of 6 kW input rating into a room. (Page 220.)

(50) Two grades of fuel are used for domestic oil fired burners; name these and explain how the fuels differ. (Page 242.)

(51) With reference to Figure 20, complete the illustration of the gravity-fed one-pipe oil supply system to show the necessary components and controls to be located on the storage tank and pipeline to enable its safe, efficient and effective use. (Page 247.)

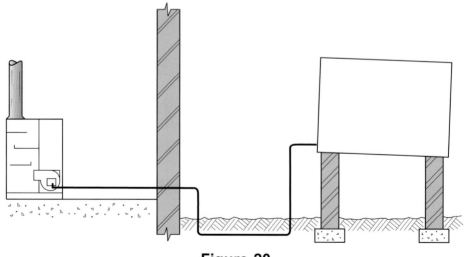

Figure 20

(52) Complete the illustration of the 'fusible link' type fire valve as shown in Figure 21, and explain its operation. (Page 249.)

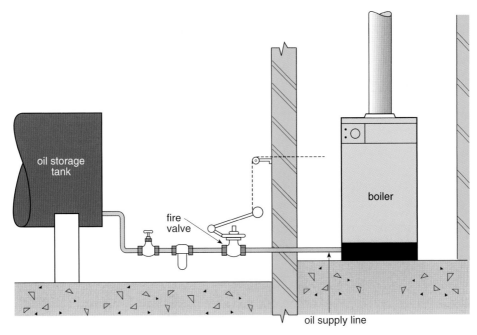

Figure 21

Operation of fire valve:

(53) State the purpose of a constant oil level control and indicate where one would be found. (Page 249.)

(54) Use the spaces provided to name the arrowed components on the pressure jet burner shown in Figure 22. (Page 253.)

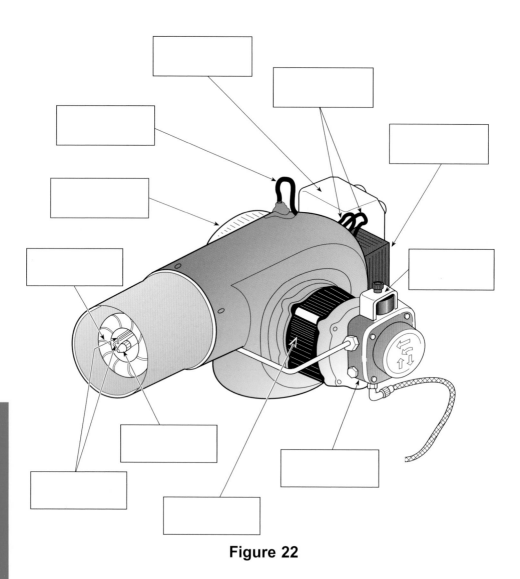

Figure 22

(55) Identify the device located in a pressure jet burner to detect light rays from within the combustion chamber to confirm ignition of the fuel and as a result making or breaking an electrical circuit. (Page 253.)

(56) Calculate the free air ventilation requirements for an open-flued pressure jet burner with an input rating of 18 kW. (Page 256.)

(57) When should a draught stabiliser be fitted to the flue way of an open-flued oil burning appliance? (Page 256.)

(58) Name the four separate tests which are carried out when completing a combustion efficiency test to an oil burning appliance. Also indicate with an 'X' the test which is not applicable to balanced flued appliances. (Page 262.)

1.	
2.	
3.	
4.	

(59) Show the correct location of the fuse to be fitted into the electrical circuit shown in Figure 23. Also calculate the size of fuse to be fitted into the 13 A 3-pin plug of an electric drill with a power rating of 550 W, designed to run on 230 V. (Page 269.)

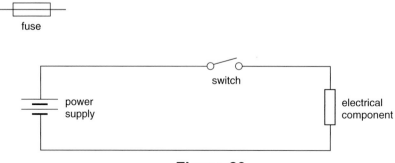

Figure 23

Fuse rating:

(60) Complete the two wiring diagrams shown in Figure 24 to produce one circuit in series and one in parallel. (Page 271.)

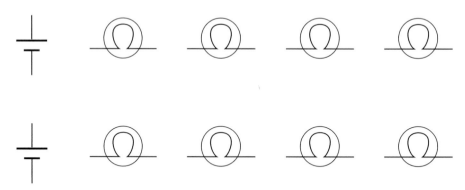

Figure 24

(61) What is the common name given to the circuit protective conductor found in domestic house wiring? (Page 272.)

(62) When would a temporary bonding wire be used? (Page 276.)

(63) When replacing a water supply main what responsibility has the plumber to the client with regard to electrical safety? (Page 276.)

(64) Figure 25 shows a completed ring circuit; show how the wiring can be altered to accommodate the connection of the spur outlet. (Page 279.)

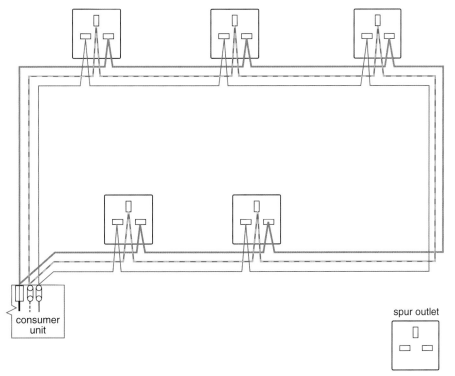

Figure 25

(65) Using coloured pencils, or the key which identifies the colour of cables, complete Figure 26 to indicate the location of the brown, blue, and green and yellow conductors into the 13 A 3-pin plug. (Page 281.)

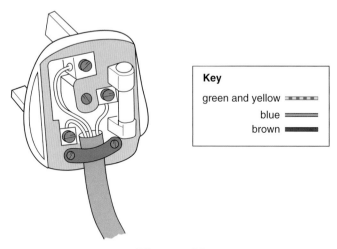

Figure 26

(66) How should cables be protected from damage when passing through metalwork? (Page 283.)

(67) Using a simple illustration explain the operation of the rod thermostat found inside an immersion heater to a dhw storage cylinder. (Page 285.)

(68) A transformer has half the number of coils on the secondary windings as those on the primary coil. Is it a step-up or step-down transformer? (Page 286.)

(69) Figure 27 shows the top of an immersion heater with the cover removed; using coloured pencils (or the key in Figure 26) complete the illustration to show the electrical connections to the heater element. Also, in the space provided, state the size and type of flex to be used. (Page 281.)

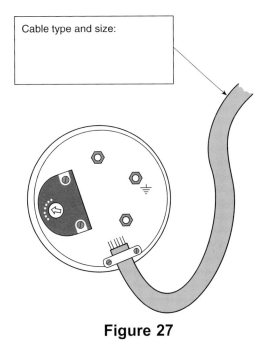

Cable type and size:

Figure 27

(70) What is a 'waterless' urinal. (Page 298.)

(71) Figure 28 shows a trap. In the space provided give the name of the trap design and indicate the depth of water seal. (Page 305.)

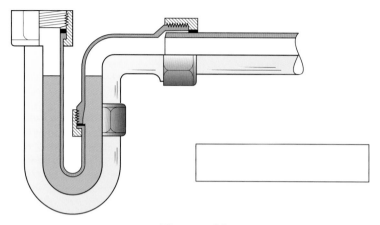

Figure 28

(72) Complete the table below to indicate the minimum size of waste fitting, trap and discharge pipe size. (Page 304.)

Type of appliance	Size of waste fitting	Discharge pipe and trap size
Wash basin		
Sink		
Bath		
Bidet		
Shower tray		
Bowl urinal		
Stall urinal		
Drinking fountain		

(73) With reference to Figure 29, in the spaces provided, give the minimum distances which need to be maintained for the proposed discharge pipes in order to avoid cross-flow of effluent from one discharge pipe into another. (Page 307.)

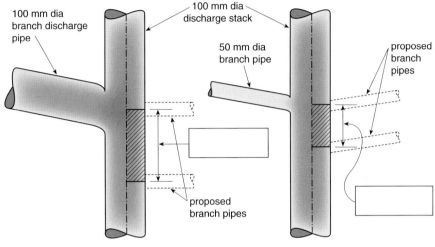

Figure 29

(74) Name the system of sanitary pipework shown in Figure 30, where no additional ventilation pipework is required. Also indicate in the space provided the maximum dimensions to be observed. (Page 309.)

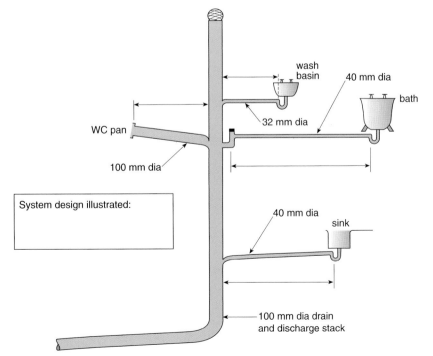

Figure 30

(75) Complete the following table to give the maximum number of appliances to be installed within a single unvented branch discharge pipe. (Page 308.)

Unvented branch discharge pipes serving more than one appliance

Appliance	Maximum no. of appliances to be fitted	Minimum pipe diameter
Wash basins		
Bowl urinals		
WC pans		

(76) What is the purpose of a resealing trap? (Page 312.)

(77) What is meant by the term 'induced siphonage'? (Page 312.)

(78) Complete the illustration in Figure 31 (over page) to show how an air test is maintained in above-ground soundness testing. Also state the minimum air test pressure. (Page 319.)

Minimum air pressure:

(79) State what is meant by the term 'performance test'. (Page 318.)

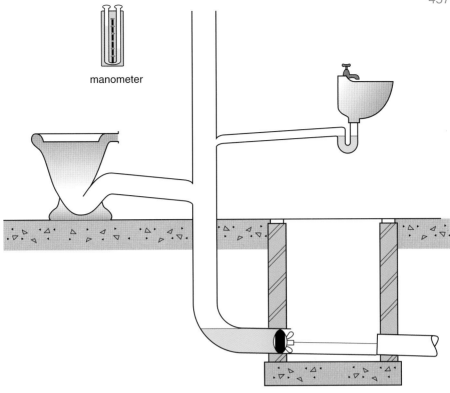

Figure 31

(80) Complete Figure 32 to show a partially separate system of drainage. (Page 325.)

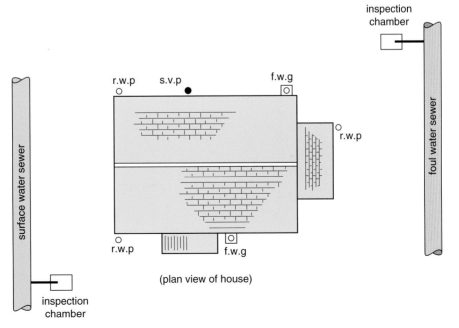

r.w.p   s.v.p   f.w.g

r.w.p

surface water sewer

foul water sewer

inspection chamber

inspection chamber

r.w.p   f.w.g

(plan view of house)

Figure 32

(81) Identify the purpose of an anti-flood gully and state the type of drainage system in which it may be found. (Page 328.)

```
┌─────────────────────────────────────────────────────────────┐
│                                                             │
│                                                             │
│                                                             │
│                                                             │
└─────────────────────────────────────────────────────────────┘
```

(82) Give four different possible locations where a point of access will be required to a system of drainage. (Page 330.)

1.	
2.	
3.	
4.	

(83) What is meant by the term 'benching'? (Page 331.)

```
┌─────────────────────────────────────────────────────────────┐
│                                                             │
│                                                             │
│                                                             │
└─────────────────────────────────────────────────────────────┘
```

(84) Show by illustration how a drainage pipe run can be made to pass through a wall, or foundation, below ground level to ensure the pipe will not fracture due to movement. (Page 327.)

```
┌─────────────────────────────────────────────────────────────┐
│                                                             │
│                                                             │
│                                                             │
│                                                             │
│                                                             │
│                                                             │
│                                                             │
└─────────────────────────────────────────────────────────────┘
```

(85) What is the name given to a piece of wood cut to an angle and used in conjunction with a spirit level for setting out the gradient to a short drainage run? (Page 335.)

```
┌─────────────────────────────────────────────────────────────┐
│                                                             │
│                                                             │
│                                                             │
└─────────────────────────────────────────────────────────────┘
```

(86) Complete Figure 33 to show the material, and indicate the depths of bedding material to be used when laying the 100 mm uPVC drainage pipe into the ground. (Page 327.)

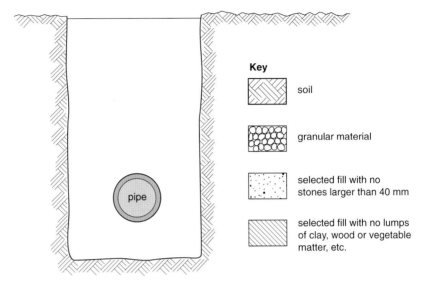

Figure 33

(87) Complete Figure 34 to show how the uPVC drain at high level can be made to connect suitably to the drain at the lower level. (Page 333.)

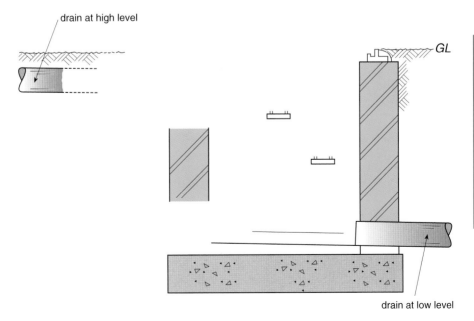

Figure 34

Figure 35

(88) Complete Figure 35 to show how a 1.5 m minimum water head pressure test can be achieved to the drainage pipe. (Page 345.)

(89) With reference to the previous question state the maximum water loss permitted from the pipe over a 30 min test period. (Page 344.)

(90) State the purpose of a soakaway. (Page 342.)

(91) Complete the following table, giving the British Standard specifications for sheet lead. (Page 348.)

BS 1178 Code No.	Colour code	Thickness Diameter
3		
4		
5		
6		

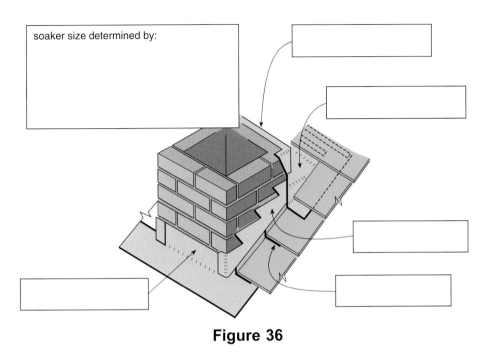

soaker size determined by:

Figure 36

(92) In the space provided name the chimney flashing details shown in Figure 36 and give the calculation used to determine the length of a soaker. (Pages 358–361.)

(93) State the maximum length to be observed for any flashing detail. (Page 358.)

(94) Figures 37 and 38 show a wood-cored roll and drip, as used for sheet lead. Complete the illustrations to show how the lead is weathered at these details, giving dimensions and explanatory notes as necessary. (Page 356.)

(95) State the name of the fixing used to secure a cover flashing into the brick course. (Page 358.)

(96) Given that code 5 sheet lead is going to be used to weather a small gutter lining, with a distance of 600 mm between the wood-cored rolls, state the maximum length to be observed between drips. (Page 354.)

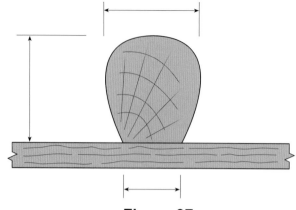

Figure 37

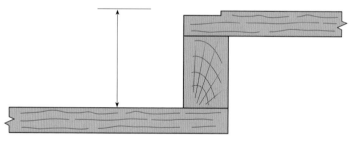

Figure 38

(97) Name the weathered detail shown in Figure 39. (Page 363.)

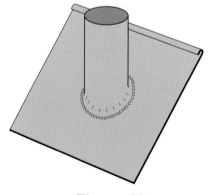

Figure 39

(98) State the minimum vertical cover that needs to be maintained with any lap joint, such as at cover flashings. (Page 358.)

(99) Name the two kinds of solar collector that can be found for a system of solar heating of domestic hot water. (Page 378)

(100) Identify the following terms (84 & 404)

Wholesome Water:	
Grey Water:	
Black Water:	

Problem Solving

Problem No. 1

The f & e cistern of the system illustrated below is continually discharging water from the overflow. The maintenance plumber who was called initially replaced the float-operated valve in the cistern, but was called back to the job to re-investigate. On this occasion the plumber turned off the service valve feeding the system and switched off the boiler, then emptied all the excess water from the cistern, but within 10 minutes the cistern started to overflow again. What is the cause?

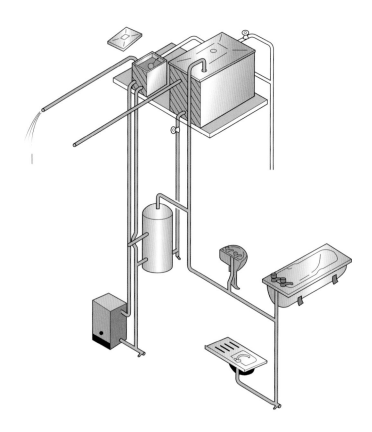

Answer:

Problem No. 2

You have been called to a job to investigate the problem of a poor and intermittent flow of water from tap 'X' when the other taps are being used. State the cause and remedy.

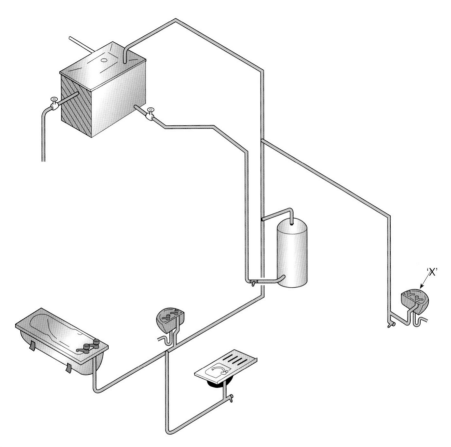

Answer:

Problem No. 3

During the summer months the fully-pumped dhw and c.h. system illustrated below is faulty in that the heat emitters at A and B heat up when the system is operating on dhw only. There is no fault with the wiring and the electrical controls are working correctly. Identify the fault.

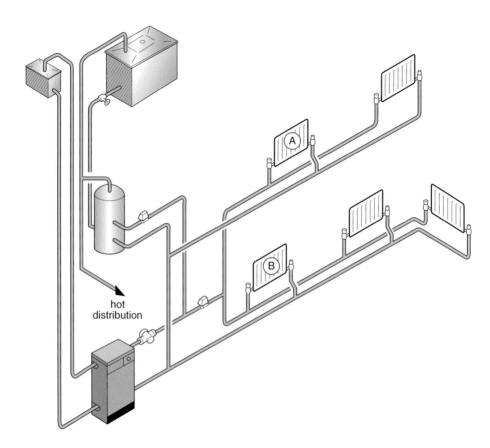

hot
distribution

Answer:

Problem No. 4

The fully pumped heating system shown below suffers the problem of air continually being drawn into the system; this invariably causes the heat emitters to become cold and as a result require frequent venting. State the cause of this problem and suggest a remedy. Also identify the long-term effects this problem will have on the system.

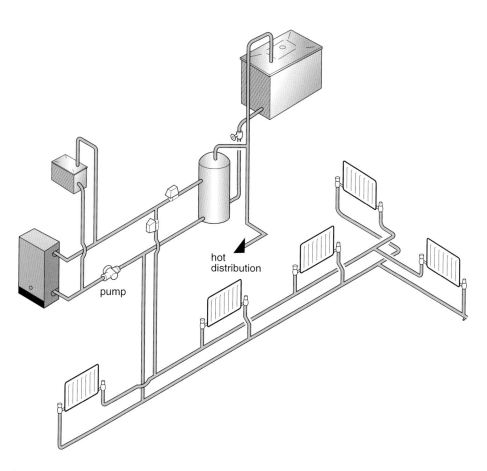

hot
distribution

pump

Answer:

Problem No. 5

You have been called to look at the dhw system illustrated below to investigate the problem of brown water which keeps appearing at the hot draw-off taps. The householder informs you that the c.h. system has recently been extended by the addition of three large radiators and that the problem did not occur before this time; the system itself was installed some 35 years ago. What is the likely cause and remedy?

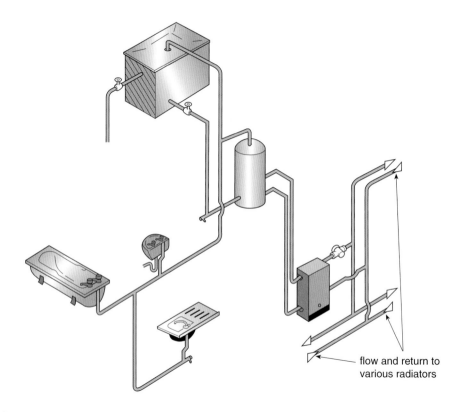

flow and return to
various radiators

Answer:

Problem No. 6

You have been called to the plumbing system illustrated below. The householder says that the dhw seems to take 6–8 hours to warm up, yet the boiler is working fine, although it is noisy at times. State the likely cause, remedy and work required to prevent this problem's recurrence.

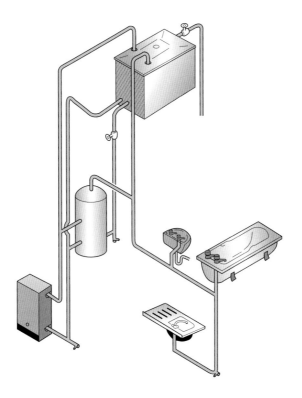

Answer:

Problem No. 7

You have been called to the gas installation shown below because the 5 kW (17 000 Btu/h) natural gas fire does not seem to operate effectively all the time and gives out very little heat; basically the flames seem to burn very low. Investigation of the flue has revealed no fault and there is a good pull to the chimney. The fire itself has no apparent defects. State the probable cause and suggest its remedy.

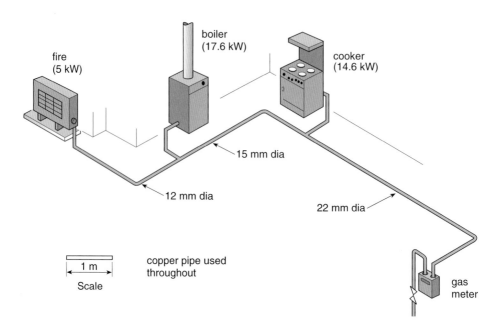

Answer:

Problem No. 8

Illustrated below is an unvented dhw unit which is experiencing intermittent discharge of water from the pressure relief valve. State the possible cause of the problem and how it can be cured. Also, the client is worried that no drain-off cock has been fitted to the base of the dhw cyclinder. Note that it is fitted to the top of the dhw storage vessel and explain how the water is removed from the vessel if necessary.

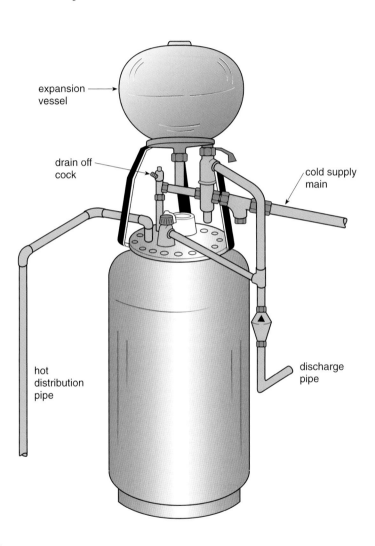

Answer:

Problem No. 9

You have been called to a dwelling to investigate the fatality of an occupant (found in the extension) owing to carbon monoxide poisoning. A $50\,cm^2$ permanent air grille is installed in the room where a $17\,kW$ open-flued natural gas burning appliance has been installed. A test on the flue system indicates no sign of spillage at the draught diverter. The appliance had been installed some 8 months previously and had been working most satisfactorily until the accident. What fault in the installation caused the death of the occupant?

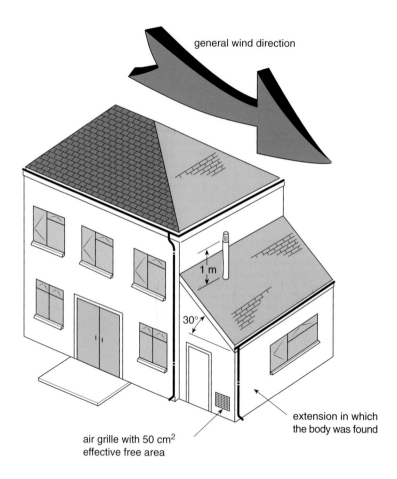

general wind direction

1 m

30°

extension in which
the body was found

air grille with 50 cm²
effective free area

Answer:

Problem No. 10

Having just completed the installation of the indirect cold water system, illustrated below, and turned on the water supply, you find that water flows freely from the bath yet only trickles out of the taps to the washbasin and WC cistern. What is the likely cause of this problem?

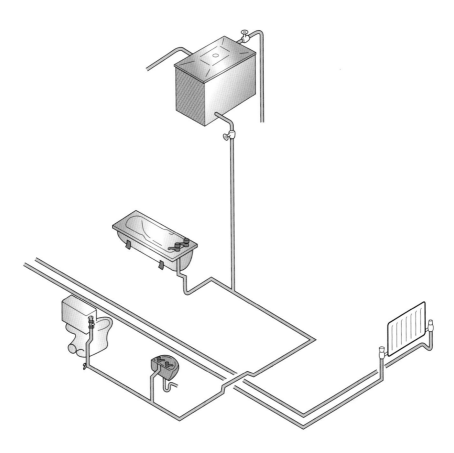

Answer:

Answers to Problems 1–10

Problem No. 1:   The heat exchanger in the dhw cylinder has developed a hole. Therefore the water from the dhw system is mixing with the water in the primary circuit. Because the water level is somewhat higher in the larger dhw feed storage cistern, the water is passing from one system through the hole into the other system, trying to maintain an equal water level. This is not possible due to the f & e cistern being installed lower, causing it to overflow as a warning.

Problem No. 2:   The problem with this system is the high level connection into the vent for the hot distribution pipe feeding tap 'X'. Due to the hydraulic gradient of water from the system (see page 110) the water ceases to flow. The problem can be overcome by re-routing the pipe from a point lower down, or by taking the connection from the distribution to the other appliances. In addition, it must be understood that the cold feed pipe to the cylinder and hot distribution pipe must be suitably sized.

Problem No. 3:   A design in which heat emitters A and B are heating up due to reversed circulation of hot water from the boiler bypassing the two-port motorised valve serving the heating circuit (see page 148).

Problem No. 4:   The fault with this system is that the cold feed and vent have not been connected to the primary heating circuit within 150 mm of each other, leading to unbalanced pressures at each point giving a negative vent connection (see page 149). A simple remedy would be to alter the pipework and connect the cold feed close to the vent. It would also be beneficial to relocate the pump so that it gives a positive pressure throughout the system, thus eliminating micro leaks through radiator glands, etc. The long term effect, if left uncorrected, would be corrosion.

Problem No. 5:   By reviewing the design of this system, which worked fine prior to the alteration, it is clear that the cylinder is of a single feed design (see page 107). Therefore, the likely cause of this problem is the air bubble being lost from the cylinder heat exchanger, resulting in the mixing of the dhw and ch waters and corrosion of the steel radiators is occurring. The remedy would be to change the single feed dhw cylinder to a double feed type, with the addition of an f & e cistern and associated pipework.

Problem No. 6:   Providing the boiler thermostat is fine, the likely cause is blocked primary pipework, either due to corrosion or, in the case of temporary hard water areas, limescale. Clearly the system should have been designed using an f & e cistern. The remedy and work to prevent reoccurrence would be to replace the primary pipework and feed this circuit via a newly installed f & e cistern, which will prevent the continued introduction of air or calcium carbonate into the primary circuit.

Problem No. 7:   By undertaking a pipe sizing calculation, as identified on page 192, it can be seen that the pipework is undersized. A calculation would suggest that an increase of the pipe size in section 'B' to 22 mm would solve the problem.

Problem No. 8:   Several problems with the system can cause the pressure relief valve to open. However, because the discharge is intermittent it is unlikely to be a faulty pressure relief valve itself or a faulty pressure reducing valve. Two possible causes are: (1) air pressure loss within the sealed expansion vessel, or (2) a faulty thermostat, which is allowing the water to expand too much and the pressure relief is opening before the temperature relief valve. Regarding the client's worries in relation to the doc located at the top of the vessel, explain that the water is simply siphoned out.

Problem No. 9:   The fault with this system is the height to which the flue terminal has been raised above the roof line. It is insufficient and should have been raised to a horizontal line of 1.5 m above the main roof level. The system appeared to work satisfactorily most of the time with the prevailing wind direction because it created a negative (suction) pressure zone at the terminal. As the wind direction changed the terminal was located in a positive pressure zone, effectively capping the flue outlet, causing the appliance to spill the products of combustion back into the room.

Problem No. 10:   Due to the cold distribution being low pressure, the most likely cause is an air lock where the pipes pass over the two central heating pipes. The problem can be resolved by passing this pipe below the heating pipework, allowing air to escape via the taps or back up into the storage cistern.

Answers to Multiple Choice Questions

(1) a	(21) a	(41) b	(61) b	(81) d
(2) c	(22) a	(42) b	(62) c	(82) c
(3) c	(23) a	(43) c	(63) d	(83) a
(4) c	(24) d	(44) b	(64) b	(84) c
(5) a	(25) c	(45) a	(65) c	(85) b
(6) a	(26) d	(46) b	(66) b	(86) b
(7) b	(27) a	(47) a	(67) a	(87) c
(8) c	(28) c	(48) a	(68) d	(88) a
(9) c	(29) a	(49) b	(69) b	(89) a
(10) a	(30) c	(50) c	(70) a	(90) b
(11) b	(31) a	(51) a	(71) c	(91) b
(12) a	(32) a	(52) d	(72) b	(92) b
(13) b	(33) b	(53) a	(73) c	(93) b
(14) c	(34) d	(54) b	(74) d	(94) a
(15) a	(35) c	(55) d	(75) a	(95) b
(16) d	(36) b	(56) c	(76) b	(96) d
(17) c	(37) c	(57) a	(77) b	(97) c
(18) b	(38) c	(58) b	(78) b	(98) a
(19) c	(39) a	(59) a	(79) d	(99) b
(20) c	(40) c	(60) a	(80) b	(100) c

Index

A

Abutment flashing, 358
Access cover, 330
Air admittance valve, 310
Air locks, 120
Air separator, 148
Air source heat pump, 386
Air test, 318, 344
Alloy, 41
Alternating current, 270
Anti-flood gully, 328
Anti-gravity valve, 140
Anti-siphon device, 247
Apron flashing, 358, 360
Areas, 34
Atmosphere sensing device, 206
Atmospheric corrosion, 46
Atmospheric pressure, 37
Automatic air release valve, 144
Automatic change over valve, 227
Automatic flow cut-off device, 302
Automatic flushing cistern, 302

B

Backdrop manhole, 332
Backgutter lead, 360
Balanced compartment, 218
Balanced flue, 216
Ballvalve, 78
Basin, 296
Basin spanner, 49
Bath, 296
Bib tap, 76
Bidet, 118, 296
Bi-metallic strip, 204, 284
Biomass, 390, 392
Blowpipe, 67
Boiler, 152
  noises, 129, 152
  sizing, 169
Boiler house, 236
Boiler interlock, 158
Boning rod, 334

Boosted water supply, 96
Bossing tools, 50
Box gutter, 364
Branched flue system, 222
Brass, 41
British Standard, 29
Bronze, 41
Bronze welding, 70
Buffer tank, 394
Building Regulations, 28
Bunded tank, 249
Burner pressure, 198
Butt weld, 72
By-pass, 162

C

Calorific value, 42
Capacitor, 286
Capacity, 34
Carbon cycle, 371
Carbon monoxide, 236
Carbon neutral, 390
Carburising flame, 69
Cast iron, 41
Catchpit, 364
Cathodic protection, 45
Centralised dhw, 102
Cesspit, 342
Cesspool, 342
Changeover valve, 227
Chase wedge, 50
Check valve, 109
Cheek, 366
Chute, 364
Circuit protective conductor, 272
Circulating head, 113
Circulating pressure, 113
Circulating pump, 144
Cladding, 354
Closure plate, 212
CO/CO_2 ratio, 228
CO detection, 236
CO_2 analyser, 228, 262
Coefficient of performance, 388

Plumbing, 4th Edition. R. D. Treloar.
© 2012 Blackwell Publishing Ltd. Published 2012 by Blackwell Publishing Ltd.

Index

Combination boiler, 154
Combined Heat & Power (CHP), 398
Combined system of drainage, 324
Combustion analysis, 228
Combustion efficiency test, 262
Combustion performance, 229
Compensation control, 164
Compression, 312
Compression fitting, 52
Compression joint, 54, 56
Condenser, 286
Condensing boiler, 156
Constant oil level control, 248
Construction team, 10
Consumer unit, 272
Convector heater, 142
Conventional flue, 206
Corrosion, 44
COSHH, 20
Coupling of storage cisterns, 92
Cover flashing, 358
CPSU, 114
Cross-flow, 307
Croydon ballvalve, 78
Cylinder thermostat, 146

D

Delayed action ballvalve, 96
Density, 36, 40
Dezincification, 44
DFE gas fire, 212, 220
Diaphragm ballvalve, 78
Direct cold water supply, 90
Direct current, 270
Direct hot water supply, 104
Discharge pipe, 306
Disconnection joint, 54
Dormer window, 366
Draeger analyser, 228
Drainage, 324
Drain cleaning, 320
Drain off cock, 76
Draught diverter, 206
Draught stabiliser, 256
Drawing symbols, 31
Draw-off tap, 76
Drip, lead, 356
Drip plate, 50
Dry riser, 98
Dual flush, 302
Ductility, 40
Dummy, 50
Durability, 40

E

Earth continuity, 276
Eaves guttering, 338
Elasticity, 40
Electricity at Work Regulations, 28
Electric storage boiler, 152
Electric water heating, 104
Electrochemical series, 44
Electrolytic corrosion, 44
Electronic analyser, 228
Employment rights, 8
Equilibrium ballvalve, 78
Equipotential bonding, 276
Evacuated tube collector, 378

F

Fan assisted flue, 222
Feed and expansion cistern, 144
Ferrous metal, 40
Filter, 248
Fire extinguisher, 26
Fire valve, 248
Flame failure device, 204
Flame rectification device, 204
Flame supervision device, 204
Flange joint, 55
Flashback arrester, 66
Flat dresser, 50
Flat plate collector, 378
Float operated valve, 78
Float switch, 96
Flue gas analyser, 228
Flue liner, 210
Flue pipe, 210
Flue terminal, 208
Flushing cistern 302
Flux, 64
Foul water, 324
Front apron lead, 360
Frost thermostat, 162
Fully pumped system, 148
Functional skills, 6
Fuse, 274
Fusibility, 40
Fusion welding, 52
Fyrite analyser, 228

G

Garage gully, 328
Gas consumption, 198
Gas cooker, 214

Gas fire, 212
Gas rate, 198
Gas regulator, 200
Gas relay valve, 202
Gas Safe Register, 3, 188
Gas Safety Regulations, 28, 188
Gas storage heater, 104
Gas supply pipework, 190
Gatevalve, 74
Globe tap, 76
Governor, 200
Gradient, 334
Gradient board, 334
Grease convertor, 328
Grease trap, 328
Greenhouse effect, 374
Greenhouse gases, 374
Grey water recovery, 404
Ground source heat pump, 386
Gully, 328
Gunmetal, 41
Gutter, 338, 362, 364

H

Hand tools, 48
Hardness, 40
Hard soldering, 64
Hard water, 82
Health & Safety at Work Act, 12
Heat emitter, 142
Heat input, 198
Heat pump, 386
Heat recovery period, 113
High tension leads, 252
Hob, 214
Hollow roll, 356
Hose reel, 96
Hot distribution pipe, 110
Hydraulic gradient, 110
Hydraulic test, 344
Hydropneumatic accumulator, 94, 129

I

Ignition systems, 284
Immersion heater, 280
Impulsive noise, 128
Indirect cold water supply, 88
Indirect dhw supply, 106
Induced siphonage, 312
Inhibitor, 45
Inspection chamber, 330
Instantaneous dhw supply, 104, 114

Instantaneous heater, 116
Intercepting trap, 328

J

Jumper, 75
Junction block, 333

K

Kettling, 129, 152
Kilowatt hour, 376

L

Ladders, 22
Lap joint, 357
Lap weld, 72
Lead bossing, 350
Lead dot, 352
Lead knife, 50
Lead sheet, 348
Lead slate, 362
Lead welding, 72
Leftward welding, 68
Lighting circuit, 278
Liquefied petroleum gas, 224
Liquid vapour device, 204
Localised dhw, 102, 116
Local Water Bylaws, 28
Low carbon steel, 54
Low pressure cut off, 202
LPG, 224

M

Magnetic valve, 286
Malleability, 40
Manhole, 330
Metric system, 32
Micro-bore, 138
Micro hydro power, 396
Mild steel, 41
Mixer tap, 76
Motorised valve, 146
Multi-function valve, 200
Multi-point, 114

N

Natural gas, 186
Neutral flame, 69
Neutral point, 148

Noise transmission in pipes, 128
Non-ferrous metals, 40

O

Obstruction test, 344
Ohm's Law, 268
Oil filter, 252
Oil level indicator, 244
Oil storage, 244
One pipe c.h. system, 140
One pipe oil supply, 244
Open flued appliance, gas, 206
Open flued appliance, oil, 256
OPSO, 225
Optimum start, 164
Oven, 214
Oxidising flame, 69
Oxygen depletion device, 202

P

Parallel circuit, 270
Partially separate system, 324
Payback, 376
Performance test, 130, 318
Permanent hardness, 82
Photocell, 252
Photovoltaic cells, 374, 384
Physical change, 43
Physical properties, 40
Piezo-electric ignition, 284
Pillar tap, 76
Pipe bending, 58
    equipment, 48
Pipe clip, 120
Pipe cutters, 48
Pipe sizing
    c.h. systems, 172
    hot and cold supply, 124
    gas supply, 192
    sanitary pipework, 316
Plant room, 236
Plasticity, 40
Plastic tube, 52
Plug cock, 74
Pneumatic storage vessel, 96
Pneumatic test, 344
Polarity, 290
Portsmouth ballvalve, 78
Pot burner, 260
Power flushing, 182
Power tools, 24
PPE, 16

Press-fitting, 56
Pressure, 37
Pressure jet burner, 250
Pressure reducing valve, 108
Pressure relieve valve, 108
Pressure switch, 96
Primatic dhw cylinder, 106
Product Characteristics Data File, 372
Programmer, 146
Pump overrun, 162
Pump sizing, 172
Purging, 197
Push-fit joint, 52, 56

Q

Quantity of heat, 42
Quarter-turn valve, 74

R

Radial circuit, 280
Radiant heating, 176
Radiator, 142
Radiator valve, 142
Rainwater harvesting, 400
Rainwater pipe, 340
Ramp, 332
Rectifier, 286
Reflection test, 344
Regulator, 66, 226
Relative density, 36
Relay, 164, 286
Renewable energy, 374
Resealing trap, 312
Reversed circulation, 148
Reversed return c.h. system, 140
RIDDOR, 28, 238
Rightward welding, 68
Ring main, 278
Risk assessment, 15
Rodding eye, 330
Room sealed appliance
    gas, 216
    oil, 258
Room thermostat, 146

S

Saddle, 361
Safety check, 228
Sanitary appliance, 296
SAP rating, 372
Screwdown valve, 74

Sealed c.h. system, 150
Sealed expansion vessel, 109, 150
Seaming pliers, 50
Secondary circulation, 110
Secret tack, 352
Se duct, 222
Self siphonage, 312
Separate system of drainage, 324
Septic tank, 342
Series electrical circuit, 270
Set-back control, 162
Setting in stick, 50
Shave hook, 50
Shower, 296
Side cheek, 366
Sight rail, 334
Single feed dhw supply, 106
Single point dhw heater, 116
Single stack system, 308
Sink, 296
Siphonic WC pan, 298
Slate, 362
Slop hopper, 298
Soakaway, 342
Soakers, 358, 361
Soft soldering, 64
Soft water, 82
Soil appliances, 298
Solar collector, 380
Solar heating, 380
Solar power, 378, 384
Solar system of dhw, 380
Solder, 41
Soldered dot, 352
Soldered joint, 56
Soldering, 64
Solenoid valve, 202, 252, 286
Solvent welding, 52
Solver flue, 222
Soundness test, 318, 344
Spanner, 48
Specific gravity, 36, 40
Specific heat, 40, 42
Spillage, 230
Sprinkler system, 98, 100
Spur outlet, 278
Stainless steel, 41
Standard Assessment Procedure, 372
Standing pressure, 198
Standing seam, 357
Stepped flashing, 358
Step turner, 50
Stillson wrench, 49
Stirling engine, 378, 398
Storage cistern, 92

Strainer, 109
Stub stack, 310
Supa tap, 76
Supplementary bonding, 276
Surface water, 324
Sustainable drainage system, 400

T

Tapered gutter, 364
Temper, 40
Temperature relief valve, 108
Temporary bonding wire, 276
Temporary hardness, 82
Tenacity, 40
Tensile strength, 40
Terminal valve, 76
Thermal expansion, 40, 43
Thermal insulation, 122
Thermal storage heater, 114, 394
Thermo-electric valve, 204
Thermoplastic, 41
Thermo-setting plastic, 41
Thermostat, 284
Thermostatic control, 163
Thermostatic radiator valve, 142
Threading equipment, 48
Three phase supply, 270
Three pipe c.h. system, 140
Tiger loop, 246
Tightness testing, 196
Transference of heat, 42
Transformer, 286
Trap, 304
Trap seal loss, 312
Traveller, 334
Tumbling bay, 332
Two pipe c.h. system, 140
Two pipe oil supply, 246
Two pipe reversed return, 140

U

U duct flue system, 222
Unvented dhw supply, 108
UPSO, 225
Urinal, 298
U valve, 166

V

Valley gutter, 362
Valve flushing cistern, 300

Index

Vaporising burner, 260
Ventilating pipe, 306
Vertex flue, 222
Viscosity, 242
Volume, 34

W

Wallflame burner, 260
Warm air heating, 180
Washdown WC pan, 298
Waste appliances, 296
Waste disposal unit, 314
Waste pipe valve, 305
Water Bylaws, 28
Water conditioner, 94
Water efficiencies, 88
Water hammer, 128
Water jacketed tube heater, 114
Water level, 355
Water power, 396

Water pressure, 38
Water softener, 94
Water Supply Regulations, 28
Water test, 344
WC macerator pump, 314
WC pan, 298
Wedge, 352
Welding equipment, 66
Welding processes, 68
Welding safety, 66
Welt, lead, 356
Wet riser, 98
Whole house boiler sizing, 170
Wind power, 396
Wood cored roll, 356
Working pressure, 198
Wrench, 48

Y

Yard gully, 328